MOLECULAR
BIOLOGY
INTELLIGENCE
UNIT

PROTEIN TRAFFICKING ALONG THE EXOCYTOTIC PATHWAY

Wanjin Hong

Institute of Molecular and Cell Biology
National University of Singapore
Singapore

CHAPMAN & HALL
I(T)P An International Thomson Publishing Company

New York • Albany • Bonn • Boston • Cincinnati • Detroit • London • Madrid • Melbourne •
Mexico City • Pacific Grove • Paris • San Francisco • Singapore • Tokyo • Toronto • Washington

R.G. LANDES COMPANY
AUSTIN

MOLECULAR BIOLOGY INTELLIGENCE UNIT
PROTEIN TRAFFICKING ALONG THE EXOCYTOTIC PATHWAY

R.G. LANDES COMPANY
Austin, Texas, U.S.A.

Please address all inquiries to the Publishers:
R.G. Landes Company, 909 Pine Street, Georgetown, Texas, U.S.A. 78626
Phone: 512/ 863 7762; FAX: 512/ 863 0081

North American distributor:
Chapman & Hall, 115 Fifth Avenue, New York, New York, U.S.A. 10003

CHAPMAN & HALL

U.S. and Canada ISBN:-0-412-11071-7

Library of Congress Cataloging-in-Publication Data
CIP Data applied for but not received by publication date.

Publisher's Note

R.G. Landes Company publishes six book series: *Medical Intelligence Unit, Molecular Biology Intelligence Unit, Neuroscience Intelligence Unit, Tissue Engineering Intelligence Unit, Biotechnology Intelligence Unit and Environmental Intelligence Unit.* The authors of our books are acknowledged leaders in their fields and the topics are unique. Almost without exception, no other similar books exist on these topics.

Our goal is to publish books in important and rapidly changing areas of bioscience and environment for sophisticated researchers and clinicians. To achieve this goal, we have accelerated our publishing program to conform to the fast pace in which information grows in bioscience. Most of our books are published within 90 to 120 days of receipt of the manuscript. We would like to thank our readers for their continuing interest and welcome any comments or suggestions they may have for future books.

Deborah Muir Molsberry
Publications Director
R.G. Landes Company

CONTENTS

Protein trafficking along the exocytotic/secretory pathway is a fundamental area of cell biology and our understanding about this pathway has experienced an exciting period during the past few years, in which many new breakthroughs were made, including the molecular identification of membrane proteins responsible for polypeptide translocation across the ER membrane, functional establishment of several soluble proteins in membrane fusion, the proposal of SNARE hypothesis for the specificity of vesicle docking/fusion, the molecular and functional elucidation of COPI and COPII protein coats of transport vesicles, functional characterization of several classes of GTPases in protein trafficking, the identification of several sorting/targeting signals for precise intracellular localization, molecular and mechanistic elucidation of neurotransmitter release and genetic as well as biochemical identification of many other proteins that participate in various aspects of protein trafficking. These exciting progresses are discussed in the context of general description of the exocytotic pathway.

Acknowledgments

I wish to acknowledge Dr. Qi Zeng, my beloved wife, and William Cheng Hong, my beloved son, for their continuous love, understanding, encouragement and support during all these years. I also like to thank Drs. Bor Luen Tang, Paramjeet Singh, Seng Hui Low for their critical reading of the entire manuscript. Thanks also to Professor Y.H. Tan and the Institute of Molecular and Cell Biology as well as Landes Bioscience Publishers for their support.

CHAPTER 1

The Structural Organization of the Exocytotic Pathway

THE EXOCYTOTIC PATHWAY

The exocytotic/secretory pathway (Fig. 1) plays an essential role in many biological processes of the living cell.[205b,303,433,477,562a,609,647,649,729,753] Its fundamental role is to maintain the dynamic structure and function of the plasma membrane and many other intracellular membranes in both proliferating and differentiated cells. Cells utilize this pathway to secrete the majority of extracellular proteins such as growth factors, peptide hormones, extracellular matrix proteins, antibodies and other serum proteins. This property is essential for all stages of development and for maintaining the normal physiology of a mature organism. Neuronal cells also use this pathway for regulated release of neuropeptides and neurotransmitters and this plays a vital role in many functions associated with the nerve system, such as learning, memory and thinking.

The general scheme of the exocytotic pathway of a mammalian cell is outlined in Figure 1. It includes several distinct membrane-enclosed compartments and intermediates that mediate dynamic connections between these compartments.[253a,303,433,477,562a,609] Proteins destined for the exocytotic pathway are initially targeted to the endoplasmic reticulum (ER) via various types of ER targeting sequences present on the nascent polypeptide chains. The topology of the targeted proteins is then established in the ER. Most of the posttranslational modifications, such as N-linked glycosylation and protein folding occur in the ER. The first vesicular transport step along the exocytotic pathway is from the ER to the cis side of the Golgi apparatus. The proteins are further transported from the cis to the trans side of the Golgi by intra-Golgi transport events. Upon reaching the trans Golgi network (TGN), proteins are sorted and packaged into distinct vesicles destined for post-Golgi structures, either the plasma membrane or the endosomal/lysosomal compartment.[303,433,477,555] For regulated secretion, the budded

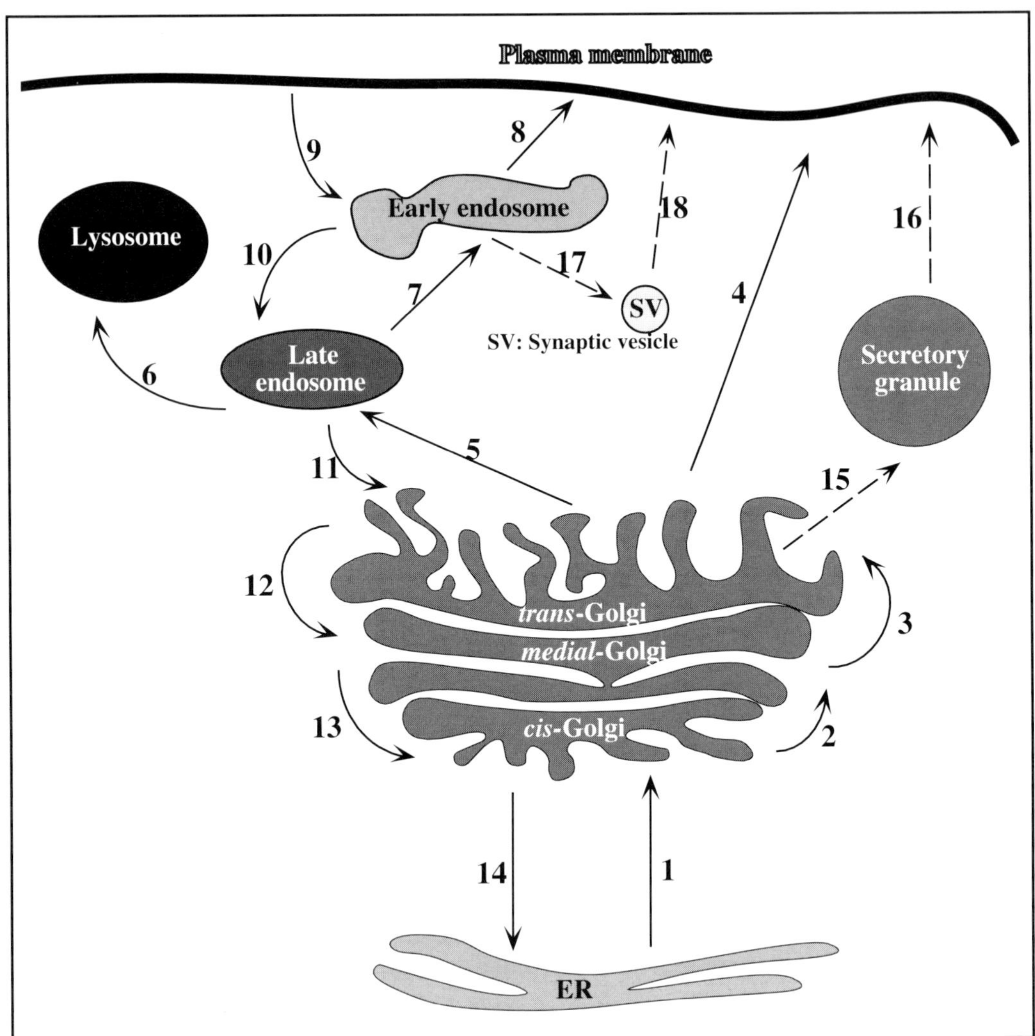

Fig. 1. Schematic illustration of the exocytotic pathway. The major organellar structures that constitute the pathway include the ER, various subcompartments of the Golgi apparatus, the endosomal/lysosomal system (the early endosome, the late endosome and the lysosomes), the plasma membrane and the membranous intermediates that mediate trafficking among these membrane structures. In neurons, there is a regulated exocytosis route mediated by synaptic vesicles which originate from an early-endosome-like compartment. In exocrine and endocrine cells, as well as peptidergic neurons, regulated exocytosis mediated by secretory granules mediate the controlled release of zymogens, peptide hormones and neuropeptides.

vesicles from the TGN undergo a maturation event and are stored as secretory granules. Upon proper stimuli, they are targeted and fused with restricted areas of the plasma membrane. Recently, retrograde trafficking from the Golgi apparatus back to the ER has been established and the major site for this pathway occurs from the cis face of the Golgi.

Morphologically, the ER consists of a network of interconnected membrane sheets, tubes, sacs and vesicles that spread throughout the entire cytoplasmic space. The outer membrane of the nuclear envelope is directly connected to the ER and could be thought of as a specialized subdomain of the ER. In addition to the nuclear envelope, the ER could be organized into other subdomains, such as the rough ER (rER) and smooth ER (sER). The rER is where nascent polypeptides are initially targeted to. It exists mainly in the form of flattened sheets with studded ribosomes on the cytoplasmic face of the membrane. The sER is more tubular and lacks ribosomes on its membrane and is mainly involved in lipid metabolism. In most cells, sER is usually scanty. However, sER is quite abundant in cells that are specialized in lipid metabolism, such as hepatocytes which are the major cell type involved in lipoprotein production.

The Golgi apparatus is primarily composed of stacked membrane cisternae and sacs which are dilated at both ends with numerous membrane vesicles present' around the Golgi stacks. The trans side of the Golgi (TGN) also has an extensive network of tubular-vesicular structures. The cis side is the entry face for vesicles derived from the ER while the trans side is the exit face of the Golgi for vesicular transport destined for post-Golgi structures. Histochemically, the cis side of the Golgi has been revealed by staining with osmium, which is reduced primarily in the cisternae and associated structures in the cis Golgi. Enzymatic activity for nucleotide diphosphatase and acid phosphatase is preferentially detected in the trans Golgi and TGN, respectively. The Golgi apparatus has a perinuclear location, usually around the microtubule organizing center (MTOC). Under the immunofluorescence microscope, the Golgi apparatus has a perinuclear crescent structure.

The endosomal/lysosomal compartment is mainly composed of the early endosomes, late endosomes and lysosomes, which are structurally and functionally distinct. Lysosomes are the major cellular degradative compartment for macromolecules. Their lumen has an acidic environment of around pH 5 and contains many acid hydrolases, including proteases, nucleases, glycosidases, lipases, phospholipases, phosphatases and sulfatases. Lysosomes are distributed throughout the cytoplasm and, although appear roughly spherical, are heterogenous in size and morphology. The late endosomes are also referred to as the prelysosomal compartment. Their major role is to mediate traffic derived from the early endosomes and the TGN to the lysosome. Late endosomes are more concentrated in the perinuclear Golgi regions as compared to lysosomes. Early endosomes are the major sorting compartment for traffic derived from the plasma membrane via endocytosis. Proteins in the early endosomes are either recycled back to the cell surface or delivered to the late endosomes. Early endosomes are heterogenous in size and morphology and are distributed throughout the cytoplasm.

LIPID BILAYER AS THE BASIC STRUCTURE OF A MEMBRANE

The basic structure of biological membranes consists of a phospholipid bilayer associated with both intrinsic (integral) and extrinsic (peripheral) membrane proteins.[302,711,715] The membrane is a dynamic, fluid structure. Lipid and protein components are free to move laterally in the plane of the membrane, unless they are restricted through various protein-protein interactions and/or aggregation. The lipid bilayer is the basic and fundamental structure of a membrane and serves as a relatively impermeable barrier to the passage of hydrophilic molecules, while the protein components confer other functional activities and, to a large extent, the specificity of the membrane. The functional and biochemical distinction among different membranes is mainly due to their different protein compositions. Membrane-associated proteins can serve as transporters that transport hydrophilic molecules across the membrane in either direction, enzymes that catalyze membrane-associated reactions, receptors that bind various ligands to transduce signals from the outside of the cell, or structural components that link the membrane to other structures (for example, the cytoskeleton or extracellular matrix). Several distinct functions could be associated with a single protein.

Phospholipids, being the most abundant lipid component of the lipid bilayer, include phosphatidylcholine, phosphatidylethanolamine, phosphatidylserine, sphingomyelin, phosphatidylinositol and phosphatidylglycerol. Phospholipids are amphipathic, having a hydrophilic head group and two hydrophobic hydrocarbon tails of varying lengths. It is this amphipathicity that allows the phospholipids to form a thermodynamically stable bilayer in an aqueous biological environment. Each layer of the lipid bilayer is referred to as a leaflet. The hydrophobic tails are embedded in the interior of the bilayer, while the polar heads are exposed to the aqueous environment (cytosol, extracellular fluid, or the lumen of an organelle). Other lipid components of the bilayer include various glycolipids and cholesterol. Glycolipids, such as galactocerebrosides and gangliosides contain sugar moieties and are associated exclusively with the noncytoplasmic leaflet of the lipid bilayer and are enriched in the plasma membrane. Like phospholipids, glycolipids and cholesterol are also amphipathic and buried in the lipid bilayer in a similar configuration as phospholipids. The lipid bilayer is, therefore, a two dimensional fluid and the lipid molecules are able to diffuse freely on its leaflet. The exact composition and the ratio of each type of lipid varies from membrane to membrane. The fluidity of a membrane is determined mainly by its lipid composition, the temperature of the environment and the cholesterol content. Cholesterol serves as a dynamic modulator of the fluidity. The amphipathic nature of the lipid bilayer enables proteins to be inserted in the lipid bilayer with their hydrophobic region(s) embedded in the interior of the membrane and their hydrophilic regions exposed on either side of the membrane.

PROTEINS ASSOCIATE WITH THE LIPID BILAYER IN DIVERSE WAYS

Proteins are associated with the membrane in two major ways (Fig. 2).[302,711,715] Peripheral proteins are attached to the membrane through noncovalent interactions with other membrane proteins on either side of the membrane. For convenience, peripheral proteins on the cytoplasmic side of the membrane are referred to as ppC, while those on the extracellular side of the plasma membrane or the luminal side of internal membranes are referred to as ppL. Since they do not extend into the hydrophobic interior of the membrane, peripheral proteins, in most cases, can be dissociated from the membrane by treatments that interfere with protein-protein interaction (for example, 1 M KCl; 0.15 M carbonate buffer, pH 11.5; and 2.5 M urea). Ankyrin is an example of ppC. In erythrocytes, ankyrin associates with the plasma membrane through its interaction with an integral membrane glycoprotein called Band 3.[302] β-COP, a component of the Golgi coatomer, could also be considered as a ppC because a fraction of it is associated with Golgi membranes via some protein-protein interactions.[189] Golgi as well as plasma membrane adaptins and components of clathrin coats are other examples of ppC. β2-microglobulin is an example of ppL. It associates with the membrane by forming a complex with the α chain of the class I major histocompatibility antigen.[302]

Integral membrane proteins are associated with the membrane by inserting their hydrophobic regions into the hydrophobic interior of the lipid bilayer (these hydrophobic regions are referred to as membrane-anchoring domains). The hydrophilic regions of integral proteins are exposed on either (or both) side of the membrane. The major type of membrane-anchoring domain is composed of a continuous sequence of hydrophobic amino acids (usually between 16 to 30 residues) termed the transmembrane domain. The other type of membrane-anchoring domain is composed of a lipid structure covalently linked to the protein. Some integral membrane proteins have both types of membrane-anchoring domains. Integral proteins anchored by their transmembrane domains can be classified into three major types. A type I integral protein has a single transmembrane domain with its amino-terminus localized in the extracellular side of the plasma membrane or in the lumen of

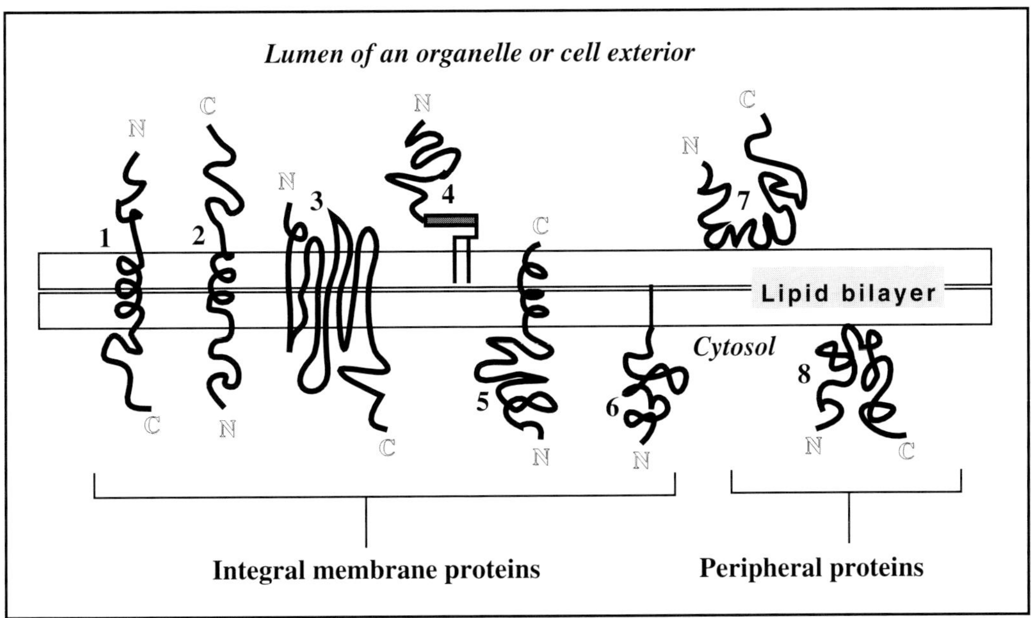

Fig. 2. *Proteins can associate with the membrane in diverse ways. Integral membrane proteins can be classified into six different types: (1) type I membrane proteins such as EGF receptor and calnexin; (2) type II membrane proteins such as dipeptidyl peptidase IV and transferrin receptor; (3) type III membrane proteins such as synaptophysin and mSec61pα; (4) GPI-anchored proteins such as Thy-1 and decay-accelerating factor; (5) C-terminal tail-anchored proteins such as synaptobrevins and syntaxins; (6)cytoplasmic lipid (prenyl groups, palmitate and myristate)-anchored proteins such as Rab proteins and ARFs. There are two types of peripheral membrane proteins: the ppL type of proteins (7) associated with the extracellular or luminal side of the membrane (such as β2-microglobulin) and the ppC type of proteins (8) associated with the cytoplasmic side of the membrane (such as the α-subunit of the SRP receptor).*

an organelle and its carboxy-terminus in the cytosol. This type I membrane topology results in three structural domains: an ectodomain or luminal domain (from the amino-terminus to the start of the transmembrane domain), a transmembrane domain and a cytoplasmic domain (from the end of the transmembrane domain to the carboxy-terminus). Ribophorin I, ribophorin II,[302] calnexin,[63] Emp24p[670] and cytochrome p450[302] are examples of type I integral proteins localized in the ER. TGN38/TGN41[446,622] and the mannose-6-phosphate (M6P) receptors[153,376] are type I integral proteins associated with the Golgi apparatus. M6P receptors are also present in the late-endosomes. Lamp1, lamp2 and lamp3 are examples of type I lysosomal integral membrane proteins.[217,376,659] Many cell surface proteins are type I integral membrane proteins, including the EGF receptor, insulin receptor, the NGF re-

ceptor, the FGF receptor, the polymeric IgA receptor, low density lipoprotein (LDL) receptor, glycophorins, the envelope glycoprotein (G protein) of vesicular stomatitis virus, hemagglutinin of influenza virus, members of the calderin family of cell surface adhesion molecules and integrins.[302] Many surface type I integral proteins can also be found in the endosomal/lysosomal system because they could be internalized into this pathway.

A type II integral membrane protein has a single transmembrane domain with its amino-terminus localized in the cytosol and the carboxy-terminus in the extracellular side of the plasma membrane or in the lumen of an organelle, resulting in a cytoplasmic domain (from the amino-terminus to the start of the transmembrane domain), a transmembrane domain and an ectodomain or luminal domain (from the end of the transmembrane domain to the carboxy-terminus). Sec12p, Sec20p and Sed4p are examples of ER localized type II membrane proteins.[172,267,758] The majority of Golgi glycan modifying enzymes, such as N-acetylglucosaminyltransferase I, β–1,4-galactosyltransferase, α–2,6-sialyltransferase and Golgi mannosidase II, are type II membrane proteins.[303,433,447] The asialoglycoprotein receptor, transferrin receptor, dipeptidyl peptidase IV, aminopeptidase N, sucrose-isomaltase and γ-glutamyl transpeptidase are examples of type II proteins localized to the plasma membrane,[302,713] some of which, such as the asialoglycoprotein receptor and the transferrin receptor are also localized in the early endosomes because they cycle between these two destinations.

A type III integral membrane protein has multiple transmembrane domains with their amino- and carboxyl-termini positioned on either side of the membrane. Sec61pα and translocating chain associating membrane (TRAM) are two type III proteins associated with the ER.[240,241] The KDEL/HDEL receptor (or Erd2p) and Erd1p are examples of Golgi-associated proteins with a type III membrane topology.[268,309,406,408,772] Synaptophysin and neurotransmitter transporters are examples of type III proteins associated with synaptic vesicles.[26,60] The α-subunit of Na+K+ ATPase, β-adrenergic receptor, glucose transporter, acetylcholine receptor and rhodopsin are examples of type III proteins present on the cell surface.[302]

Recently, a new class of integral membrane proteins has been defined and referred to as C-terminal-anchored proteins.[26,60,384a,385] A C-terminal-anchored protein has a transmembrane domain at the carboxyl-terminus, resulting in the rest of the polypeptide being oriented on the cytoplasmic side of the membrane. This class of proteins are targeted to the ER

via a mechanism different from that employed by the types of integral proteins mentioned above. A number of proteins involved in specific docking/fusion of transport vesicles have this type of membrane topology. Examples are syntaxin 1A, 1B, 2, 3, 4, 5, Sed5p, Sso1p, Sso2p, Pep12p, synaptobrevins, cellubrevin, Snc1p and Snc2p, Sec22p/Sly2p, Bet1p/Sly12p, Bos1p and a cis Golgi protein p28.

Proteins can also associate with the membrane through covalently-linked lipids. Four major types of lipid structures have been described, including glycosylphosphatidylinositol (GPI), myristate, palmitate and prenyl groups.[302] In the mature protein, the GPI-anchor is linked to the α-carboxyl group of the carboxyl-terminal amino acid by an amide linkage.[302] Many surface proteins have been identified to be anchored to the membrane by GPI-anchors, including Thy-1, Qa, alkaline phosphatase, 5'-nucleotidase, trehalase and decay accelerating factor (DAF). The polypeptide portion of a GPI-anchored protein is exposed to the extracellular side of the plasma membrane. Myristate is amide-linked to the amino-terminal glycine residues of proteins such as ADP-ribosylation factors (ARFs), pp66v-Src of Rous sarcoma virus, signal transduction proteins Gi and G_o-subunits.[346,501,625,638] Myristylated proteins are associated with the cytoplasmic side of the membrane. Palmitate is ester-linked to an internal cysteine residue and has been described for p21ras, insulin receptor and transferrin receptor.[302] A prenyl (either farnesyl or geranylgeranyl) group is linked via a thioether bond to a cysteine near the carboxy-terminus.[490] This lipid modification has been described for members of the Rab/Ypt1/Sec4 family of small GTP binding proteins, the yeast mating factors, ras protein, Src protein and nuclear lamins.

CHAPTER 2

ER Targeting
and Topogenesis

SIGNAL-MEDIATED ENTRY INTO THE EXOCYTOTIC PATHWAY

TARGETING SIGNALS AND BIOGENESIS OF MEMBRANE TOPOLOGY

Proteins destined for the exocytotic pathway are all targeted to the rER for entry into this pathway. This targeting is an active process mediated by various types of ER targeting signals present on the nascent polypeptides (Table I).[293,302,617,618,666] ER targeting signals can be roughly classified into four major types; the signal peptide, the type I signal/anchor sequence, the type II signal/anchor sequence and the C-terminal tail-anchor. The common feature among these four types of ER targeting sequences is that they are all composed of a sequence of hydrophobic amino acids flanked on both sides by hydrophilic residues. These four types of ER targeting signals are also intimately involved in topogenesis in the ER. In addition, there are two other types of primary sequences involved in topogenesis, including the stop-transfer sequence and the GPI-attachment signal (Table I). The signal peptide is found at the N-terminus of ER luminal proteins, secretory proteins, lysosomal enzymes, type I integral membrane proteins, GPI-anchored proteins and some type III integral proteins. The majority of signal peptides are proteolytically cleaved off the polypeptide during translocation into the ER. A signal peptide is composed of three regions, the central hydrophobic core region (h-region) composed of about 8 to 18 hydrophobic residues, preceded by a hydrophilic N-region of a few residues. This N-region usually has net positive charges. The hydrophobic h-region is followed by a hydrophilic c-region that contains the signal peptidase cleavage site. A signal peptide is able to initiate translocation across the ER membrane of downstream sequences unless this process is arrested by either a stop-transfer sequence or a GPI-attachment signal.

The signal peptide is the only topogenic sequence for secretory proteins, lysosomal enzymes and luminal proteins of the ER. These

Table I. Major topogenic sequences for secretory and membrane proteins					
Type of Sequence	**Location**	**Cleaved**	**ER Targeting**	**Translocation**	**Anchorage**
Signal peptide	N-terminal	Yes	Yes	Downstream	No
Signal/anchor Type I	N-terminal or internal	No	Yes	Upstream	$N_{in}C_{out}$
Signal/anchor Type II	N-terminal or internal	No	Yes	Downstream	$N_{out}C_{in}$
Stop transfer	Internal	No	No	Stop	$N_{in}C_{out}$
GPI-attachment signal	C-terminal	Yes	No	No	Replaced by GPI-anchor
C-terminal tail-anchor	C-terminal	No	Yes	Downstream	$N_{out}C_{in}$

Table I. Summary of ER targeting and/or topogenic sequences

proteins are fully translocated across the ER membrane. A stop-transfer sequence is composed of a stretch of about 16-30 hydrophobic residues flanked on both sides by hydrophilic sequences. It is located downstream of either a signal peptide or a type II signal/anchor sequence and is able to halt ongoing polypeptide translocation. Furthermore, a stop-transfer sequence is able to embed itself in the interior of the lipid bilayer and function as a stable transmembrane domain in an $N_{in}C_{out}$ orientation (the N-terminal and C-terminal ends of the transmembrane domain are on the luminal/extracellular and cytoplasmic sides of the membrane, respectively). A type I signal/anchor sequence is usually present on the N-terminal part of the nascent polypeptide and it is composed of a central hydrophobic core region of about 16-30 residues flanked on both sides by hydrophilic residues. The net charge of the 15 residues flanking the N-terminal side of a type I signal/anchor sequence is usually more negative than that flanking the C-terminal end. It is not only able to initiate translocation of a short upstream sequence across the ER membrane but also able to function as a stable transmembrane domain in an $N_{in}C_{out}$ orientation. A type II signal/anchor sequence, located either internally or in the N-terminal part of a nascent polypeptide, is capable of initiating translocation of the downstream sequence and functioning as a transmembrane domain in an $N_{out}C_{in}$ orientation. A type II signal/anchor sequence is also composed of a hydrophobic core of around 18-30 residues flanked by hydrophilic sequences on both sides. In contrast to the type I signal/anchor sequence, the net charge in the 15 residues flanking the N-terminal end of a type II signal/anchor sequence is more positive than that flanking the C-terminal end. This charge difference between the two sides of the hydrophobic core is important for the orientation of the hydrophobic domain in the membrane.[154,302] The C-terminal tail-anchor is located at the extreme C-terminus of a nascent polypeptide and consists of a hydrophobic sequence of about 18-28 residues. It is able to insert into the membrane in an $N_{out}C_{in}$ orientation, resulting in the attachment of the major portion of the polypeptide to the cytoplasmic side of the membrane. As detailed later, ER targeting mediated by a C-terminal tail-anchor is achieved posttranslationally via a mechanism that is different from that mediated by the other three types of ER targeting signals. Finally, a GPI-attachment signal consisting of a stretch of hydrophobic residues plus its upstream sequence is located exclusively at the carboxyl-terminal end. A GPI-anchor is attached to a specific residue upstream of the hydrophobic sequence and the sequence after this residue is subsequently removed. The residue to be GPI-linked holds some specificity, although many residues can be used. Thus far, the GPI-attachment signal has only been identified in proteins that have an N-terminal signal peptide.

For membrane topogenesis (Fig. 3), a ppL orientation will be established for proteins having an N-terminal signal peptide. A type I integral protein is usually established by having an N-terminal signal peptide followed by a stop-transfer sequence. It can also be established for proteins having a type I signal/anchor sequence. A type II orientation is established for proteins having a type II signal/anchor sequence. The type III orientation can be primarily achieved for proteins having alternating signal/anchor and stop-transfer sequences with or without other topogenic sequences. If the first topogenic sequence in the type III protein is a signal peptide or a type I signal/anchor sequence, the N-terminus will be located in the luminal/extracellular side of the membrane. The N-terminus will be cytoplasmic if the first topogenic sequence is a type II signal/anchor sequence. The location of the C-terminus of a type III protein is likely to be determined by the last topogenic sequence; cytoplasmic if it is a stop-transfer sequence or luminal/extracellular if it is a type II signal/anchor sequence. The GPI-anchored topology is established for proteins having an N-terminal signal peptide and a C-terminal GPI-attachment signal (Fig. 3). ppC proteins are synthesized in the cytosol and then attached to the target membrane via interactions with the integral components of the membrane. For proteins anchored on the cytoplasmic side of the membrane via lipid linkers (myristate, palmitate and/or prenyl groups), they are synthesized and lipidated in the cytosol and then become associated with the membrane. Lipidation

is a specific modification and only proteins having proper modification motifs are lipidated. The prenyl group is added to the Cys residue of the C-terminal CAAX sequence with the proteolytic removal of the last three residues.

SIGNAL RECOGNITION PARTICLE (SRP) MEDIATES ER TARGETING

Detailed studies of ER targeting signals, targeting and translocation mechanisms are made possible by the reconstitution in vitro of ER targeting signal-dependent translocation across the ER membrane.[617,618,826,827] Among all the events associated with the exocytotic pathway, ER targeting/translocation is the first that was reconstituted in vitro. This system combines an in vitro translation system prepared from either rabbit reticulocyte lysate or wheat germ extract with purified dog pancreatic microsomal membranes. It reconstitutes faithfully ER targeting, membrane translocation, signal peptide cleavage, N-linked glycosylation and other associated events. Several important components of the ER targeting/translocation machinery have thus been identified and extensively studied using this in vitro system.

Three of the four types of ER targeting signals (with the exception of C-terminal tail-anchor) are recognized by an 11S cytosolic factor upon emerging from the ribosome. This cytosolic factor, referred to as the signal recognition particle (SRP), is a ribonucleoprotein complex (Fig. 4).[823,825] SRP binding also causes a specific arrest of polypeptide chain elongation.[824] The ribosome-nascent polypeptide-SRP complex is then targeted to the membrane of the rER via a heterodimeric integral protein complex (composed of SRα and SRβ subunits) referred to as the SRP receptor or docking protein.[484,763] The interaction with the SRP receptor also releases the chain elongation arrest of the nascent polypeptide and leads to the dissociation of SRP from this complex for subsequent use. SRP therefore cycles between the ER membrane and the cytosol.

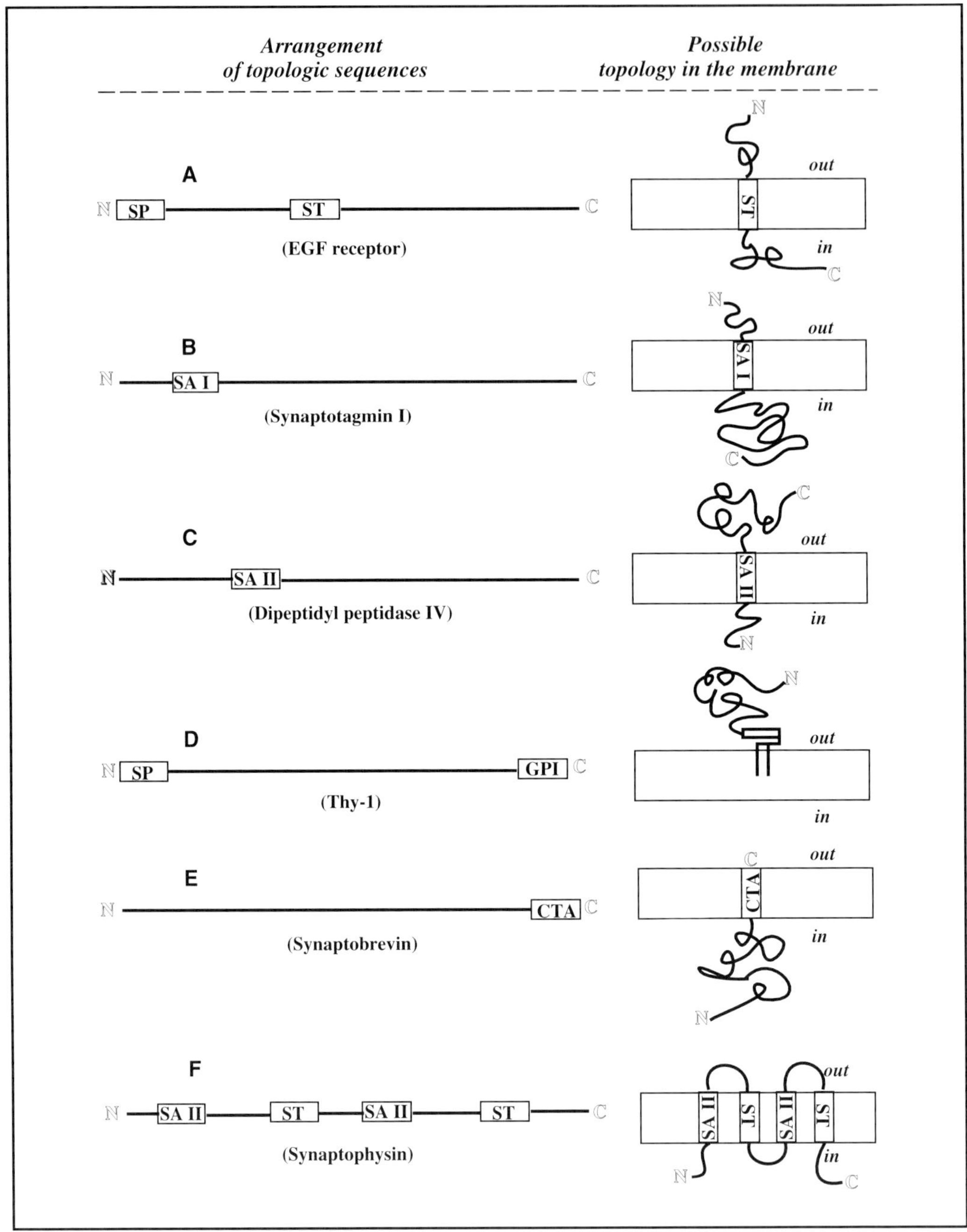

Fig. 3. Topogenesis of membrane topology. A type I topology will be generated for proteins with (A) an N-terminal signal peptide (SP) followed by a stop-transfer (ST) or with (B) a type I signal/anchor (SA I) sequence. C, Proteins with a type II signal/anchor (SA II) sequence will have a type II topology. D, GPI-anchored topology will be established for proteins possessing an N-terminal signal peptide followed by a C-terminal GPI-attachment signal (GPI). E, C-terminal tail-anchored topology is achieved for proteins with a C-terminal tail-anchor (CTA). F, type III topology will be generated for proteins with alternating type II signal/anchor sequences, stop-transfer sequences and other topogenic signals. The N-terminus will be cytoplasmic if the first topogenic signal is a type II signal/anchor but will be oriented to the extracellular/luminal side if the first topogenic signal is a signal peptide or a type I signal/anchor sequence. The C-terminus will be cytoplasmic if the last topogenic signal is a stop-transfer sequence or it will be luminal/extracellular if the last topogenic signal is a type II signal/anchor sequence.

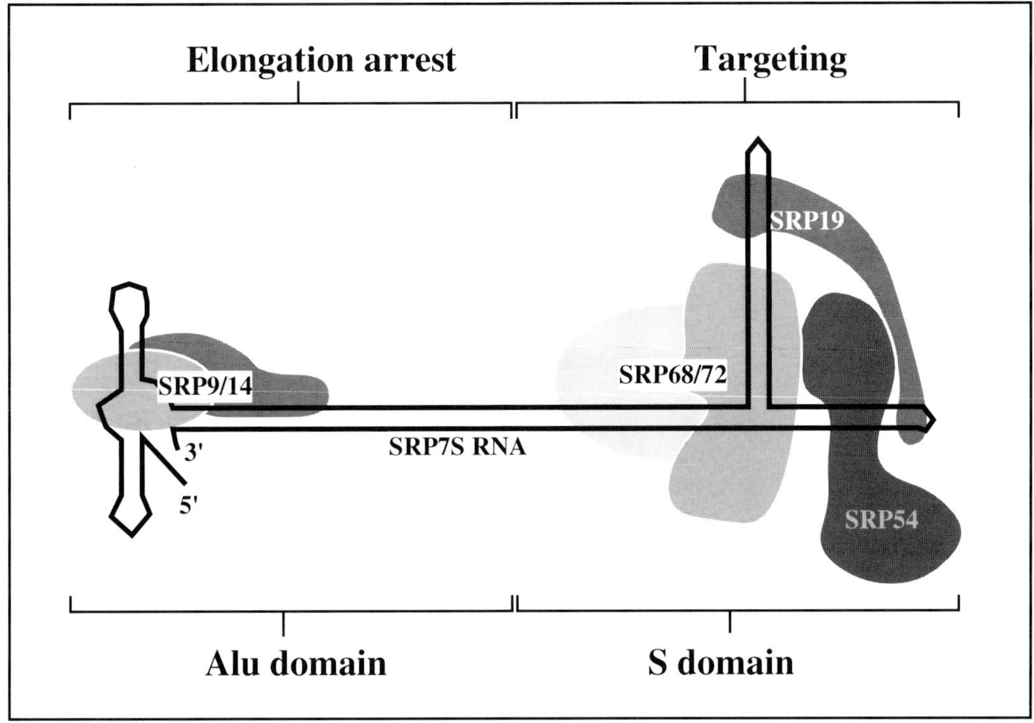

Fig. 4. The signal recognition particle is composed of six SRP proteins and an SRP 7S RNA that organizes into Alu and S domains. SRP9 and SRP14 bind to the Alu domain as a dimer and are involved in elongation arrest. SRP68 and SRP72 form a dimer and, together with SRP9, SRP19 and SRP54, bind to the S domain and are involved in recognizing ER targeting signals and targeting to the ER.

Purification of the SRP was achieved using an in vitro translation system of wheat germ extract which does not contain functional SRP activity, in combination with dog pancreatic microsomal membrane which has SRP associated on the cytoplasmic side of the membrane. ER targeting and translocation are faithfully reconstituted in this system. However, when the microsomal membrane is washed with high salt, which removes SRP from the membrane, this system is rendered incompetent for ER targeting and translocation. Supplementing the salt-extract restores the ER targeting and translocation. Biochemical fractionation of the salt-extract for restoring activity resulted in the purification of SRP. It is composed of six distinct polypeptides and an SRP7S RNA.[823,825,826,827] The six polypeptides have molecular masses of 9, 14, 19, 54, 68 and 72 kDa and are referred to as SRP9, SRP14, SRP19, SRP54, SRP68 and SRP72, respectively (Fig. 4).

The SRP7S RNA not only provides a structural framework for the binding of SRP proteins in the assembly of functional SRP but is also functionally involved in the association of SRP with the ribosome and the SRP receptor.[826,827] The binding regions of SRP7S RNA for the SRP proteins have been mapped. The 300-nucleotide mammalian SRP7S RNA can be structurally divided into two domains. A region of about a 100-nucleotides at the 5' end and another of about a 50-nucleotides from the 3' end are homologous to the highly repetitive Alu DNA family and form the Alu domain by incomplete base-pairing of the 3' end 50 nucleotides with nucleotides 47 to 97 at the 5' end, resulting in the formation of a contiguous domain. The central 150 nucleotide sequence forms another domain referred to as the S domain. SRP9 and SRP14 bind to SRP7S RNA as a stable heterodimer (SRP9/14) and this dimerization is essential for binding to the RNA. Individual subunits of SRP9 and 14 produced by in vitro translation do not bind the SRP7S RNA unless they are in the dimerized form. Similarly, SRP68 and SRP72 also form a stable heterodimer (SRP68/72) for binding to the SRP7S RNA. SRP68 can bind the RNA weakly in the absence of SRP72. Association of SRP72 with SRP68 significantly enhances the binding affinity for the RNA. In contrast, SRP19 and SRP54 bind to the RNA individually. Stable binding of SRP54 to the RNA requires the presence of prebound SRP19. Prebinding of SRP19 removes a destabilizing influence of a specific region in domain S on the SRP54 binding. SRP9/14 binds to the Alu domain of SRP7S RNA and is involved in elongation arrest of nascent polypeptides. SRP68/72, SRP19 and SRP54 bind to distinct regions of the S domain. SRP68/72 binding is essential for interaction with the SRP receptor on the ER membrane and is therefore important for ER targeting and subsequent translocation. SRP54 is a subunit of the SRP that interacts directly with ER targeting signals.

Although cDNA cloning has revealed the primary amino acid sequence for all the subunits of mammalian SRP, SRP54 has been the one whose primary sequence gives the most significant information on its structure and function and has been studied most extensively.[826,827] The SRP 54 sequence is evolutionally highly conserved. Mammalian SRP54 is composed of 504 residues and three structural domains have been defined.[826] The 100-residue sequence from the N-terminus is referred to as the NH_2-terminal domain (N-domain) followed by a GTP binding domain (G-domain) from residue 101 to residue 297. The C-terminal part (from residue 298 to the C-terminal residue 504) is termed the methionine-rich domain (M-domain). Residues 298-300 form a flexible hinge region that is protease-sensitive. This property of SRP54 has allowed

the functional dissection of the individual M-domain and the rest of the polypeptide (the N/G-domain). The G-domain of SRP54 is characterized by regions with short sequences that are conserved in most of the known GTPases and is most closely related to the GTP binding domain of the SRP receptor α-subunit (SRα). The M-domain is highly enriched in methionine residues (about 12%) and these residues are predicted to be located on one side of three potential α helices of the M-domain. This structure has been proposed to form a flexible hydrophobic pocket for recognizing the hydrophobic ER targeting signal. The flexible side chains of methionines may be particularly suited for accomodating ER targeting signals of different structures. Proteolytically-released M-domain possesses the binding capacity for SRP7S RNA, establishing that the M-domain is also responsible for SRP54 interaction with the RNA.[641,871] Furthermore, the M-domain, particularly, its C-terminal 6 kDa fragment has been shown to be responsible for recognizing ER targeting signals.[295,871] The G-domain is not required for SRP54 binding to either RNA or to the ER targeting signal. SRP54 is stabilized in a nucleotide-free (empty) state by the ER targeting signal in the ribosome-SRP-nascent polypeptide complex.[483] ER targeting of this complex does not require GTP. However, upon the interaction of SRP with the SRP receptor, the affinity of SRP54 for GTP is increased and its GTPase activated. The G-domain of SRP54 is therefore involved in events after targeting of the ribosome-SRP-nascent polypeptide complex to the ER. Nucleotide-mediated conformational change of SRP54 may regulate the release of the ER targeting signal and docking of the ribosome-nascent polypeptide complex at the ER membrane.

SRP-like components have been evolutionally conserved in both structure and function.[9,91,826,855] The yeast *S. cerevisiae* Sec65 gene was identified in genetic screens for genes essential for ER targeting and translocation. The derived amino acid sequence established that Sec65p is the yeast homologue of SRP19 (Sec65p/Srp19). The yeast ScR1 RNA is equivalent to SRP7S RNA and it exists in a 16S ribonucleoprotein complex containing Sec65p/Srp19 and yeast SRP54 (Srp54). Recent purification of this ribonucleoprotein complex has revealed five additional subunits of this complex.[91] Amino acid sequencing and gene cloning for four of these proteins established that yeast contains yet more homologues of mammalian SRP subunits. The newly identified Srp14p, Srp68p and Srp72p correspond to the mammalian SRP14, SRP68 and SRP72 subunits, respectively. One component that has not been molecularly-cloned has a size of 7 kDa and may correspond to the mammalian SRP9. One cloned component (Srp21) does not show homology to any of the mammalian SRP subunits and may be a unique component of the yeast SRP. Among the yeast SRP proteins, Srp54 is most identical (47% sequence identity) to the mammalian counterpart. Therefore, SRP of the *S. cerevisiae* contains one ScR1 RNA and six distinct protein subunits, five of which correspond to mammalian SRP subunits. Mutational studies have established that ScR1, Srp14p, Srp19/Sec65p, Srp21p, Srp45p, Srp68p and Srp72p are all required for the formation of functional SRP. SRP-like components have also been identified in bacteria. The bacterial 4.5S-RNA and a 48 kDa protein named FfH are structurally related to SRP7S RNA and SRP54, respectively. The sequence similarity (31% amino acid identity) between SRP54 and FfH was discovered when the primary sequence of SRP54 was derived from the cloned cDNA sequence. It is this sequence similarity that has allowed the cloning of yeast SRP54 gene and initiated the structural and functional characterization of the yeast SRP. The 4.5S-RNA and FfH also form a ribonucleoprotein complex that specifically recognizes nascent polypeptides containing a targeting signal for the translocation machinery of the bacterial membrane. The existence of structurally and functionally related SRP-like complex in mammals, yeast and bacteria highlights the importance of SRP or SRP-like complexes in membrane targeting and translocation.[855] Components of SRP have also been identified in other organisms.

A cytoplasmic heterodimeric protein complex has been recently purified and named the nascent-polypeptide associated complex (NAC).[842] NAC binds nascent polypeptides emerging from the ribosomes unless an ER targeting signal is fully exposed, upon which SRP displaces NAC and binds to the targeting signal. It has been speculated that the physiological function for NAC is to prevent nonspecific binding of SRP to nascent polypeptides that do not contain ER targeting signals. This prevents nonspecific ER targeting and translocation of polypeptides that do not contain ER targeting signals. Therefore, NAC may assure the fidelity of ER targeting and subsequent translocation.

SRP RECEPTOR MEDIATES ER DOCKING OF RIBOSOME-SRP-NASCENT POLYPEPTIDE COMPLEX

The SRP receptor defines specific docking sites on the ER membrane for the ribosome-SRP-nascent polypeptide targeting complex.[617,618] It was purified by two independent biochemical approaches. The first approach was based on the observation that protease (elastase)-treated dog pancreatic microsomal membranes is rendered inactive in ER targeting and translocation. More importantly, ER targeting and translocation activities could be restored by adding back the protease-treated extract. Biochemical fractionations of the protease-treated extract led to the purification of a 52 kDa protein fragment that restored the ER targeting and translocation activity of the protease-treated microsomal membrane. This 52 kDa polypeptide is a proteolytic fragment of the α-subunit of the SRP receptor (SRα). Independently, SRα was purified based on the observation that binding of SRP to the nascent polypeptide-ribosome complex causes a chain elongation arrest which is released upon targeting of the ribosome-SRP-nascent polypeptide complex to the ER membrane. The putative elongation-arrest-releasing activity is assumed to be a protein component of the ER membrane. Biochemical fractionation of a detergent extract of the ER membrane resulted in the purification of a 69 kDa protein that possesses this elongation-arrest-releasing activity. This protein was established to be identical to the SRα identified by the first approach. The SRP receptor is now known to be a heterodimeric protein complex, containing another subunit (SRβ) of about 30 kDa.[484,763] The primary sequence derived from cDNA sequencing for SRα contains 638 amino acids.[392] Comparing the N-terminal amino acid sequence of the purified 52 kDa proteolytic fragment established that it begins with residue 152. Although two regions (residues 1-22 and 64-79) are relatively hydrophobic, the presence of positively charged residues (Lys 10 for the first region and Lys 72 for the second region) in the middle of these two regions renders them unlikely to be transmembrane domains. Furthermore, unlike integral membrane proteins, SRα is extractable by chaotropic agents. These suggest that SRα is a peripheral membrane protein on the cytoplasmic side of the ER membrane (ppC orientation). The cloning of mouse SRβ established that it is a type I integral membrane protein of 269 amino acids.[484] The insertion of SRβ is mediated by a type I signal/anchor sequence, the hydrophobic core of which is composed of residues 33 to 52 and functions also to anchor the protein in the ER membrane. The N-terminal 33-residue sequence preceding this hydrophobic core is luminally-oriented, while the C-terminal 217-residue sequence (residues 53-269) is cytoplasmically oriented. Database search using the mouse SRβ sequence has identified a yeast protein that is 23% identical and also has the same membrane topology, indicating that it is a yeast counterpart of the mammalian

SRβ.[484] Association of SRα with the membrane results mainly from its dimerization with SRβ. Since the purified 52 kDa fragment of SRα could restore the ER targeting and translocation activities of protease-treated microsomal membranes, this fragment may contain sufficient structure for association with the SRβ.

After it was discovered that the post-targeting events are dependent on GTP, re-examination of the SRα amino acid sequence identified a GTP binding domain at its C-terminal portion.[139,826] This GTP binding region is most closely related to that of SRP54 and this sequence similarity extends through the N-domain of SRP54. The cytoplasmic domain of SRβ also contains a GTP binding domain that is not closely related to those of SRP54 and SRα.[484] The presence of three distinct but directly-interacting GTP binding proteins in the ER targeting and translocation system complicated studies aimed at understanding how these three GTP binding proteins were involved in posttargeting events. As discussed in the previous section, SRP54 is in an empty (nucleotide-free) state when its M-domain is occupied by an ER targeting signal of the nascent polypeptide. Its GTP binding and GTPase activity are then turned on upon association with the SRP receptor. GTP binding and the GTPase activity of SRP54 may play a role in posttargeting events, such as the dissociation of SRP from the ER targeting signal, proper docking of the ribosome-nascent polypeptide complex onto the translocation machinery of the ER membrane and/or dissociation of SRP from the SRP receptor. Although both subunits of the SRP receptor are also GTP binding proteins, their exact role and regulation of GTP binding and GTPase activities in posttargeting events have been a puzzle. By supplementing wild-type and several point-mutated SRα subunits individually to the SRα-depleted microsomal membranes, it was firmly established that the GTP binding domain of SRα is essential for polypeptide translocation. More defined studies revealed that GTP binding to the SRα is necessary for the dissociation of SRP54 from the ER targeting signal as well as for the insertion of the nascent polypeptide into the translocation site.[138,139,140,615,616] GTP binding and GTPase activities of SRα may therefore proceed in parallel with those of SRP54 in these posttargeting events. Whether the GTP binding domain of SRβ is essential for any of these posttargeting events has not been reported. It seems possible that these three GTP binding proteins may coordinate with each other to regulate the fidelity and precise order of these posttargeting events so that the nascent polypeptide will be faithfully translocated and the SRP will be dissociated in a timely manner for subsequent cycles of targeting. The SRP receptor thus plays a fundamental role in the ER targeting of ribosome-SRP-nascent polypeptide complex, in the dissociation of SRP from the ER targeting signal and the SRP receptor and in initiating the docking of ribosome-nascent polypeptide on the translocation machinery, leading to the formation of the ribosome-nascent polypeptide-translocation complex.

The derived amino acid sequence of SRP54 not only suggests that FfH may be the bacterial homologue of SRP54 but also reveals a striking sequence similarity between SRα and the bacterial FtsY. This has led to the proposal that FtsY could play an SRP receptor-like function in bacteria. Evidence for an SRα-like role for FtsY has been recently established. In much the same way as SRP binds to the SRP receptor, FfH/4.5S ribonucleoprotein complex binds tightly to FtsY in a GTP-dependent manner.[482,855] More significantly, nucleotide triphosphate hydrolysis by FfH and FtsY occurs in reciprocally coupled reactions in which two interacting GTPases act as regulatory proteins for each other. This interaction results in GTP hydrolysis that can be inhibited by synthetic targeting signals.[603a] SRα-like proteins may also exist in yeast, implying that the docking mechanism is also evolutionally conserved.

TRANSMEMBRANE PROTEINS MEDIATE TRANSLOCATION ACROSS THE ER MEMBRANE

Upon docking of the ribosome-SRP-nascent polypeptide complex on the ER membrane, SRP is dissociated from the complex. The ribosome-nascent polypeptide complex is then positioned with the translocation machinery of the ER membrane and the ER targeting signal of the nascent polypeptide is inserted into the translocation machinery. This process may involve a specific interaction of the ER targeting signal with a component(s) of the translocation machinery, leading to the assembly of active translocation machinery. The translocation is believed to be mediated by an aqueous protein-conducting channel,[232,479,617,618] which is supported by recent electrophysiological data.[148] Although the SRP and SRP receptor were identified and purified around 1980, the key protein components of the translocation machinery have been established only recently. This was achieved by the combination of molecular identification of several candidates for the translocation machinery and the development of a protein translocation system using proteoliposomes reconstituted from purified components. Genetic screens have led to the identification of a number of yeast genes that are involved in signal-mediated ER targeting and translocation. Three of these genes (Sec61, Sec62 and Sec63) are integral membrane proteins of the ER and are potential candidates for the protein-conducting channel of the translocation machinery.[826] The derived amino acid sequence of Sec61p is related to bacterial SecYp, an essential component of the bacterial translocation apparatus. Biochemical approaches have identified several mammalian proteins in the vicinity of the translocating polypeptide, including the signal peptidase, the oligosaccharide transferase, the translocon-associated protein

(TRAP) or signal sequence receptor (SSR) complex, the translocating chain associating membrane (TRAM) protein and the mammalian homologues of the yeast Sec61p (mSec61p). Only TRAM protein and Sec61p have multiple membrane spanning domains expected for a protein-conducting channel.

The TRAM protein was identified by a crosslinking approach to be a glycoprotein that is in close proximity to nascent chains of secretory proteins early in translocation.[240,296] Its derived 376 amino acid sequence revealed that TRAM protein has multiple (most likely 8) membrane spanning domains, some of which contain a number of hydrophilic and charged residues, suggesting that the TRAM protein has the potential to form a hydrophilic environment in the ER membrane required for polypeptide translocation. The C-terminal sequence of about 60 amino acids is exposed to the cytosol and this tail contains an ER retrieval signal ($KXKXX_{COOH}$) (see chapter 4). The TRAM protein in the ER membrane is as abundant as membrane-bound ribosomes, indicating that it may be present in each translocation machinery bound by the ribosome.

The yeast Sec61p was biochemically determined to be the major protein crosslinked to the translocating polypeptide at a late stage.[504] Furthermore, Sec61p is tightly associated with ribosomes on the ER membrane, suggesting that it may be the ribosome receptor. These observations led to a search for a similar component in the mammalian ER membrane.[241] A 40 kDa nonglycoprotein has thus been identified and purified. The N-terminal amino acid sequence was used to isolate the cDNA of this protein. The derived 476-residue protein sequence has 56% identity with the yeast Sec61p. Like the yeast Sec61p, the mSec61p is located in the immediate vicinity of nascent translocating polypeptides.[241,294,296] Furthermore, it is also tightly associated with membrane-bound ribosomes.[344] The derived mSec61p sequence also predicts 10 transmembrane domains with internal hydrophilic residues

that may form a hydrophilic environment in the ER membrane. Similar to the yeast Sec61p, mSec61p also bears significant similarity in its amino acid sequence as well as transmembrane topology to SecYp of bacteria. The structural similarity of mSec61p with yeast Sec61p and bacterial SecYp indicates that mSec61p may be important for protein translocation in the ER. The tight association of Sec61p with membrane-bound ribosomes also indicates that Sec61p may function as a ribosome receptor. In this way, ribosomes may contact directly with the protein-conducting channel and the nascent polypeptide passes directly from the ribosome into the Sec61p-containing translocation machinery. Biochemical purification of mSec61p under nondenaturing conditions revealed that mSec61p is a component of a protein complex containing two additional subunits with sizes of about 14 and 8 kDa, respectively.[242,273] The 40, 14 and 8 kDa subunits have been referred to as the mSec61pα, mSec61pβ and mSec61pγ, respectively. mSec61pγ is the mammalian homologue of the yeast SSS1 gene product,[205,273] a suppresser of yeast sec61 mutant. SSS1 is an essential yeast gene. Depletion of Sss1p in yeast led to a block in polypeptide translocation, suggesting that Sss1p plays an essential role in translocation. Furthermore, mSec61pγ and Sss1p are both structurally related to bacterial SecEp, another essential component of the bacterial translocation machinery. The Sec61pβ is encoded by SBH1 gene. Furthermore, Sbh1p/Sec61pβ is structurally related to mSec61pβ.[273] These results strengthen the speculation that the mSec61p complex is a key component of the protein-conducting channel in the ER membrane.[460] Recent biochemical data have established that mSec61p is indeed the receptor for translating ribosomes on the ER membrane.[344,464]

Identification of the key components of the ER protein-conducting channel was made possible by demonstrating that translocation-competent vesicles could be reconstituted from an unfractionated detergent extract of dog pancreatic microsomal membranes.[481,525] This observation has led to the development of translocation-competent proteoliposomes reconstituted from purified components.[242] This system has confirmed that the SRP receptor is necessary and sufficient for the signal-mediated ER membrane targeting of ribosome-SRP-nascent polypeptide complex but lacks translocation activity. The mSec61p complex is the key component of the protein-conducting channel required for all SRP-mediated translocations.[296] The TRAM protein may be another component of the ER translocation apparatus required only for some translocations. When crosslinking experiments were performed, the TRAM protein seemed to interact with the amino acid residues preceding the hydrophobic core of the ER targeting signal whereas the Sec61pα seemed to interact primarily with the hydrophobic core as well as residues succeeding it. Furthermore, the TRAM protein could be crosslinked during the early stage of translocation, while Sec61pα could be crosslinked to the translocating polypeptide at all stages. The TRAM protein may play a role in mediating the transfer of the ER targeting sequence of the nascent polypeptide from the SRP to the mSec61p complex. This mediating effect of the TRAM protein could be bypassed for strong ER targeting signals so that translocation of some proteins could be achieved in reconstituted proteoliposomes containing only the SRP receptor and the mSec61p complex. For weak signals, the mediating effect of the TRAM protein may be necessary so that their translocation in the reconstituted proteoliposomes requires the TRAM protein in addition to the SRP receptor and the mSec61p. The TRAM protein may be involved in all translocations in vivo to increase the specificity and efficiency of translocation across the ER membrane, which has many other ongoing protein-protein interactions that are not related to polypeptide translocation.

ACCESSORY COMPLEXES OF THE CORE PROTEIN-CONDUCTING CHANNEL

In addition to the mSec61p complex and the TRAM protein, the holotranslocation machinery is expected to contain other proteins or protein complexes. This is suggested by the observation that other ER integral membrane proteins are crosslinked to the translocating polypeptide and the fact that the cleavage of the signal peptide, N-linked glycosylation and folding of the translocated portion of the polypeptide are essentially cotranslocational events. Proteins involved in these events should be in the vicinity of the core-protein conducting channel composed of the mSec61p complex and the TRAM protein as well as the ribosome bound to the mSec61p complex. Therefore, the holotranslocation machinery may also contain the translocon-associated protein (TRAP) or signal sequence receptor (SSR) complex, the signal peptidase complex, the oligosaccharyl transferase complex and luminal ER proteins, such as Bip and protein disulfide isomerase involved in polypeptide folding.

The TRAP complex is composed of four integral membrane proteins (α, β, γ and δ subunits) with molecular weights of about 34, 22, 20 and 18 kDa, respectively.[212] The α and β subunits have a type I membrane orientation, while the δ subunit has a type II membrane orientation. The γ subunit probably spans the ER membrane four times with both N- and C-termini in the lumen of the ER. The α subunit was originally identified by its property of being crosslinked with the nascent translocating polypeptide and was named signal sequence receptor (SSR). It was later demonstrated that SSR is not required for polypeptide translocation because proteoliposomes reconstituted from an ER membrane detergent extract immunodepleted of SSR showed unimpaired translocation activity. SSR was renamed TRAP and additional subunits were identified. Although not essential for translocation, other evidence also suggest that the TRAP complex may be in the vicinity of mSec61p complex, in addition to the fact that TRAPα can be crosslinked to the translocating polypeptide. TRAPα (and probably the entire complex) is segregated to the rER and could be crosslinked with membrane-bound ribosomes. The association of TRAP with ribosomes has also been detected after solubilization of the rER membrane with detergent. Furthermore, the TRAP complex is abundant in the rER and antibodies against TRAPα could inhibit protein translocation, most likely via a sterical effect on the core-translocation apparatus. The exact role of TRAP has not been established and it could play an unknown function associated with translocation.

Since the signal peptide is cotranslationally cleaved, the signal peptidase must be located in the vicinity of the core protein-conducting channel. The signal peptidase complex purified from a detergent extract of dog pancreatic microsomal membranes consists of five subunits of 25, 22/23, 21, 18 and 12 kDa, referred to as SPC25, 22/23, 21, 18 and 12, respectively.[250] The purified yeast signal peptidase complex has four subunits of 25, 20, 18 and 13 kDa, referred to as ySPC25, 20, 18 and 13, respectively.[859] ySPC18 was found to be the product of SEC11 gene, which was identified in genetic screens for genes involved in the yeast secretory pathway. SEC11 is necessary for signal peptidase activity and therefore Sec11p/ySPC18 is an essential component of the yeast signal peptidase complex.[71,859] The canine SPC21 and 18 sequences are 81% identical to each other and are both about 50% identical to the Sec11p/ySPC18 sequence,[697] suggesting that canine SPC21 and 18 are both mammalian homologues of Sec11p/ySPC18. Three subunits (SPC18, SPC21, SPC22/23) of the signal peptidase complex have been shown to have a type II membrane orientation with the bulk of the polypeptide oriented in the ER lumen.[698] Since the signal peptidase complex is functionally associated with polypeptide translocation and in the vicinity of the core protein-conducting channel, it may be reasonable to consider the

signal peptidase complex as another part of the holotranslocation machinery.

Another protein complex that can be considered as part of the holotranslocation machinery is the oligosaccharyl transferase complex. It is responsible for transferring the oligosaccharyl moiety from a dolichol intermediate to Asn residues (in the sequence context of Asn-X-Ser/Thr, where X is any residue except for Pro) in the luminal portion of the nascent translocating polypeptide. The oligosaccharyl transferase complex purified from dog pancreatic microsomal membranes consists of three subunits of 66, 63 and 48 kDa, respectively.[353] The 66 and 63 kDa subunits were found to be ribophorins I and II, respectively, two previously characterized abundant ER integral membrane proteins of unknown function. The purified yeast oligosaccharyltransferase complex has six subunits of about 60-64, 45, 34, 30, 16 and 9 kDa, referred to as α, β, γ, δ, ε and ζ-subunit, respectively.[352] The 45 kDa β-subunit and 30 kDa δ-subunit correspond to products of Wbp1 and Swp1 genes, respectively. The Wbp1p and Swp1p are essential for Asn-linked glycosylation in vivo. The 48 kDa subunit of canine oligosaccharyl transferase complex is structurally related (about 25% sequence identity) to the Wbp1p (the yeast 45 kDa β-subunit).[708] The sequence similarity of the Swp1p (the yeast 30 kDa δ-subunit) with the C-terminal half of ribophorin II suggests that they are related gene products. Cloning the gene (OST1) encoding the yeast α-subunit established that the protein is about 28% identical to human ribophorin I, providing further evidence to the similarity of oligosaccharyl transferase complex in mammals and the yeast.[707] The 16 and 34 kDa subunits are encoded by OST2 and OST3 genes, respectively.[347,707a] Ribophorin I, II and the 48 kDa subunit are all type I integral membrane proteins. The oligosaccharyl transferase complex is tightly associated with membrane-bound ribosomes even after membrane solubilization by detergent. Although ribophorins are not required for translocation in reconstituted proteoliposomes, antibodies against ribophorins inhibit protein translocation. This is supportive of the proposal that the oligosaccharyl transferase complex is in close proximity to the core-translocation apparatus as indicated by the cotranslocational N-linked glycosylation of the nascent translocating polypeptide.[375,528]

The oligosaccharyl transferase complex, the signal peptidase complex, the TRAP complex and some other unidentified components may interact and/or coordinate with the core-protein conducting channel composed of mSec61p complex and the TRAP protein to form a highly organized structure (the holotranslocation machinery) that will mediate translocation and translocation-associated events. The major components involved in ER targeting and translocation are summarized in Table II and the membrane topologies for SRP receptor, TRAM and mSec61p complex are shown in Figure 5.

A Model for ER Targeting and Translocation

A model for the ER targeting and translocation is described below.[617,618,826] Stage 1 (SRP binding): the short nascent polypeptide emerging from the ribosome is bound by the NAC complex. Upon full appearance of the ER targeting signal from the ribosome, the SRP binds to the targeting signal via the M-domain of SRP54, leading to the dissociation of GDP from the G-domain of SRP54 and stabilization of SRP54 in a nucleotide-free state. The SRP9/14 associated with the Alu domain of SRP7S RNA causes a chain elongation arrest via some kind of interaction with the ribosome. Stage 2 (ER targeting): the elongation-arrested ribosome-nascent polypeptide-SRP complex is then targeted to the rER membrane through an interaction of SRP68/72 (associated with the S-domain of SRP7S RNA) with the SRP receptor. Stage 3 (GTP loading): the interaction of SRP with the SRP receptor leads to GTP loading of the G-domain

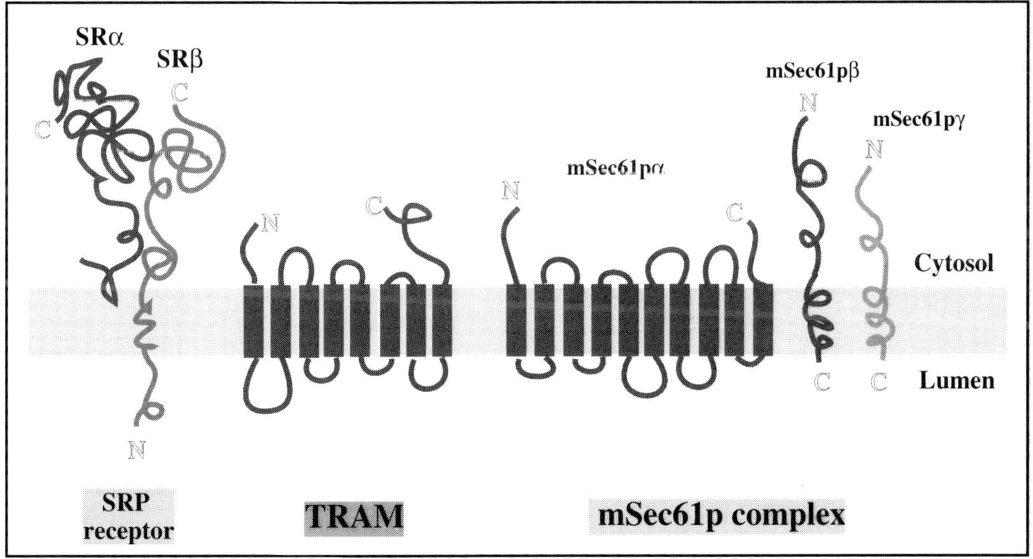

Fig. 5. Topologies for proteins involved in ER targeting and translocation. The β-subunit of the SRP receptor (SRβ) is a type I protein generated by its type I signal/anchor sequence. The α-subunit (SRα) is a peripheral protein (ppC topology) that is associated with the ER membrane via its interaction with SRβ. The TRAM protein and mSec61pα are type III membrane proteins with both their N- and C-termini oriented on the cytoplasmic side of the membrane. mSec61pβ and mSec61γ are both C-terminal tail-anchored proteins. mSec61pα, mSec61pβ and mSec61γ form the mSec61p complex, which is the key component of the core protein-conducting channel.

Table II. Components involved in ER targeting and translocation

Name	Composition	Size
SRP	SRP7S RNA	300 nt
	SRP9	9 kDa
	SRP14	14 kDa
	SRP19	19 kDa
	SRP54	54 kDa
	SRP68	68 kDa
	SRP72	72 kDa
SRP receptor	SRα	69 kDa
	SRβ	30 kDa
mSec61p complex	mSec61α	40 kDa
	mSes61β	14 kDa
	mSec61γ	8 kDa
TRAM	TRAM	40 kDa
Signal peptidase complex*	SPC12	12 kDa
	SPC18	18 kDa
	SPC21	21 kDa
	SPC22/23	22-23 kDa
	SPC25	25 kDa
Oligosaccharyl transferase complex*	ribophorin I	66 kDa
	ribophorin II	63 kDa
	OST48	48 kDa
TRAP complex*	TRAPα	34 kDa
	TRAPβ	22 kDa
	TRAPγ	20 kDa
	TRAPδ	18 kDa

TRAM: translocating chain-associating membrane protein
TRAP: translocon-associated protein
* : accessory complexes

Table II. Summary of proteins involved in ER targeting, translocation and events associated with translocation

of SRP54 and probably nucleotide exchange of both SRα and SRβ of the SRP receptor from the GDP- to the GTP-bound forms. Stage 4 (Transfer of the ER targeting signal): the GTP loading of SRP54 and probably nucleotide exchange of one or both subunits of the SRP receptor result in the dissociation of the ER targeting signal from the SRP and the transfer of the ER targeting signal to a translocation component such as the TRAM protein and/or mSec61p complex. During this transfer, SRP is dissociated from the ribosome-nascent polypeptide and becomes bound to the SRP receptor. GTP hydrolysis by SRP54 results in the dissociation of SRP from the SRP receptor and its return to the cytosol for subsequent cycles of use. Whether GTP hydrolysis of either or both subunits of the SRP receptor is involved in the SRP dissociation is unknown. Stage 5 (Assembly of holotranslocation machinery): the transfer of the ER targeting signal and the conformational change of either or both subunits of the SRP receptor associated with their nucleotide exchange and/or GTP hydrolysis triggers the functional docking of the ribosome-nascent polypeptide complex onto the mSec61p complex and recruitment of accessory complexes, leading to the assembly of the holotranslocation machinery. The chain elongation arrest is released and the nascent polypeptide is transferred across the ER membrane via the core-protein conducting channel composed of mSec61p complex. The accessory complexes, including the ER luminal chaperones of the holotranslocation machinery mediate other events associated with the translocation, such as cleavage of the signal peptide, N-linked glycosylation, polypeptide folding and dissociation of the stop-transfer sequence from the protein-conducting channel and subsequent association with the lipid bilayer.

Posttranslational Translocation in Yeast

Posttranslational translocation also occurs in yeast. Since ribosomes are no longer associated with the translocating polypeptide, posttranslational translocation may involve additional components that may fulfill the functions conferred by ribosomes. Significant progress was made by reconstituting this mode of translocation using purified components. Genetic screens have identified other genes in addition to the Sec61p complex involved in ER translocation, including Sec62p, Sec63p, Sec71p and Sec72p.[89,161,162,206,826] In detergent extracts from yeast microsomes, a protein complex composed of Sec61p, Sbh1p, Sss1p, Sec62p, Sec63p, Sec71p and Sec72p has been purified.[161,565] This 7-protein complex, when reconstituted in proteoliposomes, can posttranslationally translocate prepro-α factor.[565] The translocation of prepro-α factor is significantly stimulated by ATP and yeast Bip (Kar2p). Posttranslational translocation of pro-OmpA, a bacterial protein known to be translocated posttranslationally into yeast microsomes, into the proteoliposome is absolutely dependent on Kar2p and ATP. Kar2p has been shown to interact with the luminal domain of Sec63p,[89] suggesting that Kar2p interacts with both the translocating polypeptide as well as this 7-protein complex. Posttranslational translocation in yeast microsomes therefore requires this 7-protein complex and luminal Kar2p in an ATP-dependent manner. Furthermore, the Sec61p complex also exists in a different form that is tightly associated with ribosomes and does not contain Sec62p, Sec63p, Sec71p and Sec72p, suggesting that this form is involved in cotranslational translocation in yeast and cotranslational translocation may be independent of Sec62p, Sec63p, Sec71p and Sec72p. Consistent with this, the 7-protein complex could be dissociated into the Sec61p complex as well as another complex composed of Sec62p, Sec63p, Sec71p and Sec72p.

POSTTRANSLATIONAL INSERTION OF C-TERMINAL TAIL-ANCHORED PROTEINS

As mentioned earlier, rER targeting mediated by the C-terminal tail-anchor of a novel class of proteins is not dependent on the SRP pathway described for the other three types of ER targeting signals. The discovery of a novel class of proteins that are anchored by a C-terminal hydrophobic tail, as well as the demonstration of their importance in mediating specific vesicular transport, have led to recent studies on the pathway of their membrane insertion and intracellular transport. The best studied protein of this class is synaptobrevin which is found on synaptic vesicles and plays a role in vesicle docking and fusion with the presynaptic membrane.[385] It was shown that synaptobrevin is initially targeted to the ER and then transported via the Golgi apparatus to the target membrane. In contrast to cotranslational ER targeting mediated by the SRP-dependent signals, SRP-independent ER targeting of synaptobrevin occurs posttranslationally in an ATP-dependent manner. Proteoliposomes reconstituted from an unfractionated detergent extract of dog pancreatic microsomes have a similar activity as does the intact microsomal membranes. When the SRP receptor or mSec61p complex was immunodepleted from the reconstituted proteoliposomes, the ER insertion of synaptobrevin was not affected. Furthermore, proteoliposomes reconstituted from purified SRP receptor, the TRAM protein and the mSec61p complex were incompetent for synaptobrevin insertion, although they were fully active for SRP-dependent translocation. Microsomal membranes pretreated with protease are rendered inactive for synaptobrevin insertion. The posttranslational ER targeting of synaptobrevin is therefore not dependent on SRP, SRP receptor, the TRAM protein or the mSec61p complex. Rather, it requires other ER membrane proteins. The components and the mechanism responsible for posttranslational ER targeting of C-terminal tail-anchored proteins remain to be established.

N-LINKED GLYCOSYLATION AND ITS SUBSEQUENT MODIFICATIONS

Proteins that are targeted to the ER are usually glycosylated in the region exposed to the lumen if there are proper acceptor sites for the modification.[204,375] N-linked glycosylation is well studied (Fig. 6), and the carbohydrate chain is attached to the Asn residue if it is in the context of -Asn-X-Ser/Thr- (where X cannot be proline) and accessible to the glycosylation machinery localized in the lumen of the ER. The first step in this process involves the transfer of a large oligosaccharide from a dolichol pyrophosphate oligosaccharide precursor to the Asn residue. This oligosaccharide precursor is synthesized on the ER membrane through many steps of reactions that occur both outside (early steps) and in the lumen (later steps) of the ER. The transferred sugar chain has three glucose (Glc), nine mannose (Man) and two N-acetylglucosamine (GlcNac) sugar residues in the configuration of Glc3Man9GlcNac2-asn (Fig. 6).

When the proteins are transported along the exocytotic pathway, their oligosaccharide chains are modified sequentially by enzymes present throughout the pathway. The terminal three glucose residues are removed by glucosidase I and II in the ER. An ER mannosidase removes one of the nine mannose residues. When transported to the Golgi apparatus, the Golgi mannosidase I in the cis Golgi compartment removes three more mannose residues, resulting in a sugar chain (Man5GlcNac2-) that is available for the addition of an N-acetylglucosamine residue by the N-acetylglucosaminyl transferase I in the medial Golgi compartment, resulting in the structure GlcNac1Man5GlcNac2-. The Golgi mannosidase II then removes two more mannose residues, resulting in a structure (GlcNac1Man3GlcNac2-) to which one

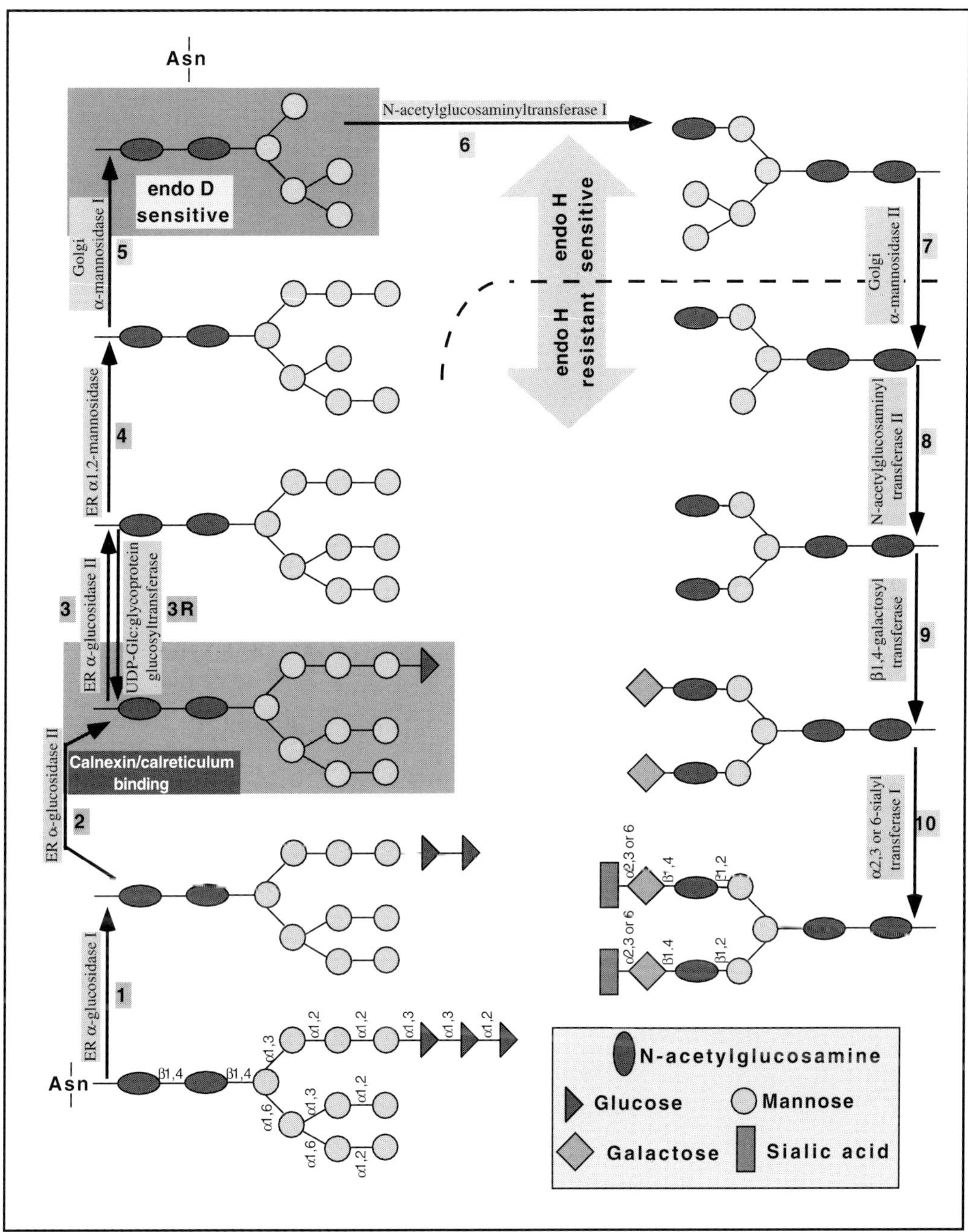

Fig. 6. Biogenesis of the N-linked glycan. Steps 1-4 occur in the ER, while step 5 is in the cis Golgi. Steps 6-7 occur in the medial Golgi of CHO, NRK and other cells, but also occur in the cis Golgi of NRK cells. Steps 8-10 occur in medial, trans Golgi and the TGN. The glycan structure generated by step 2 and step 3R is recognized by calnexin and calreticulin. The glycan structure generated by step 5 is the only one that is sensitive to endoglycosidase D. All glycan structures before step 7 are sensitive to endoglycosidase H while that generated by step 7 and beyond are resistant to endoglycosidase H.

more GlcNac residue is added (resulting in the GlcNac2Man3GlcNac2- structure). The GlcNac1Man3GlcNac2- structure and its subsequent derivatives are resistant to endoglycosidase H (endo H), while all preceeding structures are sensitive to endo H. The concerted actions of N-acetyl-glucosaminyl transferase I and Golgi mannosidase II is thus a key step to confer endo H resistance, a property that has been used widely to measure transport to the medial Golgi cisternae. However, recent studies suggest that endo H resistant glycans can be generated in the cis Golgi of NRK cells. The GlcNac2Man3GlcNac2- structure is the core structure for all N-linked glycans. Its subsequent modification by a variety of sugar transferases such as other types of N-acetylglucosaminyl transferases, various types of galactosyl-transferases, sialyltransferases and fucosyl-transferases will result in a variety of glycans with diverse structures. These further modifications occur in the trans side of the Golgi apparatus. In particular, galactose residues are added at the trans cisternae, while sialic acids are added to the galactose residues in the trans cisternae and TGN. The sequential and highly organized modification of N-linked sugar chains has been used extensively to monitor the transport of proteins along the exocytotic pathway. Among all the glycan intermediates and final products, the Man5GlcNac2- structure derived after modification by Golgi mannosidase I is the only species that is sensitive to the endoglycosidase D (endo D). Since this structure is efficiently modified by the Golgi N-acetylglucosaminyl transferase I to give rise to the GlcNac1Man3GlcNac2- structure, endo D sensitive glycans are usually undetectable unless in cells that are deficient in Golgi N-acetylglucosaminyl transferase I activity (such as CHO 15B cells). Since lectins are capable of recognizing specific structures, they are also useful for monitoring the status of glycan modifications as a measure of transport along the exocytotic pathway. Erythrina cristagalli (ECA) recognizes Gal-β1,

4-GlcNac of N-linked glycans and can be used to measure the modification by trans Golgi β1,4-galactosyltransferase. ECA binding is lost upon further addition of sialic acid to the terminal Gal. Sambucus nigra (SNA) and Maackia amurensis (MAA) recognize sialic acids α2,6- and α2,3-linked to Gal, respectively. They are useful in measuring the extent of modifications by α2,6- and α2,3-sialyltransferase, respectively. Wheat germ agglutinin (WGA) recognizes terminal sialic acid and N-acetylglucosamine while succinylated WGA (sWGA) binds only N-acetylglucosamine. The combined use of WGA and sWGA is also very useful for monitoring the glycan modifications.

PROTEIN FOLDING AND ER QUALITY CONTROL

Polypeptide folding starts when the nascent polypeptide is being translocated and continues posttranslationally. Furthermore, most proteins are assembled into either homo- or hetero-oligomers before being exported from the ER. The ER thus plays a fundamental role in polypeptide folding and oligomer assembly.[231,284,317,644] These processes are intimately dependent on the luminal microenvironment and assisted by a set of ER proteins that function as molecular chaperones and/or folding enzymes. The proper folding is also dependent on the cleavage of the signal peptide and N-linked glycosylation, because proper folding for some proteins is prevented when their signal peptides are not cleaved or when N-linked glycosylation is inhibited. The signal peptidase complex and oligosaccharyl transferase complex are therefore functionally associated with protein folding/assembly.

The luminal environment of the ER has a relatively high concentration of Ca^{2+} (in the millimolar range). Polypeptide folding and oligomer assembly of some proteins depend on this Ca^{2+} concentration because Ca^{2+} ionophores such as A23187 can inhibit the folding pathway of some proteins.[386,429] Similarly, the ER Ca^{2+} pump blocker thapsigargin also inhibits the folding process of some proteins.[782] The lumen of the ER is also about 10 times more oxidized than the cytosol, mainly due to the higher ratio (1/10) of oxidized to reduced glutathione in the ER as compared to the cytosol (1/100).[320] This relatively oxidized ER environment facilitates disulfide bond formation, which occurs at higher frequencies for proteins destined for the exocytotic pathway as compared to cytosolic proteins.[82] The ER redox states can be modulated by adding either reducing agents, such as 2-mercaptoethanol or dithiothreitol (DTT), or oxidizing agents such as diamide.[82,428,430,776] Luminal ATP is also important for polypeptide folding in the ER and some ER proteins facilitating the folding are able to bind ATP and require ATP to function.

Most of the ER luminal proteins are involved in protein folding and oligomer assembly. Bip/GRP78 and GRP94 are two of the most abundant protein chaperones of the ER and are structurally related to HSP70 and HSP90, respectively.[75,610,724] A chaperone is defined as a protein that is able to bind transiently to unfolded, partially folded and unassembled intermediates and then facilitate their folding and/or assembly. A chaperone usually does not interact with fully folded/assembled proteins. One mechanism of chaperone function is to prevent premature misfolding and aggregation of intermediates, allowing productive folding/assembly. A chaperone may also play an active role in polypeptide folding and assembly. Bip/GRP78 was originally identified as a binding protein for the immunoglobulin heavy chain (Bip) and glucose-regulated protein (GRP78), respectively.[367] It possesses an ATPase domain. Although binding of nascent polypeptides to Bip/GRP78 is ATP-independent, subsequent release of these polypeptides from Bip/GRP78 is associated with ATP hydrolysis by the ATPase domain.[68] Bip/GRP78 comprises about 5% of ER luminal proteins at steady state and binds transiently with a variety of nascent polypeptides, facilitating their folding and oligomer assembly. When

assisting in the folding of the translocated portion of nascent polypeptides, Bip/GRP78 may be involved in the unidirectional polypeptide translocation. GRP94 may play a similar chaperone function. Recently, it was shown that Bip/GRP78 and GRP94 interact sequentially with immunoglobulin chains,[478] indicating that GRP94 is a chaperone which acts after Bip/GRP78 for proper polypeptide folding and assembly.

The role of an ER membrane protein termed calnexin as a chaperone has been recently established.[63,261,283,394,560,817,818] Calnexin was independently identified by several approaches. A murine 88 kDa ER membrane protein (p88) was identified that associates transiently with class I major histocompatibility molecules when they are in the early ER stage of synthesis in various murine lymphoma cell lines.[6] A human protein designated IP90 was identified that transiently associates with a variety of newly synthesized proteins, including the T cell receptor (TCR), the B cell antigen receptor (membrane Ig) and class I molecules.[300] Incompletely folded/unassembled molecules are more stably associated with p88/IP90. p88 and IP90 are identical to calnexin,[6] an ER Ca^{2+}-binding integral membrane phosphoprotein. Calnexin was purified as a protein complex containing three additional polypeptides, TRAPα, TRAPβ and a novel protein named gp25L.[818] The association of the two subunits of the TRAP complex with calnexin indicates that the TRAP complex may also be involved in protein folding/assembly. Calnexin is a type I integral membrane protein of 573 residues. Its luminal domain is structurally related to calreticulin, the major ER luminal Ca^{2+}-binding protein. Calnexin functions as a molecular chaperone for diverse soluble secretory and membrane glycoproteins, recognizing glycoproteins only when they are incompletely folded. Glycosylation of nascent polypeptides is required for their association with calnexin. Inhibition of glycosylation with tunicamycin or inhibition of glycan modification by ER glucosidase I and II with castanospermine and 1-deoxynojirimycin abolish association of nascent polypeptides with calnexin. Monoglucosylated N-linked glycans of nascent polypeptides are most likely the major species recognized by calnexin, although polypeptide domains of the nascent glycoproteins may also play an important role for calnexin recognition.[126,283] In this regard, UDP-Glc:glycoprotein glucosyltransferase is also involved in protein folding.[731] Calnexin is evolutionarily conserved and its primary sequence has been established for diverse species,[787] including yeast *S. cerevisiae* and *Schizo. pombe*.[328,566,567] Recent results have revealed that Bip/GRP78 and calnexin function sequentially during the folding of VSV G-protein and thyroglobulin in that Bip/GRP78 binds early folding intermediates while calnexin chaperones bind later folding species.[167,262,356] It seems that various ER chaperones may act synergistically and/or sequentially during the folding/assembly of nascent proteins.

In addition to chaperones such as Bip/GRP78, GRP94 and calnexin, the ER also possesses a number of enzymes that are involved in polypeptide folding and assembly. Protein disulfide isomerase (PDI) is an abundant component of the ER lumen and its most well established role is in protein folding.[218] PDI catalyzes thio:disulfide interchange reactions, which in appropriate conditions can lead to the shuffling or isomerization of protein disulfide bonds, an essential process in the folding of nascent polypeptides. Molecularly, PDI is a homodimer of about 60 kDa. Its primary sequence shows the presence of two internal homologous domains which are closely related to thioredoxin, a small redox protein with an active site disulfide formed by vicinal Cys residues in the sequence context of Trp-Cys-Gly-Pro-Cys-Lys (WCGPCK). The corresponding regions of PDI are also the active sites involved in disulfide interchange reactions. PDI is identical to ERp59. In addition, two other ER luminal proteins referred to as ERp61 and ERp72 are structurally related to PDI/ERp59. ERp61 contains two thioredoxin-like domains found in the same relative positions as in PDI, although the rest of its sequence is different from PDI. ERp72 contains three thioredoxin-like domains, two of which are spaced as in PDI, embedded in an otherwise unrelated sequence. It seems likely that ERp61 and ERp72 may also be involved in native disulfide bond formation and protein folding/assembly for proteins targeted to the exocytotic pathway. Another folding enzyme in the ER lumen is peptidyl prolyl cis trans isomerase (PPI/ER).[761] PPI catalyzes interconversion of the alternative cis and trans conformers of the peptide backbone around the planar imide bond in the sequence -X-Pro-. A cytosolic form of PPI was identified as cyclophilin. Cyclophilin binds with high affinity to cyclosporin A, a cyclic fungal undecapeptide that inhibits T cell activation, particularly the induction of interleukin 2 (IL2) synthesis, and has been used as an immunosuppressive drug. Another distinct cytosolic PPI is the FK506-binding protein (FKBP), which does not bind cyclosporin A but rather binds FK506 and rapamycin, two other widely used immunosuppressive drugs. FKBP is structurally related to cyclophilin. Like cyclosporin A, FK506 also inhibits the activation of T cells, while rapamycin inhibits the response of T cells to lymphokine IL2. Human cyclophilin B is an ER luminal PPI (PPI/ER), targeted to the ER via an N-terminal signal peptide. The Drosophila ninaA protein is also a PPI/ER that is required for the folding of some rhodopsin isoforms. Evidence suggests that PPI/ER may also be involved in folding/assembly of collagen and transferrin.[427] It seems that PPI/ER may also play a role in folding/assembly of other proteins targeted to the exocytotic pathway. Prolyl 4-hydroxylase (P4H) is also an ER luminal enzyme catalyzing co- and posttranslational hydroxylation of Pro residues in the sequence context of -X-Pro-Gly- of collagens and proteins with collagen-like sequences.[332] The presence of

4-hydroxyproline residues in the newly synthesized procollagen polypeptide is essential for the folding/assembly of the polypeptide to form triple-helical collagen. P4H is a tetramer ($\alpha2\beta2$) with subunits of 64 kDa (α) and 60 kDa (β). The 64 kDa α-subunit is the catalytically more important subunit, while the 60 kDa β-subunit has been shown to be identical to PDI. The main role of the β-subunit (PDI) is to prevent misfolding and aggregation of the catalytic α-subunit and to retain the active $\alpha2\beta2$ complex in the ER lumen. Another enzyme important for folding/assembly of collagen or collagen-like proteins is lysyl hydroxylase that catalyzes the hydroxylation of Lys residues.[602,863] The importance of lysyl hydroxylase is manifested by the demonstration that deficiency in lysyl hydroxylase is the major cause of the type IV variant of the Ehlers-Danlos syndrome (EDS), a recessively inherited connective-tissue disorder.

Polypeptide folding and oligomer assembly are not only required for the achievement of normal conformation of proteins but also play an essential role in their export from the ER. Misfolded and/or unassembled proteins are prevented from export from the ER. They are retained in the ER and subsequently degraded by an ER degradative pathway. Chaperones, such as Bip/GRP78, GRP94 and calnexin may play an important role in retaining these misfolded/unassembled proteins in the ER. The term "ER quality control" was coined to describe the intriguing phenomenon that only properly folded/assembled proteins are exported.[263,317,644] In the event that misfolded/unassembled proteins are accumulated in the ER, the transcription of Bip/GRP78, GRP94, PDI/ERp59 and ERp72 is enhanced, resulting in increased levels of these proteins.[379] Accumulation of misfolded/unassembled proteins in the ER can be induced by several ways, such as preventing N-linked glycosylation with reagents such as tunicamycin or deoxyglucose, starving cells for glucose, preventing disulfide-bond formation with reducing agents such as 2-mercaptoethanol, depleting the ER of Ca^{2+} with Ca^{2+} ionophore A23187, Ca^{2+} pump blocker thapsigargin, or overexpression of mutant proteins that cannot fold/assemble properly and thus are accumulated in the ER.

The phenomenon that accumulation of misfolded/unassembled proteins in the ER triggers the increased production of ER chaperones and folding enzymes has been referred to as the unfolded protein response pathway.[379,694] This pathway is similarly triggered in yeast cells by tunicamycin, deoxyglucose, 2-mercaptoethanol and overproduction of mutant proteins that accumulate in the ER. In addition, temperature sensitive mutations in SEC53 or SEC11 gene cause the accumulation of misfolded/unassembled proteins in the ER and trigger this pathway at the restrictive temperature. As mentioned earlier, SEC11 encodes the 18 kDa subunit of the yeast signal peptidase complex,[71] suggesting that cleavage of the signal peptide is essential for protein folding/assembly. At the restrictive temperature, sec53 mutants lack phosphomannomutase activity and are defective in protein glycosylation.[694] The enhanced transcription of the target genes of this pathway, such as Bip/GRP74, GRP94 and PDI, has been shown to be dependent on a promoter element in both mammalian and yeast cells. This 22 bp promoter element in yeast was referred to as unfolded protein response element (UPRE) and is conserved in several target genes, including KAR2 (encoding yeast Bip), PDI1 (encoding yeast PDI), EUG1 (encoding Eug1p that is related to PDI) and FKB2 (encoding peptidyl-prolyl cis trans isomerase). Furthermore, UPRE is sufficient for inducible gene transcription upon its transfer to reporter genes. Genetic screens in yeast have identified one gene that is necessary for this unfolded protein response pathway (named IRE1 and ERN1, respectively, in two independent studies).[145,495] Mutations in IRE1/ERN1 abolish the induced transcription of endogenous KAR2, PDI1 or EUG1 gene in response to the ER accumulation of misfolded/unassembled proteins. The IRE1 gene was previously indentified as a gene required for inositol prototroph and its mutation results in inositol auxotrophs.

The IRE1/ERN1 gene encodes a protein of 1115 amino acids. The N-terminal 26 residues forms the signal peptide which is cleaved during translocation. Residues 27 to 526 is the luminal domain followed by a transmembrane domain (residues 527-556). In the C-terminal 559-residue (residues 557-1115) cytoplasmic domain, there is a serine/threonine kinase domain (residues 677-985) related to the Cdk (cell division cycle kinases) class such as the cdc2/CDC28 kinase. The type I transmembrane topology of IRE1p/ERN1p indicates that it belongs to a small but growing class of serine/threonine transmembrane kinases, such as the activin receptor and TGF-β type II growth factor receptor. Furthermore, this topology also suggests an intriguing signal transduction pathway for the unfolded protein response. The luminal domain of IRE1p/ERN1p may function as a sensing domain to monitor the levels of misfolded/unassembled proteins in the ER. Upon accumulation of misfolded/unassembled proteins in the ER, the luminal domain transduces a signal to the cytoplasmic kinase domain, resulting in the activation of the kinase activity. The kinase domain of IRE1p/ERN1p has been shown to be

essential for the unfolded protein response pathway, since the mutation of a specific Lys into Ala or Arg residue in the catalytic domain abolished its function in the unfolded protein response pathway. Cytoplasmic truncated versions of IRE1p/ERN1p lacking the kinase domain behaved like dominant-negative mutants, probably by interfering with the normal sensing function of the luminal domain in the wild-type protein. The luminal domain of IRE1p/ERN1p may sense the concentration of free Bip (Bip not bound to misfolded/unassembled proteins) as a measure of the amount of misfolded/unassembled proteins. During low levels of misfolded/unassembled proteins in the ER, the level of free Bip is high and IRE1p/ERN1p remains in the inactive form (possibly as a monomer). Upon accumulation of misfolded/unassembled proteins, the concentration of free Bip falls, resulting in the activation of IRE1p/ERN1p (probably via dimerization of the inactive monomer). In support of this, overexpression of Bip in the yeast diminishes the unfolded protein response pathway, most likely due to the higher levels of free Bip in the ER.[694] Similarly, when the free Bip is artificially reduced by expressing a mutant Bip that is poorly retained in the ER, the transcription of KAR2 gene is induced. IRE1p/ERN1p homologues may also be expressed in the mammalian cells and involved in a similar unfolded protein response pathway.[111]

IDENTIFICATION, PURIFICATION AND CLONING OF IMPORTANT FACTORS REQUIRED FOR TRANSPORT

In Vitro

The development of various in vitro transport systems has not only allowed the identification and purification of novel components involved in vesicular transport but also provided a working system for detailed dissections of the functional aspects and working mechanisms for these components and other proteins identified by different approaches. The intra-Golgi cis to medial transport system has been used to identify and isolate several important molecules, including the N-ethylmaleimide-sensitive factor (NSF), soluble NSF attachment proteins (SNAP), p115/TAP and coat components (COP, for coat proteins) of the Golgi-derived vesicles.

NSF

Several important early findings are essential for the isolation of NSF. The first observation was that mild treatment of both the donor and acceptor Golgi membranes with N-ethylmaleimide (1 mM for 15 min at 0°C) inhibited the intra-Golgi transport, even in the presence of untreated cytosol prepared under standard conditions (50 mM KCl in the absence of ATP). Normal transport could be achieved only when either the donor or the acceptor but not both membranes were pretreated, suggesting that this N-ethylmaleimide-sensitive factor (NSF) could be provided by either the donor or the acceptor Golgi membrane.[233] The second observation was that transport could be restored by the addition of untreated Golgi membranes or more importantly by an ATP-extracted supernatant of untreated Golgi[67] membranes. Thirdly, cytosol prepared under standard conditions actually had some NSF activity that was lost completely within an hour on ice, while cytosol prepared under a modified condition (100 mM KCl in the presence of 5 mM ATP) had a high NSF activity due to the ability of ATP

to extract NSF from the Golgi membrane and to stabilize the extracted NSF. It was also found that 1% polyethylene glycol (PEG) 4000 or 10% glycerol and the reducing agent DTT could further stabilize NSF.[67] These observations led to the design of an assay for NSF purification using the pretreated donor and acceptor membranes in the presence of NSF-free cytosol. The NSF-containing cytosol prepared under the modified condition was biochemically fractionated for an activity that could restore the transport, leading to the purification of a single polypeptide of about 76 kDa.[67] Kinetic studies of intra-Golgi transport suggest that NSF functions after vesicle budding from the donor membrane and transfer to the acceptor membrane. Morphological studies demonstrated that in the absence of NSF (using N-ethylmaleimide treated membranes and NSF-free cytosol), vesicles accumulated on the acceptor membrane, which could be consumed upon addition of functional NSF. These results suggest that NSF functions at a stage that is necessary for the fusion of the transferred vesicles with the acceptor membrane. The hamster (from CHO cells) cDNA encoding the full-length NSF was cloned and encodes a 744 amino acid polypeptide (numbering from the initiation Met residue 1 which was originally numbered as residue 9)[852] that is 48% identical to a protein (Sec18p) encoded by the yeast SEC18 gene, which was identified in genetic screens for genes involved in vesicular transport.[195] Sec18p has been shown to be essential for ER-Golgi transport in intact yeast cells. Biochemical studies also established the existence of NSF-like activity in yeast cytosol that is sensitive to N-ethylmaleimide and becomes inactivated in the absence of ATP. Furthermore, cytosol prepared from yeast cells that overproduce Sec18p has many-fold higher NSF activity than the control cytosol, while cytosol prepared from yeast cells harboring a mutant SEC18 gene has no NSF activity, confirming that Sec18p is the yeast functional counterpart of mammalian NSF. Consistent with a role for NSF in vesicle fusion, yeast cells harboring the mutant SEC18 gene accumulate ER derived 50 nm transport vesicles. It is now established that NSF is important for many other vesicular transport or membrane fusion events in mammalian cells, including endosomal fusion, transcytosis and neurotransmitter release, in addition to its role in ER Golgi and intra-Golgi transport events.[54,452,838,839] Sec18p has also been shown to be involved in many membrane fusion events in yeast.[246] NSF is thus a general fusion factor involved in many diverse intracellular vesicular transport and membrane fusion events. A second form of NSF has recently been described and cloned.[78,563]

The 744-residue NSF (Fig. 8) contains two internal homologous ATPase regions (residues 206 to 477 form the D1 domain and residues 478 to 744 form the D2 domain) that are about 23.5% identical to each other. Similarly, two ATPase domains also exist in Sec18p. When compared to Sec18p, the D1 domain (about 64% identity) of NSF is evolutionally more conserved than the D2 domain (about 42% identity). The consensus Walker A box G/A-X-X-X-X-G-K-T/S for nucleotide binding is present in both D1 (residues 260 to 267: G-P-P-G-C-G-K-T) and D2 (residues 543 to 550: G-P-P-H-S-G-K-T) domains. The conserved Lys residues 266 in the D1 and 549 in the D2 domain are key residues for nucleotide binding. Similarly, consensus Walker B box (DEAE box) for ATP hydrolysis is also present in the D1 domain (residues 328-331: DEID) and D2-domain (residues 603-606: DDIE). The N-terminal 205 residues have been referred to as the N-domain. The domain structure of NSF is supported biochemically by limited proteolysis. When incubated with increasing concentrations of trypsin, the 76 kDa NSF first gives rise to a 60 kDa fragment and then to a 31 kDa fragment. The 60 kDa fragment is due to a cleavage between the N-domain and D1-domain (between Arg_{205}-Gln_{206}), while the 31 kDa fragment is derived from the 60 kDa fragment by a cleavage between the D1 and D2 domains (between Arg_{477}-Gly_{478}).[762,838]

Originally suggested to be homotetramer, more accurate studies have shown that functional NSF is a homotrimer.[839] The functional aspects of the two ATPase domains have been examined by site-directed mutagenesis or domain rearrangements.[756,839] The D2 domain is required for the 76 kDa subunit to oligomerize into the functional homotrimer. The three subunits in the trimerized NSF may function in a coordinated and concerted way because a trimer composed of two wild-type and a mutant subunit has no NSF activity. Mutations of the D1-domain by substituting Lys 266 with Ala (K266A) or Glu329 with Gln (E329Q), affecting ATP-binding or hydrolysis, respectively, com-

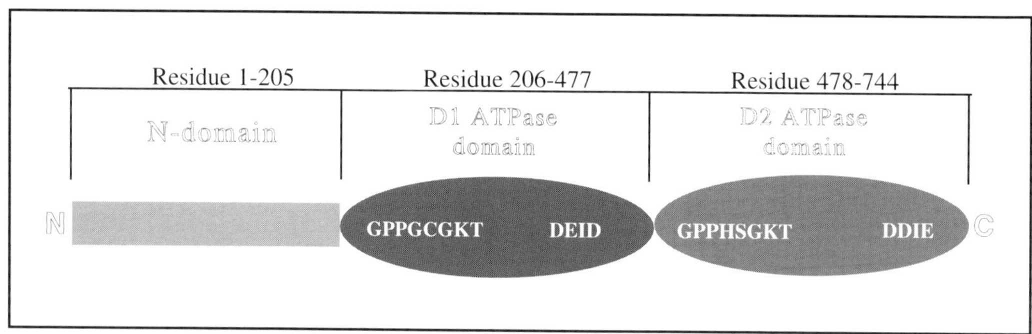

Fig. 8. Domain-structure of NSF. Functional NSF is a homotrimer composed of three 76 kDa polypeptides, each of which is organized into the N-domain, D1 ATPase domain and D2 ATPase domain.

pletely abolished NSF activity. As compared to wild-type NSF, K266A or E329Q mutation caused an 82% or 69% decrease in ATPase activity, respectively, suggesting that ATP binding and subsequent hydrolysis in the D1 domain is essential for NSF activity. Furthermore, these NSF mutants also inhibit intra-Golgi transport, most likely by competing with endogenous NSF for functional interacting proteins. Mutations in the D2 domain by substituting Lys 549 with Ala (K549A) or Asp604 with Gln (D604Q) reduce the NSF activity to about 14 or 42%, of the wild-type protein, respectively. The ATPase activity of NSF has recently been shown to be modulated by its interacting protein (SNAP). Using purified NSF and SNAP, it was revealed that the ATPase activity of NSF is enhanced in a dose-dependent manner by immobilized (plastic-absorbed) but not soluble SNAP. In the presence of immobilized SNAP, the enzymatic kinetics for ATPase activity of NSF is consistent with the presence of two functional ATPase domains with different affinities for ATP.[492] The enhanced ATPase activity of NSF is most likely mediated by decreasing the Km of its low affinity sites by about 100-fold.

Recently, it was shown that an NSF-like ATPase termed p97,[583] together with NSF, is involved in the cisternal regrowth from mitotic Golgi fragments.[476,611] Similarly, re-formation of Golgi stack from ilimaquinone-vesiculated Golgi membranes also requires p97 in addition to NSF.[4] The membrane fusion mediated by p97 is mechanistically different from that mediated by NSF in that it does not require soluble SNAPs or Rab proteins. A yeast protein (Cdc48p) is about 76% identical to p97 and also related to NSF and Sec18p. It was discovered that fusion of ER membranes is independent of Sec18p or Sec17p but is actually mediated by Cdc48p.[391] Fusion of ER/nuclear membrane is an essential process required for normal progression of the nuclear cell cycle, karyogamy. It is possible that p97 may also mediate ER/nuclear membrane fusion in mammalian cells and that Cdc48p may also be involved in events similar to the regrowth of Golgi in yeast.

SNAP

Since NSF functions at the stage of vesicle fusion with the target membrane, the functional site of NSF should be at the membrane, consistent with the fact that NSF was originally identified as a membrane-associated factor that was extracted by ATP. To study how NSF functions, the reassociation of NSF with NSF-free Golgi membrane was examined. It was found that binding of purified NSF to the Golgi membrane required the presence of NSF-free cytosol.[831] Furthermore, NSF binding was linearly dependent on the concentration of cytosol present, suggesting that a cytosolic factor(s) distinct from NSF was stoichiometrically involved in the membrane association of NSF. In the presence of cytosol, binding of NSF to the Golgi membrane is saturable and requires an integral membrane protein(s) of the Golgi membrane that is protease and heat sensitive. The cytosolic activity required for NSF binding to the Golgi membrane was named soluble NSF attachment factor (SNAP) and is both protease and heat sensitive. SNAP can be separated from the bulk of the other cytosolic proteins. The requirement of SNAP and integral membrane components for NSF binding provided the foundation for the molecular identification of SNAP and the integral membrane components involved in vesicle docking/fusion.[132,831]

A modified intra-Golgi transport assay was designed for the identification of peripheral proteins other than NSF that are involved in intra-Golgi transport.[133] In this modified assay, both the donor and acceptor Golgi membranes were stripped of peripheral proteins (such as NSF) with 1M KCl (referred to as K donor or acceptor Golgi membrane). When purified NSF and NSF-free CHO cytosol were present, normal transport was achieved with K Golgi membranes, suggesting that stripped peripheral proteins other than

NSF could be provided by the NSF-free cytosol. However, when the transport was assayed with K Golgi membrane, purified NSF and yeast cytosol prepared in the absence of ATP (to inactivate the endogenous Sec18p), no transport was detected. Yeast cytosol was shown to be able to substitute for CHO cytosol during transport when untreated Golgi membranes were used. These observations suggest that 1M KCl removed some peripheral protein(s) other than NSF from the Golgi membranes that could not be provided by the yeast cytosol. The K Golgi membranes, purified NSF and yeast cytosol were used to identify factors that were extracted from the Golgi membranes by 1 M KCl and could be provided by mammalian but not yeast cytosol. From bovine brain cytosol, two fractions Fr1 and Fr2 of an average native size of about 500 and 40 kDa, respectively, were shown to be required for restoring intra-Golgi transport using the yeast cytosol/K Golgi membranes/purified NSF assay. In the presence of the Fr1 fraction, this assay was used to purify active components from the Fr2 fraction. Three proteins of 35, 36 and 39 kDa were purified that could each complement this assay for intra-Golgi transport.[133] The 35 kDa protein had the highest specific activity, whereas the 36 kDa protein was somewhat lower and the 39 kDa protein had the lowest specific activity in restoring intra-Golgi transport using this assay. Detailed studies revealed that these three proteins were not required for vesicle budding and attachment to the acceptor Golgi membrane. Rather, they acted in parallel with NSF at the fusion stage of transport. More importantly, it was found that each of these proteins could support NSF binding to Golgi membranes and that each of them could form a protein complex with NSF when immobilized on a plastic surface. These results led to the conclusion that they are SNAPs originally defined in assaying NSF binding to the Golgi membrane. The 35, 36 and 39 kDa proteins were termed α-SNAP, β-SNAP and γ-SNAP, respectively.

Since the function of NSF is dependent on SNAP and yeast cells contain a functional homologue of NSF, it was speculated that a SNAP-like protein should also be present in yeast cytosol. However, yeast SNAP-like activity was not functional in the Fr1/yeast cytosol/K Golgi membranes/purified NSF assay, which enabled the purification of mammalian SNAPs. One interpretation is that although Sec18p (the yeast NSF) can functionally substitute for CHO NSF in intra-Golgi transport, the yeast SNAP-like protein may be functionally less-conserved and is not able to function along with the CHO NSF. This has been shown to be the case because when Sec18p was used as the source of NSF activity, intra-Golgi transport was observed with the Fr1/yeast cytosol/K Golgi membrane/Sec18p assay system. These results confirm that a yeast SNAP-like activity, in the context of Sec18p but not CHO NSF, can complement the requirement of SNAP for intra-Golgi transport. This observation led to an examination as to whether any of the existing yeast secretory pathway (sec) mutants had a deficiency in SNAP activity. For this purpose, the intra-Golgi transport assay consisted of K Golgi membranes, Fr1 derived from bovine brain cytosol and yeast cytosol prepared from wild-type and various yeast mutants under modified conditions to extract and stabilize endogenous yeast NSF (Sec18p) activity. Cytosol prepared from wild-type yeast cells and all the mutants was active in this intra-Golgi transport assay with the exception of cytosol prepared from sec17 mutant cells under various condtions. Even cytosol prepared from a sec17 mutant that overproduced Sec18p was still inactive, suggesting that the yeast SNAP activity is deficient in sec17 mutant cells and that the SEC17 gene product (Sec17p) may be the yeast SNAP.[132,251] This was confirmed by the demonstration that cytosol prepared from sec17 mutant cells could be restored to normal activity when purified α-SNAP was added. However, β-SNAP and γ-SNAP could not restore the SNAP activity of cytosol prepared from sec17 mutant cells. Therefore, yeast NSF (Sec18p) can functionally interact with both yeast SNAP (Sec17p) and mammalian α-SNAP, while yeast Sec17p can only functionally interact with Sec18p but not with mammalian NSF. Consistent with this, the CHO NSF gene could not complement the SEC18 gene when introduced into sec18 yeast cells and the mammalian α-SNAP but not β- or γ-SNAP could functionally interact with Sec18p. Amino acid sequences derived from cDNA cloning revealed that all three forms of mammalian SNAPs [837] are structurally related to Sec17p (291 aa). α-SNAP (295 aa) and γ-SNAP (328 aa) are widely expressed and may act synergistically in intra-Golgi transport, while β-SNAP (298 aa) is a brain-specific isoform of α-SNAP. α-SNAP and β-SNAP are 83% identical to each other and 34% and 33% identical to Sec17p, respectively. γ-SNAP is 25% identical to α-SNAP and 23% identical to β-SNAP and Sec17p. Three potential phosphorylation sites (Ser residues at positions 4, 35 and 201) for protein kinase C and seven putative casein kinase II phosphorylation sites (Ser residue at position 4, 36, 77, 126, 201, 258 and 270) are present in α-SNAP. α-SNAP is also predicted to have three regions (residue 5-28, 127-150, 166-188) that may form coiled-coil domains. β-SNAP contains five potential phosphorylation sites for casein kinase II (Thr-36, Ser-77, S-84, T-186 and S-199) and two regions (residue 127-150 and 166-188) that can potentially form coiled-coil domains. One potential site for tyrosine phosphorylation (Tyr-292) and one potential coiled-coil domain (residue 185-205) are predicted for γ-SNAP.

The physical and functional interaction between NSF and SNAP is also conserved in yeast.[251] Genetic evidence suggests that Sec17p and Sec18p interact functionally. Temperature sensitive mutant sec17 or sec18 cells grow normally at the permissive but not the restrictive temperature. However, a sec17 and sec18 double mutant does not grow at the permissive temperature, a phenomenon referred to as synthetic lethal interaction. Purified Sec17p can physically interact with Sec18p with a 1:1 stoichiometry. The interaction between Sec17p and Sec18p occurs only in the presence of microsomal membranes. In normal cells, the majority of Sec17p is associated with membrane structures as a peripheral protein. However, Sec17p is mostly soluble when overexpressed, suggesting the presence of a saturable membrane receptor for Sec17p. The dependence on microsomal membranes for Sec17p and Sec18p interaction is consistent with the observation that interaction of NSF and SNAP occurs only in the presence of Golgi membrane or when SNAP is immobilized. It seems that when SNAP/Sec17p is bound to its membrane receptor (or immobilized), a conformational change of SNAP/Sec17p exposes a high affinity binding site for NSF/Sec18p, leading to the recruitment of NSF/Sec18p to the membrane. SNAP/Sec17p can be considered as an adapter, mediating membrane association of NSF/Sec18p and at the same time is also actively involved in membrane fusion by working together with NSF/Sec18p. This possibility is consistent with the observation that only immobilized but not soluble SNAP is able to enhance the ATPase activity of NSF/Sec18p. The N-domain of NSF may be involved in the interaction with SNAP.

SNAP receptors

The existence of a membrane receptor(s) for SNAP that is involved in membrane recruitment of NSF triggered studies on the identity of the SNAP receptor (SNARE).[722,836,850] When binding studies were performed with [35S]-Met labeled α-SNAP with 1M KCl-stripped Golgi membranes, it was shown that α-SNAP binds to the membrane in a saturable fashion when a fixed amount of the membrane is used. Increased binding was achieved with increasing amounts of the membrane. Binding of α-SNAP is independent of NSF and occurs in the absence of Mg^{2+}/ATP. Pretreatment of membranes with proteinase K abolished this specific binding. Furthermore, unlabeled α-SNAP and β-SNAP, but not γ-SNAP compete with labeled α-SNAP for binding. This suggests that the binding site for α-SNAP may be the same for β-SNAP but distinct from that for γ-SNAP, since γ-SNAP can bind the membrane and this γ-SNAP binding is reduced when the membrane is pretreated with protease K. It was shown that the distinct binding sites for α/β-SNAP and γ-SNAP may exist in the same protein complex. Crosslinking experiments show that α-SNAP can interact with a Golgi protein of about 30 kDa.[836] The membrane receptor was thought to function as an assembly factor for the NSF-α-SNAP-receptor complex. Although ^{35}S-Met labeled α-SNAP does not interact with NSF in solution, specific interaction between the labeled α-SNAP and NSF occurs when Golgi membranes or its detergent extract is present. Furthermore, in the presence of myc-epitope-tagged NSF (NSFmyc) and Golgi detergent extract, the labeled α-SNAP is incorporated into a protein complex with a sedimentation coefficient of 20S.[850] This 20S α-SNAP-containing complex can be specifically immunoprecipitated with a monoclonal antibody against the myc tag on the NSFmyc molecule. The α-SNAP receptor (assembly factor) is also incorporated into this 20S complex, suggesting that this 20S complex is composed of NSFmyc, α-SNAP and the α-SNAP receptor(s). The 20S complex is stable under conditions that prevent ATP hydrolysis (in the presence of ATP/EDTA or ATP-γ-S) but becomes disassembled when ATP is hydrolysed (in the presence of ATP and Mg^{2+}). Since NSF is a Mg^{2+}-dependent ATPase, a working model for membrane recruitment of NSF has been proposed. Cytosolic free NSF homotrimer is unable to interact with cytosolic free SNAP or integral membrane proteins. Membrane binding of free SNAP to the SNARE causes a conformational change, leading to the appearance of a high affinity site for NSF. Binding to SNAP

in the SNAP-SNARE complex results in the membrane recruitment of NSF, the formation of the 20S NSF-SNAP-SNARE complex and activation of the ATPase activity of NSF in the protein complex. Hydrolysis of ATP by NSF leads to the disassembly of this protein complex; NSF dissociates from SNAP and returns to the cytosol for subsequent use. SNAP may also dissociate from the SNAREs. The assembly and subsequent ATP hydrolysis-driven disassembly of the protein complex is intimately associated with vesicle docking and fusion with the target membrane. This 20S protein complex has also been referred to as the 20S fusion particle.

The specific incorporation of putative SNAREs into the 20S fusion particle was employed to purify SNAREs from crude bovine brain total membranes. A detergent extract of bovine brain membranes was incubated with purified α-SNAP, γ-SNAP and NSFmyc (the C-terminus of NSF is tagged with a myc epitope) under conditions (2 mM EDTA/0.5 mM ATP or 0.5 mM ATP-γ-S) that facilitate formation of the 20S particles. The 20 S particles are selectively immunoprecipitated with myc-specific monoclonal antibodies immobilized on protein G-sepharose. After removing nonspecifically bound proteins, the immunoprecipitated 20S particles are disassembled by ATP hydrolysis catalysed by NSF under conditions (4 mM free Mg^{2+}/0.5 mM ATP) that facilitate the disassembly process, releasing potential SNAREs, α-SNAP and γ-SNAP, while NSF remaining bound to the beads. Upon characterization of the released proteins by microsequencing, syntaxin 1B, syntaxin 1A, SNAP-25 (for a 25 kDa synaptosome-associated protein) and synaptobrevin-2/VAMP 2 were identified in addition to the α- and γ-SNAP.[722] It has been suggested that syntaxins, SNAP-25 and synaptobrevin 2 are involved in synaptic vesicle docking and fusion with the presynaptic membrane. Their specific incorporation into the 20S fusion particle suggests that they are the most abundant SNAREs in the total brain membrane and that NSF and SNAP may be involved in synaptic vesicle docking and fusion. Since synaptobrevin 2 is specifically associated with synaptic vesicles while syntaxins and SNAP-25 are confined to the presynaptic membrane, this finding led to the SNARE hypothesis for the specificity of vesicle docking/fusion (discussed later).[207,649,650,722]

p115/TAP

Using the yeast cytosol/K Golgi membrane/purified NSF assay for essential peripheral Golgi membrane proteins that are extracted by 1M KCl but can be rescued by mammalian but not yeast cytosol, the high molecular weight Fr1 fraction is shown to be required in addition to the low molecular weight Fr2 fraction composed of SNAPs. Based on this observation, a modified intra-Golgi transport assay was developed for monitoring high molecular weight cytosolic components. This assay (K Golgi membranes/NSF/Fr2 assay) consists of 1M KCl extracted donor and acceptor Golgi membranes (K Golgi membranes), purified CHO NSF, the crude Fr2 fraction of bovine brain cytosol and other necessary common constituents. When crude bovine brain cytosol was fractionated by an FPLC MonoQ anion exchanger, three high molecular weight fractions eluted at 210, 270 and 380 mM KCl were required for optimal transport. The 210, 270 and 380 mM KCl eluted fractions were termed Fr1α, Fr1β and Fr1γ, respectively, leading to the design of an assay for the purification of components in the Fr1γ fraction. From bovine liver cytosol, a unique protein (based on partial amino acid sequence) of about 115 kDa (p115) was purified that constitutes the Fr1γ activity.[829] Using monoclonal antibodies specific for p115, it was established that p115 is a peripheral membrane protein enriched in the Golgi apparatus. p115 is a homodimer and has an elongated structure with two globular heads (about 10 nM in diameter) and an extended rod-like tail (about 40 nM long) reminiscent of myosin II. When bovine and rat cDNAs for p115 were

cloned and sequenced, p115 was found to be identical to the transcytosis-associated protein (TAP) cloned from rat.[50,660] TAP was originally identified as a protein associated with transcytotic vesicles and required for their fusion with the apical plasma membrane of epithelial cells. The rat and bovine p115/TAP molecules are about 92% identical. Furthermore, p115/TAP is structurally related to a yeast protein, Uso1p (1790 amino acids long). The 959 amino acid sequence of p115/TAP can be divided into three regions: an N-terminal domain (residue 1-651); an internal region with several potential coiled-coil regions (residue 652-930); and a 28-residue C-terminal acidic region (residue 931-958). In the p115/TAP homodimer, the two N-domains form the two globular heads while the coiled-coil domains may interact with each other to form the rod-like tail. Uso1p is required in yeast for ER Golgi vesicular transport and also has a similar three domain structure.[508] The significantly larger size (1790 residues) of Uso1p is mainly due to its longer coiled-coil domain. p115/TAP may be involved in vesicle docking to the target membrane, since it was identified in two different transport events. Like NSF and SNAP, p115/TAP and Uso1p may function also as a general transport factor involved in vesicle docking/fusion.

Coat proteins

Morphological examination of the in vitro intra-Golgi transport has established that the VSV G-protein is transported to the acceptor membrane via vesicles that bud from the donor membrane. These vesicles have a size of about 70 nm in diameter, similar to that of Golgi vesicles observed in cells. Furthermore, all the buds/vesicles in the process of budding as well as most of the budded vesicles have a distinct coat of about 18 nm thick on the cytoplasmic face, which resembles but is distinct from the clathrin coat of the endocytotic vesicles.[550,551] As discussed previously, NSF is required for vesicle fusion and its inactivation by

N-ethylmaleimide treatment leads to the accumulation of vesicles on the acceptor membrane. These accumulated vesicles are not coated, suggesting that the vesicles must be uncoated before fusion with the acceptor membrane via the NSF-dependent process. When the yeast SEC4 gene (which is involved in Golgi to surface transport in yeast) was cloned, it was found that Sec4p is structurally-related to the ras-like small GTPases.[656] This observation triggered an investigation into whether GTP hydrolysis was involved in intra-Golgi transport. It was found that nonhydrolysable analogues of GTP such as GTP-γ-S effectively inhibited transport.[475] Electron microscopy reveals that, in the presence of GTP-γ-S, vesicle budding is not affected, but vesicle consumption is, resulting in the accumulation of vesicles on the acceptor membrane. In contrast to N-ethylmaleimide treatment, the vesicles accumulated in the presence of GTP-γ-S are coated. Treatment with both GTP-γ-S and N-ethylmaleimide resulted in the accumulation of only coated vesicles, suggesting that GTP-γ-S blocks a step preceding that affected by N-ethylmaleimide treatment and GTP-γ-S may actually inhibit the uncoating process. Furthermore, it suggests that GTP hydrolysis by a GTPase is involved in the uncoating process before the vesicles can undergo the NSF-dependent fusion step. The accumulation of coated vesicles due to the selective block of uncoating by GTP-γ-S led to the biochemical purification of the coat and its protein components.[453] Eight distinct proteins have been identified as components of the coat (Table III).[200,647,649,690,830] One of them is a small GTPase identified earlier called ADP-ribosylation factor (ARF), some of which is associated with the Golgi apparatus while the majority exists in the cytosol as a monomer. The remaining seven proteins are defined as coat proteins (COPs) according to their sizes, namely α- (160 kDa), β- (110 kDa), β'- (102 kDa), γ- (98 kDa), δ- (61 kDa), ϵ- (36 kDa) and ξ-COP (20 kDa). An important fortuitous finding during the purification of p115/

TAP was that these seven COPs exist in the cytosol as a protein complex referred to as the coatomer which can be easily purified.[830]

When the 110 kDa β-COP purified from Golgi derived vesicles was sequenced,[691] it was found to be identical to a 110 kDa Golgi peripheral protein (110K protein) identified by a monoclonal antibody (M3A5).[189] The 953 amino acid sequence of rat β-COP derived from cDNA cloning shows significant homology to β-adaptin (a component of clathrin-coated vesicles), especially in the 500-residue N-terminal region.[189] β'-COP was initially not detected due to its comigration with β-COP in the SDS-PAGE gel but became detectable when purified coatomer was resolved by a modified gel system containing urea and an increased ratio of acrylamide to bisacrylamide.[270,742] β'-COP was cloned by two independent approaches. Internal peptide sequences derived from β'-COP of purified bovine brain coatomer were used to design an oligonucleotide probe for screening a bovine brain cDNA library, leading to cDNA cloning of bovine β'-COP. Independently, the human β'-COP was cloned by screening human cDNA expression libraries with a monoclonal antibody (23C). Among seven monoclonal antibodies generated against the C-terminal half of mouse TCP-1, 23C and 72A are two monoclonal antibodies that do not detect TCP-1 but crossreact fortuitously with a unique protein of 102 kDa (p102) in primate cells. Cloning the human cDNA revealed that p102 is β'-COP and 23C and 72A crossreact with the C-terminal tail of β'-COP. The derived 906 amino acid sequence of β'-COP contains six WD-40 repeats in its N-terminal portion (residue 2-44, 45-86, 87-128, 129-172, 173-216 and 217-258) which are similar to those found in the β-subunit of trimeric G proteins, yeast CDC4 gene product, yeast Sec13p and other regulatory proteins.[515] The significance and function of these WD-40 repeats in β'-COP remain to be established. The partial amino acid sequences derived

Table III. Comparison of known coat complexes

Name	Composition	Size
Surface clathrin (AP2) coat	Clathrin heavy chain	180 kDa
	Clathrin light chain	30 kDa
	α-adaptin	115 kDa
	β-adaptin	110 kDa
	AP50	50 kDa
	AP17	17 kDa
TGN clathrin (AP1) coat	Clathrin heavy chain	180 kDa
	Clathrin light chain	30 kDa
	γ-adaptin	100 kDa
	β'-adaptin	100 kDa
	AP47	47 kDa
	AP19	19 kDa
COPI coat	α-COP	160 kDa
	β-COP	110 kDa
	β'-COP	102 kDa
	γ-COP	98 kDa
	δ-COP	61 kDa
	ε-COP	36 kDa
	ζ-COP	20 kDa
	ARF1	21 kDa
COPII coat	Sec13p	33 kDa
	Sec31p	150 kDa
	Sec16p	240 kDa
	Sec23p	84 kDa
	Sec24p	105 kDa
	Sar1p	21 kDa

Table III. Summary of three well-characterized protein coats for transport vesicle

from purified 20 kDa ξ-COP were used to design oligonucleotides for a polymerase chain reaction to clone a cDNA fragment for ξ-COP, leading to the elucidation of the entire amino acid sequence for ξ-COP.[383] The derived 177 amino acid sequence of ξ-COP is 24% identical to AP17 and AP19, two distinct small chains of clathrin adapter complexes. Polyclonal antibodies directed against ξ-COP block the binding of coatomer to Golgi membranes and inhibit the assembly of COP-coated vesicles. Unlike other characterized coatomer subunits (β-, β'-, γ-, ε-COP) that exist primarily in the cytoplasmic coatomer complex, ξ-COP exists in both coatomer associated and cytosolic free pools. The existence of free ξ-COP raised the question as to whether association and dissociation of ξ-COP with the coatomer represents a regulatory event for the coatomer function. The bovine cDNA for the 36 kDa ε-COP has also been cloned and predicts a novel protein of 308 amino acids.[266] When a His6-tagged ε-COP was overexpressed in stably-transfected CHO cells, radiolabeled coatomer complex can be easily affinity-purified by Ni-NTA agarose after an initial enrichment by Mono Q chromatography. This permitted the radiolabeled coatomer to be used to establish that all the subunits of the coatomer encase the coated vesicles as an intact protein complex.[266] The hamster cDNA for ε-COP was also cloned by its ability to correct the temperature sensitive phenotype of CHO idlF mutant,[256] which is defective in Golgi structure and ER-Golgi transport at the nonpermissive temperature, providing in vivo evidence that coatomer may be essential for normal Golgi structure and ER to Golgi membrane traffic. The 98 kDa γ-COP[743] is related to yeast Sec21p.[308] The primary amino acid sequence of the 61 kDa ξ-COP remains to be established.

The yeast SEC21 gene is essential for ER-Golgi transport in the yeast secretory pathway. The SEC21 gene encodes a 105 kDa protein of 935 amino acids (Sec21p).[308] Sec21p exists in both membrane-bound and soluble forms. The membrane-bound Sec21p behaves like a peripheral protein while the soluble Sec21p is present in a protein complex of about 700-800 kDa as assessed by gel filtration chromatography. When the Sec21p-containing complex was purified, six distinct subunits of about 150, 110, 105, 73, 35 and 25 kDa were detected. This subunit composition resembles that of mammalian coatomer complex. The 105 kDa subunit is Sec21p and represents the yeast γ-COP.[308] The 110 kDa subunit represents the yeast β-COP as it can be recognized by a peptide-specific antibody against mammalian β-COP and this antibody-re-

acting species copurifies with the Sec21p-containing complex. Partial amino acid sequences of this 110 kDa subunit was used to clone the yeast gene. The derived amino acid sequence predicts a protein of 973 amino acids that is 43.8% identical to rat β-COP and 40.6% identical to Drosophila β-COP.[190] Rat and Drosophila β-COP molecules are themselves 65% identical. Similar to rat β-COP, the N-terminal 470 amino acid sequence of this yeast 110 kDa protein is related to mammalian β-adaptin (about 22.5% identical). These results firmly establish that the 110 kDa subunit of yeast Sec21p-containing protein complex is the yeast β-COP and has been referred to as Sec26p (The corresponding gene is named SEC26). Similar to Sec21p, Sec26p is essential for growth and protein transport from the ER to the Golgi. In addition to being present in the yeast coatomer complex, Sec26p also interacts genetically with Sec21p and Sec27p, as overexpression of Sec26p suppresses the sec21-1 and sec27-1 mutations. The sec27-1 mutation is a recessive temperature-sensitive mutation originally isolated in a screen for mutants defective in mitotic spindle formation and its suppression by Sec26p suggests that it may correspond to another subunit of the yeast coatomer. The Sec27p is involved in ER-Golgi transport as sec27-1 mutant cells are defective in this transport step at the nonpermissive temperature. By complementing the sec27-1 mutation, the SEC27 gene was cloned and it encodes a protein of 889 amino acids which is about 45% identical to bovine and human β'-COP (the bovine and human β'-COP are themselves 98.5% identical).[190] Similar to β'-COP, there exist five WD-40 repeats[515] in the N-terminal 280 amino acids of Sec27p and this N-terminal region is most similar to β'-COP (about 60% amino acid identity). The remaining carboxyl two-thirds of Sec27p is about 35% identical to the corresponding region of β'-COP. Sec27p is present in the yeast coatomer complex and one N-terminal amino acid sequence derived from the

110 kDa fraction of purified yeast coatomer matches perfectly with that of the derived Sec27p sequence, suggesting that, like the mammalian coatomer, the 110 kDa fraction of the purified yeast coatomer contains two distinct proteins: the Sec26p and Sec27p. These mean that Sec27p is the yeast homologue of mammalian β'-COP. The yeast α-COP encoding gene was identified in a genetic screen for yeast mutants defective in Golgi retrieval of ER membrane proteins with a cytoplasmic dilysine motif (discussed later) and is named RET1.[403] The RET1 gene was cloned by complementing yeast ret1-1 mutant and it encodes a protein of 1201 amino acids. Its identity as the yeast α-COP was established by the perfect match of its predicted N-terminal amino acid sequence with the N-terminal 20-residue sequence derived from microsequencing of the purified yeast α-COP (the 150 kDa protein of yeast coatomer). Similar to β'-COP and Sec27p (yeast β'-COP), the N-terminal portion of yeast α-COP also contains six WD-40 repeats (residue 4-45, 46-87, 88-129, 130-171, 202-243 and 246-287). Using partial amino acid sequences derived from purified protein, the yeast α-COP was cloned and sequenced independently.[229] Partial amino acid sequences of mammalian (bovine) α-COP are highly homologous to the corresponding regions of yeast α-COP, suggesting that this protein is evolutionarily conserved.[229] Since Sec21p (yeast γ-COP), Sec26p (yeast β-COP) and Sec27p (yeast β'-COP) are all involved in yeast ER-Golgi transport and mutation of ε-COP in CHO idlF mutant cells results in defective Golgi structure and ER-Golgi transport, the coatomer may be an essential protein complex for ER-Golgi transport in both yeast and mammalian cells. The functional involvement of coatomer in ER-Golgi transport of mammalian cells has also been supported by the demonstration that microinjection of antibodies against β-COP inhibits transport of VSV-G protein from the ER to the Golgi.[576] Similarly, transport of VSV G-protein in vitro is also

inhibited by antibodies against β-COP.[581] Immunocytochemical studies have established that components of coatomer are confined mainly to the cis Golgi and pre-Golgi structures implicated in trafficking between the ER and the Golgi.[253,549] Since components of the coatomer were initially identified in the intra-Golgi transport system and are essential for COP-coated vesicle transport between the cis and medial cisternae of the Golgi, the combined evidence strongly suggests that coatomer may play an essential role in intra-Golgi transport in vivo, in addition to its role in ER to cis Golgi transport. This conclusion is further supported by the observation that ARF is another key component of COP-coated vesicles and is functionally important for ER-Golgi transport and intra-Golgi transport both in vivo and in vitro.[188,253,256,609,647,649,766]

GENETIC IDENTIFICATION AND CLONING OF GENES ESSENTIAL FOR IN VIVO TRANSPORT

Genetic screens for yeast (*S. cerevisiae*) mutants defective in the secretory pathway have been most fruitful in identifying and cloning genes involved in the exocytotic pathway (Fig. 9).[609] Most of them are temperature-sensitive (ts) mutants. In 1979, the first two sec (secretion) mutants (sec1-1 and sec2-1) were isolated by screening an unselected collection of 87 temperature-sensitive strains for mutants that accumulate precursor forms of invertase and acid phosphatase.[538] At the nonpermissive temperature (37°C), Sec1 mutant cells stop dividing, while protein and phospholipid synthesis continue for at least 3 hr, resulting in dense cells that can be separated from other cells by a Ludox density gradient. This observation led to the design of a procedure that would enrich for ts sec mutants. Yeast cells are mutagenized with ethyl methanesulfonate (EMS) or nitrous acid and allowed to recover by growth at 25°C for 16 hr. Cells are then allowed to grow for 3 hr at 37°C for potential ts sec mutants to grow into dense cells, which are then

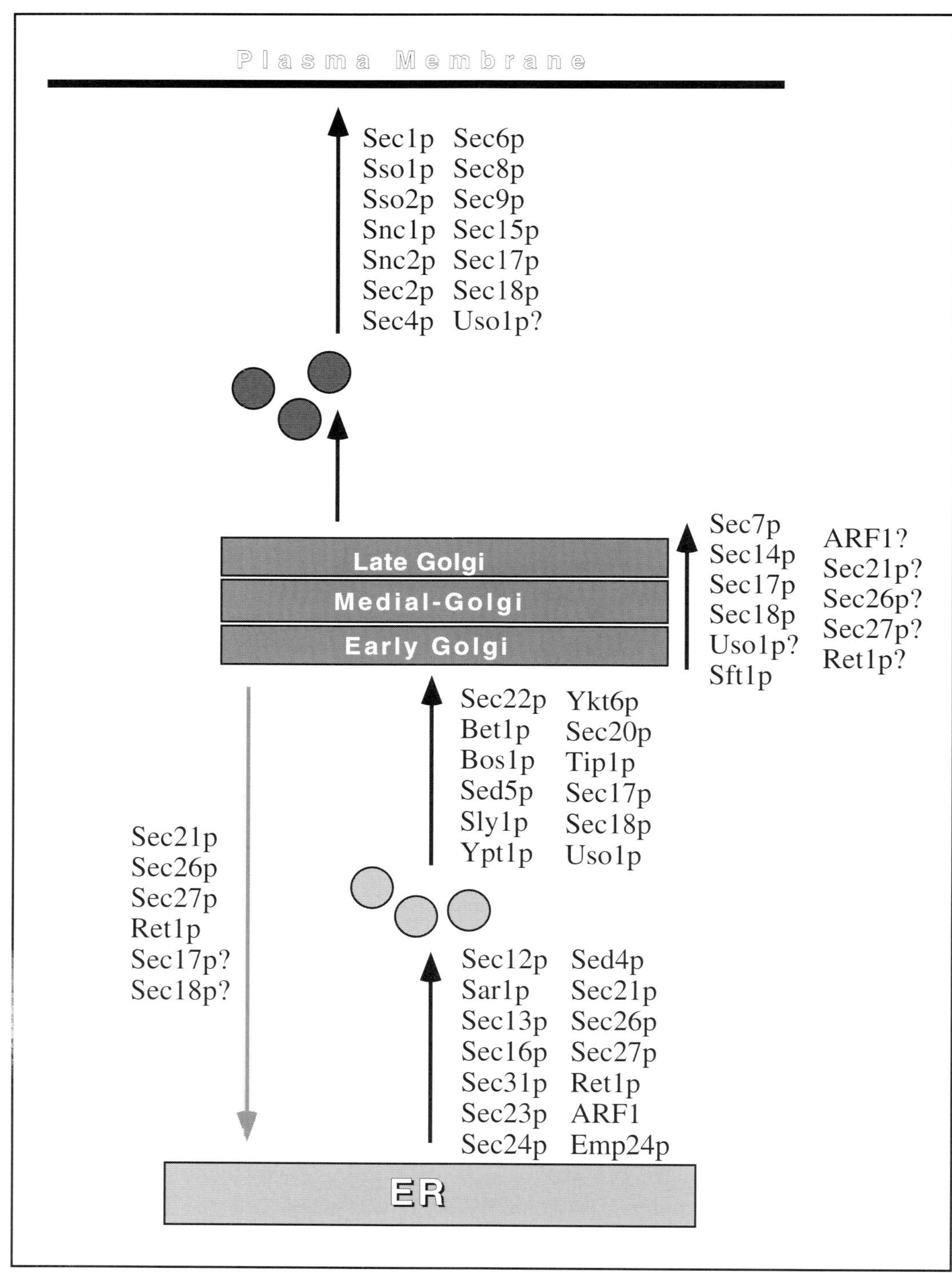

Fig. 9. Proteins that play a role in the yeast secretory/exocytotic pathway.

enriched by a Ludox density gradient. The enriched dense cells are then spread on agar culture plates and incubated at 20-25°C for 2 days to allow individual colonies to grow up. The colonies are replica plated onto two fresh plates; one replica is incubated at room temperature and the other at 37°C. ts mutants for growth are picked from the master plate after comparing the two replica plates. These ts growth mutants are then assayed for ts defects in secretion of invertase and acid phosphatase. This screening procedure yielded to a collection of 188 ts mutant strains that accumulate intracellular pools of invertase at 37°C and fall into 23 complementation groups (sec1-23).[204,536,537] Two additional ts mutants bet1 and bet2 (blocked early in transport) were isolated based on a ^{3}H-mannose suicide selection in which cells with defective protein transport to the Golgi do not receive lethal ^{3}H-mannose addition to their glycoproteins.[519] A large number of vps (vacuolar protein sorting) mutants have been isolated by the mislocalization of vacuolar resident proteins to the plasma membrane (see chapter 6). They are defective in protein sorting to the vacuole and/or in vacuole biogenesis. ret1, as mentioned earlier, was isolated by its defect in retrieval from the Golgi of cytoplasmic dilysine-based ER membrane proteins. Genes involved in the yeast exocytotic pathway were also identified by other genetic strategies, such as isolating suppressers for existing ts mutants or loss of a particular gene. SAR1 was cloned as a suppresser for sec12;[510] BOS1 as a suppresser for bet1;[700] SED1, SED2, SED3, SED4 and SED5 as multicopy suppressers for the loss of the ERD2 gene (to be discussed later);[267,269] SSO1 and SSO2 as suppressers of sec1 mutations,[2] and SLY2, SLY12, SLY41 and a mutant allele of SLY1 (Sly1-20) as suppressers for the functional loss of the YPT1 gene.[160,558] SED2 is identical to SEC12. SLY2 is identical to SEC22 while SLY12 is identical to BET1. Additional approaches have also identified other genes involved in the yeast

secretory pathway. For example, TIP1 (SEC twenty interacting protein) was identified by its ability to suppress the lethal effect of overexpressing the cytoplasmic domain of Sec20p and interacts physically with Sec20p.[759] A 105 kDa protein (p105) was identified by its interaction and copurification with Sec23p.[292] The gene encoding p105 (or Sec24p) has been named SEC24. The complex composed of Sec23p and Sec24p has been referred to as the Sec23p complex. Similarly, a 150 kDa protein (p150) was identified by its association with Sec13p.[608] p150 has been termed Sec31p and its gene SEC31. Sec13p and Sec31p form the Sec13p complex. Subsequent studies established that SEC12/SED2, SEC13, SEC16, SEC17, SEC18, SEC20, SEC21, SEC22/SLY2, SEC23, SEC24, SEC26, SEC27, USO1, BET1/SLY12, BET2, BOS1, SLY1, SEC31 and SED5 are required for ER-Golgi transport.[609] Four small GTPases, SAR1, YPT1, ARF1 and ARF2, are also involved in ER-Golgi transport.[609] Among these genes, SEC12/SED2, SEC13, SEC16, SEC23, SEC24, SEC31 and SAR1 have been determined to be involved in vesicle formation, while SEC17, SEC18, SEC22/SLY2, BET1/SLY12, BOS1, SED5, YPT1 and SLY1 are involved in vesicle docking and/or fusion. BET2, SEC17 and SEC18 are also involved in other transport events. In addition to SEC17 and 18, SEC7 and SEC14 are implicated in intra-Golgi transport and/or Golgi export. Genes involved in transport from the Golgi to the plasma membrane include SEC1, SEC2, SEC3, SEC4, SEC5, SEC6, SEC8, SEC9, SEC10, SEC15, SNC1, SNC2, SSO1 and SSO2. Some of the gene products are discussed in more detail below, while others, such as Sec21p, Sec26p, Sec27p, Ret1p and Uso1p have been discussed earlier.

Sec1p

The SEC1 gene is required for docking and fusion of Golgi derived vesicles with the plasma membrane and was cloned by complementation of the sec1-1 mutant.[3] The SEC1 gene encodes a hydrophilic protein of 724 amino acids. When expressed from a high copy number plasmid, SEC1 can suppress the sec3-2, sec15-1 and sec10-2 mutations, indicating interaction of Sec1p with these gene products. Two high copy suppressers (SSO1 and SSO2) for sec1 have been isolated that encode proteins related to syntaxins, suggesting that Sec1p can interact with Sso1p and Sso2p on the plasma membrane and may be involved in vesicle docking and fusion.[2] In yeast, Sec1p, Sly1p, Vps33p/Slp1 and Vps45p form a family of proteins involved in vesicle docking and fusion.[1,288+V?S45] The Drosophila Rop and C. elegans UNC 8 proteins are homologues of Sec1p. Based on the conserved sequences among Sec1p, Rop and UNC 8, a mammalian Sec1p homologue was identified and has been referred to as n-Sec1 or rbSec1.[223,587] This mammalian Sec1p-like protein was also identified by a biochemical approach for syntaxin-binding proteins and has been referred to as munc-18.[275,301] munc-18/n-Sec1/rbSec1 is expressed primarily in the brain and binds syntaxin 1, 2 and 3, but not 4. munc-18/n-Sec1/rbSec1 is about 65%, 59% and 27% identical to Rop, UNC-18 and Sec1p, respectively. A ubiquitously expressed mammalian homologue has also been cloned from mouse (muSec1/Munc-18b)[348,779] and rat (Munc-18-2),[276] and is about 63% identical to the brain-specific munc-18/n-Sec1/rbSec1. In addition, another ubiquitously-expressed form (Munc-18c) that is about 51% identical to the brain-specific form has also been cloned.[779]

Sso1p and Sso2p

SSO1 and SSO2 were each isolated as a high copy suppresser for the sec1 ts mutation and encode highly related proteins of 290 and 295 amino acids, respectively.[2] Sso1p and Sso2p are 72% identical and structurally related to syntaxins and other syntaxin-like proteins, including yeast Sed5p and Pep12p. Pep12p is also known as Vp16p, Vpt13p or Vps6p. Like syntaxins, both Sso1p and Sso2p are anchored by a C-terminal hydrophobic tail anchor to the cytoplasmic side of the membrane. Although SSO1 and SSO2 can suppress the sec1 mutation, they cannot rescue SEC1 disruption, suggesting that Sso1p and Sso2p interact with Sec1p functionally. Genetically, SSO1 and SSO2 can also suppress other sec mutations, including sec3-2 and sec15-1, sec9-4. SSO1 and SSO2 themselves together perform an essential function. Cells depleted of Sso1p and Sso2p accumulate a large number of 100 nm Golgi-derived vesicles, suggesting that they are involved in docking and fusion of these vesicles. Biochemical studies establish that Sso proteins can indeed form protein complexes with Sec9p and Snc proteins. The assembly of this complex may be regulated by Sec4p and plays a critical role in vesicle docking and fusion.

Snc1p and Snc2p

The SNC1 gene was isolated by its ability to suppress the loss of C-terminal functions of CAP in S. cerevisiae strains possessing an activated allele of yeast RAS2.[230] CAP is a component of the RAS-responsive S. cerevisiae adenylyl cyclase complex. SNC1 encodes a membrane protein of 117 amino acids. The C-terminal 24-residue sequence functions as a C-terminal tail anchor, resulting in the cytoplasmic exposure of the main part of the polypeptide (residue 1-93). Most importantly, the Snc1p sequence is about 40% identical to synaptobrevin 1 and 2, which are associated with synaptic vesicles. Gene disruption suggests that SNC1 itself is not essential. Since there are two highly related synaptobrevins, oligonucleotides based on the conserved regions between Snc1p and synaptobrevins were used in polymerase chain reaction (PCR) to isolate an SNC1 homologue termed SNC2.[606] The encoded protein (Snc2p) consists of 115 amino acids and is about 79% identical to Snc1p. Snc2p has a cytoplasmic 92-residue sequence (residue 1-92) followed by a 23-residue C-terminal hydrophobic tail anchor (residue 93-115). Gene disruption of SNC2 alone has no detectable phenotype. However, disruption of both SNC1 and SNC2 resulted in conditional lethal phenotypes. Depletion of cellular Snc proteins led to a defect in protein secretion and accumulation of 100 nm Golgi-derived vesicles. Immunolocalization has established that Snc proteins are associated mainly with Golgi-derived vesicles. Protein interaction has been observed among Snc proteins, Sec9p and Sso proteins.[144] Snc proteins may thus play an important role in Golgi-derived vesicle docking and fusion by interacting with a protein complex composed of Sso proteins, Sec1p and Sec9p on the plasma membrane.

Sec2p

Sec2p plays an essential role in vesicle docking and/or fusion with the plasma membrane.[506] Cloned by complementation of a sec2 ts mutant, SEC2 encodes a hydrophilic protein of 759 amino acids. Antibodies raised against Sec2p revealed that Sec2p has an apparent size of about 105 kDa on the SDS polyacrylamide gel. The N-terminal domain of Sec2p is essential for its function and contains a potential coiled-coil domain. Deletion of the C-terminal 251 amino acids of Sec2p had no apparent effect on its role in vesicle docking/fusion. However, truncation of the C-terminal 368 amino acids results in a ts phenotype. Sec2p is found predominantly in the soluble fraction and has a native molecular mass of more than 500 kDa, suggesting that Sec2p exists either in a homooligomeric or a heterooligomeric form. The sec2 ts phenotype can be partially suppressed by duplication of SEC4 gene, indicating a functional interaction between Sec2p and Sec4p.

Sec4p

Mutations in SEC4 gene block docking and/or fusion of Golgi-derived vesicles with the plasma membrane.[656,824] The SEC4 gene was cloned by its ability to partially complement the sec15-1 growth defect through the genetic interaction between SEC4 and SEC15. The amino

acid sequence of Sec4p derived from its gene sequence is very informative as the 215-residue sequence of Sec4p bears amino acid sequence similarity with members of the ras transforming proteins. The Sec4p is about 30% identical to human H-ras and is thus a member of the ras-like small GTPase superfamily. The major similarity between Sec4p and ras is confined to four domains involved in the binding and hydrolysis of GTP. This discovery was the first evidence that small GTPases may be involved in the regulation of vesicular transport. Furthermore, Sec4p is about 45% identical to yeast YPT1gene product (Ypt1p). YPT1 was cloned in 1983 and shown to be an essential gene, although its exact role was not known.[221,672,673] The high degree of amino acid identity between Sec4p and Ypt1p and the involvement of Sec4p in vesicle docking and/or fusion suggested that Ypt1p may also be involved in the secretory pathway. Ypt1p was subsequently shown to be involved in docking and/or fusion of ER derived vesicles with the Golgi apparatus.[53,339,414,687,688] Since SEC4 gene was isolated by its ability to partially suppress the sec15-1 mutation, the genetic interaction of SEC4 gene with others was investigated. The duplication of SEC4 gene was found to partially suppress sec2-41 and sec8-9, in addition to sec15-1. Synthetic lethality between sec4 and sec2, sec3, sec5, sec8, sec10, sec15 or sec19 was observed, suggesting that Sec4p may interact functionally with diverse gene products involved in Golgi to surface transport. In wild-type cells, Sec4p is present on the cytoplasmic side of both the plasma membrane and vesicles in transit to the cell surface,[245] while Sec4p is predominantly associated with the vesicles accumulated in sec mutants blocked in docking/fusion with the plasma membrane. Pulse-chase experiments suggest that Sec4p is synthesized as a soluble form that rapidly and tightly associates with vesicles and the plasma membrane. The membrane association of Sec4p is due to the addition of a geranylgeranyl group(s), a 20-carbon hydrophobic lipid, to its C-terminal CysCys residues. Since Sec4p and Ypt1p are highly related and required specifically for Golgi-surface and ER-Golgi transport, respectively, chimeric proteins between these two proteins have allowed the identification of three domains that specificity the localization and function of Sec4p and Ypt1p.[86,91] These include the effector domain (residues 44-55: KFNPSFITTIGI for Sec4p), the ras-equivalent loop 7 (L7) domain (residues 120-128: EHANDEAQL for Sec4p) and the C-terminal hypervariable domain (residues 172 to 215 for Sec4p). The effector domain is involved in the interaction of Sec4p with downstream effector proteins, while the hypervariable domain is involved in specific cellular localization.[121] When the L7 and the hypervariable domains of Sec4p are replaced with the corresponding regions of Ypt1p, the resulting chimeric protein has both Sec4p and Ypt1p functions and can complement the simultaneous deletion of both SEC4 and YPT1 genes at 25°C. Although this chimeric protein can functionally replace Sec4p at both 14°C and 37°C, it could not replace Ypt1p function at 14°C and 37°C, suggesting that other regions of Ypt1p are required for full Ypt1p function. When the effector domain of this chimeric protein is also replaced with that of Ypt1p, the resulting chimeric protein possesses only Ypt1p function. This suggests that the effector domain of Sec4p can not be replaced by that of Ypt1p and is thus uniquely required for its function, while Ypt1p can use its own as well as Sec4p's effector domain for its function. In contrast, a chimeric protein using Ypt1p as the framework in which all these three domains of Ypt1p are replaced with those of Sec4p has neither Sec4p nor Ypt1p function, suggesting that the effector domain of Sec4p is not sufficient but requires the Sec4p framework for its function. The importance of the effector domain of Sec4p is further revealed by the observation that Sec4p lost its function when a four-residue sequence (46-49: NPSF) of this do-

main is replaced by the corresponding sequence (TNDY) of Ypt1p. A single mutation of the Ser residue at position 48 to Pro (sec4-P48) results in cold sensitivity in that yeast strains containing this sec4-P48 allele as the only copy of SEC4 show extremely slow growth at 14°C, although they grow normally at 25°C. At 14°C, the sec4-P48 mutant accumulates a large number of 100 nm Golgi-derived vesicles and has a pronounced defect in invertase secretion, suggesting that sec4-P48 protein cannot function properly at this temperature. The sec4-P48 mutant was used to isolate high copy suppressers that support normal growth at 14°C. Of the eight independent suppressers isolated, HSS47 (high copy suppresser of sec4) is most interesting as it not only suppresses the sec4-P48 mutation at 14°C but also the sec9-4 mutation at 37°C. These observations led to the conclusion that HSS47 is identical to SEC9, suggesting strongly that Sec9p can be one of the downstream effector proteins of Sec4p.[94]

Sec6p

SEC6 gene is required for docking and/or fusion of Golgi derived vesicles with the plasma membrane.[603] By complementation of a sec6 ts mutation, SEC6 was cloned and predicts a hydrophilic protein of 733 amino acids. Sec6p is fould predominantly in the soluble fraction and exists in a high molecular weight (Sec6/8/15) complex that also contains Sec8p and Sec15p.[780] The Sec6/8/15 complex contains five other uncharacterized proteins (of about 144, 107, 100, 91 and 70 kDa) in addition to Sec6p (88 kDa), Sec8p (121 kDa) and Sec15p (113 kDa). The Sec6/8/15 complex is disrupted in sec3-2, sec5-24 and sec10-2 strains, suggesting that these gene products may directly or indirectly participate in the formation of this complex. The Sec6/8/15 complex is believed to be involved in docking/fusion of Golgi-derived vesicles. Furthermore, based on the localization of Sec8p, the Sec6/8/15 complex is probably localized to small bud tips and may be involved in selective docking/fusion of vesicles with the growing bud.

Sec7p

Yeast sec7 mutants are pleiotropically deficient in protein transport within the Golgi apparatus and cause the proliferation of a large array of Golgi cisternae at 37°C.[5] SEC7 encodes a large hydrophilic protein of 2008 amino acids with a predicted size of 230 kDa. The N-terminal portion of Sec7p contains a 125-residue region (residue 89-213) that is highly acidic. This acidic region consists of 47% acidic residues (29% glutamate and 18% aspartate) and is rich in Ser residues (21%). The function of this acidic Ser-rich domain is unknown but has been proposed to interact with lipids or proteins on the cytoplasmic side of the Golgi membrane, since Sec7p is associated with the cytoplasmic surface of the Golgi membrane. Sec7p has been suggested to play a role in protein transport between distinct compartments of the yeast Golgi apparatus.

Sec8p

SEC8 encodes a 121 kDa hydrophilic protein composed of 1065 amino acids.[80] The N-terminal domain of Sec8p contains a region (residues 30-230) that is leucine rich and bears sequence similarity with that of yeast adenylate cyclase encoded by the CYR1 gene. This domain of yeast adenylate cyclase is required for its activation by ras. By analogy, this leucine rich domain of Sec8p may be involved in some protein-protein interaction. Genetic analysis indicates that Sec8p interacts functionally with Sec4p. Furthermore, Sec8p exists in the Sec6/8/15 complex and is confined to the bud tip.[780]

Sec9p

SEC9 gene, a high copy suppresser for sec4-P48 cold sensitive mutant, encodes a hydrophilic protein of 651 amino acids.[94] Although the calculated size of Sec9p is 74 kDa, the apparent size of Sec9p as determined by SDS-PAGE is about 106 kDa. In the 651-residue sequence of Sec9p, a 17-residue Gln-rich (Q-rich) region (residue 373-389) (Q_6-X_5-Q_6) divides Sec9p into two domains: an N-terminal 372-residue (residue 1 to 372) domain and a C-terminal domain of 262 residues (residue 390-651). The most striking feature of Sec9p is that the C-terminal 231-residue sequence bears homology (about 19% identity and 60% similarity in a 205-residue overlap) with SNAP-25 protein that is associated with the presynaptic membrane and involved in docking and fusion of synaptic vesicles. The functional importance of the SNAP-25-like domain of Sec9p is revealed by the fact that the sec9-4 mutant allele results from a mutation of the evolutionally conserved Gly residue 458 to an Asp residue. Furthermore, a fragment of Sec9p from residue 416 to the C-terminus can fulfill all essential functions of Sec9p, suggesting that the SNAP-25-like domain of Sec9p represents the sole essential domain and the N-terminal region (residue 1-415) is dispensable for Sec9p function. Similar to SNAP-25, Sec9p is localized peripherally to the cytoplasmic side of the plasma membrane. Sec9p can associate with both Sso and Snc proteins to form a protein complex that is similar to the SNAP receptor complex composed of SNAP-25, syntaxins 1A/B and synaptobrevin. Sec9p may be a downstream effector of Sec4p and Sec4p may regulate directly or indirectly the assembly of the Sec9p-Ssop-Sncp SNARE complex, which plays a fundamental role in docking and fusion of Golgi-derived vesicles with the plasma membrane.

Sec12p and Sar1p

SEC12 encodes a 70 kDa glycoprotein of 471 amino acids.[172,509] There is a hydrophobic region of about 19 residues (residues 355-373). Preceding this hydrophobic region, there are five potential sites for N-linked glycosylation in the N-terminal 354-residue sequence, while two potential sites are present in the downstream C-terminal 98-residue sequence. Sec12p is also rich in Ser and Thr residues that are potential sites for O-linked glycosylation. Sec12p is a type II integral protein with the N-terminal 354-residue sequence and the C-terminal 98-residue sequence oriented in the cytoplasmic and luminal side, respectively, of the membrane.[70,172] The luminal 98-residue domain is heavily glycosylated by both N- and O-linked glycans. The C-terminal 71-residue sequence of Sec12p is not essential for ER-Golgi transport, suggesting that the majority of the luminal domain is dispensable. The integrity and membrane association of the cytoplasmic domain are essential for its function. When expressed alone as a soluble form, the N-terminal 354-residue cytoplasmic domain has an inhibitory effect on ER-Golgi transport, probably by functioning as a dominant negative mutant. A gene (SAR1) encoding a small GTP binding protein Sar1p was isolated as a multicopy suppresser of a sec12 ts mutant.[46,510,544,545] Sar1p is about 21 kDa and consists of 191 amino acid residues. The inhibitory effect of the soluble cytoplasmic domain of Sec12p can be suppressed by overexpression of Sar1p, similar to the suppression of sec12 mutant by overproduced Sar1p. Furthermore, membrane association of Sar1p can be enhanced by elevated levels of Sec12p, suggesting that the cytoplasmic domain of Sec12p may interact with Sar1p.[174] Since Sar1p is required for ER-Golgi transport, the role of Sec12p in ER-Golgi transport may be to recruit Sar1p to the membrane. Sec12p functions as a guanine-nucleotide exchange factor for Sar1p to enhance the conversion of GDP-bound Sar1p to the GTP-bound form, resulting in the membrane recruitment of GTP-bound Sar1p.[48] However, GTP hydrolysis by Sar1p is not enhanced by Sec12p. Sec12p and Sar1p are mainly involved in vesicle budding from the yeast ER membrane, probably by mediating the recruitment of other components of the coat proteins.[510,544,545] Sar1p itself is a component of the coat protein for yeast ER derived vesicles. Both Sec12p and Sar1p genes have been isolated from the fission yeast (*Schizosaccharo-*

myces pombe) and a plant (*Arabidopsis thaliana*).[173] Two highly related mammalian Sar1p cDNAs have also been cloned and the mammalian Sar1p has been shown to participate in ER-Golgi transport by promoting vesicle budding from the ER.[382,699]

Sed4p is structurally related to Sec12p in that it is also a type II membrane protein with an N-terminal cytoplasmic domain that is homologous to the corresponding region (45% identity) of Sec12p. Recent studies suggest that Sed4p interacts with Sec16p and participates in vesicle formation in the ER.[231a] Sec16p is a 240 kDa protein (2194 amino acids) and function as a new component of COPII coat because it interact with Sec23p.[205a]

Sec13p-Sec31p complex

SEC13 encodes a 33 kDa protein of 297 amino acids.[608,655] The majority of the 297-residue sequence can be divided into six WD40 repeats,[515] each of about 40 residues. The importance of the WD40 repeats is revealed by noting that three of the ts alleles (sec13-1, sec13-4, sec13-5) of SEC13 are due to mutations of conserved residues in the WD-40 repeats. In wild-type cells, the majority of Sec13p behaves like a peripheral membrane protein in that it sediments mainly in a particulate fraction and can be extracted by pH 11.5 carbonate buffer. When Sec13p is overexpressed most of the over-expressed protein is in the soluble fraction, suggesting that Sec13p binding to the putative membrane association sites is saturable. Functional Sec13p-dihydrofolate reductase (Sec13p:DHFR) fusion protein can be produced in which the entire coding region of mouse DHFR is fused to the C-terminus of Sec13p. Due to the high affinity (even at 1M KCl) of DHFR for its inhibitor, methotrexate, Sec13p:DHFR can be easily purified by affinity chromatography using methotrexate-agarose. Two forms of Sec13p:DHFR exist: a monomeric species and a high molecular weight form. A protein of 150 kDa (p150) copurifies with the high molecular weight form. Using a Sec13p-dependent ER vesicle budding assay, it was established that the Sec13p complex is essential for vesicle budding from the ER. Furthermore, Sec13p and p150 are components of a novel protein coat (COPII) associated with yeast ER derived transport vesicles.[47,608,655] The gene encoding p150 has been referred to as SEC31 and p150 as Sec31p. SEC31 was recently cloned (as WEB1) and the predicted Sec31p is composed of 1273 amino acids. The N-terminal portion of Sec31p contains nine WD-40 repeats, while its C-terminal portion contains a region rich in Pro. A human cDNA has been reported that encodes a protein (designated SEC13Rp) of 322 amino acids.[757] The derived amino acid sequence of this human protein is about 53% identical to Sec13p. Similarly, there exists six WD40 repeats in SEC13Rp. Antibodies against SEC13Rp revealed that it is confined to structures involved in ER-Golgi transport.[696] A chimeric protein composed of the N-terminal half of Sec13p and the C-terminal half of the SEC13Rp is able to complement a SEC13 gene deletion in yeast, establishing that SEC13Rp is indeed the human counterpart of Sec13p.[696]

Sec14p

SEC14 is required for protein export from the Golgi and/or transport within the Golgi. The cloned SEC14 gene encodes a 37 kDa hydrophilic protein of 304 amino acids.[38] Sec14p resides primarily in the cytosolic fraction and is evolutionally well conserved. A significant amount of Sec14p is peripherally associated with the Golgi membrane. Sec14p is identical to the PIT1 gene product, which is a phosphatidylinositol/phosphatidylcholine transfer protein.[37] Phospholipid transfer proteins are a class of cytosolic proteins that are ubiquitous among enkaryotic cells and are distinguished by their ability to catalyse the exchange of phospholipids between membranes in vitro. Sec14p has been proposed to maintain a reduced phosphatidylcholine content in the Golgi membrane and to function as a sensor of Golgi membrane phospholipid composition,[471] thus maintaining a phosphatidylcholine content compatible with vesicular transport.

Sec15p

SEC15 encodes a hydrophilic protein of 911 amino acids. Although the calculated size is about 105 kDa, the apparent size of Sec15p detected by antibodies is about 113 kDa.[81,657] Overexpression of Sec15p interferes with the secretory pathway, resulting in the formation of a cluster of vesicles and a patch of Sec15p revealed by immunofluorescence. Furthermore, the Sec15p patch formation upon its overexpression requires the functional Sec2p and Sec4p. As mentioned earlier, Sec15p exists in the Sec6/8/15 complex.[780]

Sec17p

As discussed earlier, Sec17p is a 33 kDa protein composed of 291 amino acids and is the functional yeast homologue of mammalian SNAP.[251] By binding to membrane SNAP receptors, Sec17p plays an essential role in membrane recruitment of Sec18p (the yeast NSF) and in vesicle fusion.

Sec18p

By complementation of sec18 ts mutation. SEC18 was cloned before the establishment of the primary structure of NSF.[195] The cloning of hamster NSF and functional analysis of Sec18p established that Sec18p is the yeast counterpart of mammalian NSF. Sec18p is composed of 757 amino acids. Sec18p exists both as an 84 kDa species as well as an 82 kDa form in the cells. Similarly these two forms of Sec18p can be produced by in vitro transcription/translation of the cloned SEC18 gene. The 82 kDa species is most likely due to translational initiation at the internal Met residue at position 19 and/or 22.

Sec20p

Sec20 mutants are defective in ER-Golgi transport. However, morphological analysis did not reveal which step (budding vs fusion) of transport is affected because cells accumulate both ER membrane and some clusters of vesicles at the nonpermissive temperature. The number of accumulated vesicles is much less than that observed in sec18 mutants, in which accumulation of ER membrane was not observed. SEC20 encodes a type II membrane protein of 383 amino acids.[758] The N-terminal 275-residue sequence represents the cytoplasmic domain followed by a transmembrane domain of 17 residues (residue 276-292). The C-terminal 91-residue (residue 293-383) sequence forms the luminal domain, which is glycosylated by O-linked glycans. Sec20p is about 50 kDa as revealed by SDS-PAGE and is associated predominantly with the ER membrane. Most interestingly, Sec20p terminates with the HDEL sequence, which has been established as the ER localization signal for luminal proteins and is recognized by a membrane receptor encoded by the ERD2 gene (see later). Since the Sec20p C-terminal HDEL sequence is located in the lumen, it can be recognized by the receptor. Deletion analysis suggests that the Sec20p C-terminal HDEL sequence plays a role in maintaining intracellular levels of Sec20p, although it is dispensable for Sec20p function. Similar to sec20 mutants, cells depleted of Sec20p also accumulate ER membrane and some clusters of vesicles. The potential step for sec20p function has not been revealed, although it is suggested that Sec20p may play a role in vesicle docking and fusion. This is because sec20 and sec18 double mutants accumulate vesicles to the same extent as in sec18 cells, more than in sec20 cells, indicating that the function of Sec20p may be downstream of Sec18p.

Tip1p

TIP1 was cloned by its ability to suppress the toxic effect of the Sec20p cytoplasmic domain upon overexpression and encodes an 80 kDa hydrophilic protein of 701 amino acids.[759] Tip1p can interact with the cytoplasmic domain of Sec20p,

suggesting that the Sec20p-Tip1p complex may be important for the function of both proteins. Depletion of cellular Tip1p led to an inhibition of ER-Golgi transport and accumulation of extensive ER network and some vesicles, establishing that Tip1p is a component involved in ER-Golgi transport.

Sec22p/Sly2p

SEC22 was identical to SLY2, which is a suppresser for the functional loss of Ypt1p.[160,520,558] Sec22p/Sly2p is a 24 kDa membrane protein composed of 214 amino acid residues anchored by its C-terminal 26-residue tail anchor (residue 189-214). The rest of the polypeptide is exposed on the cytoplasmic side of the membrane. The region consisting of residues 125 to 188 is structurally related to synaptobrevin and may form two consecutive coiled-coil domains (residue 133-163 and residue 164-188). Sec22p/Sly2p is required for ER-Golgi transport and is a membrane component of the ER derived transport vesicle, possibly as a SNARE. Sec22p/Sly2p interacts functionally with Bet1p/Sly12 and Bos1p in ER-Golgi transport.

Sec23p-Sec24p complex

SEC23 encodes an 84 kDa protein of 768 amino acid residues.[291] Similar to Sec13p, Sec23p is mainly associated with a particulate fraction. The majority of Sec23p can be released into a soluble form by incubation at pH 11.5 carbonate buffer or 2.5 M urea, suggesting that Sec23p is also a peripheral membrane protein. When perforated spheroblasts and cytosol prepared from the sec23 cells grown at the permissive temperature are used, in vitro ER-Golgi transport is not observed at 30°C. The defective transport at 30°C can be rescued when wild-type cytosol replaces the sec23 cytosol, providing a biochemical assay for purification of Sec23p. Biochemical fractionation of wild-type cytosol revealed that functional Sec23p has an apparent size of about 300 kDa. When overexpressed, Sec23p exists in two forms: a monomeric species and the 300 kDa form. When purified, the 300 kDa form contains Sec23p and another protein of about 105 kDa (p105). Antibodies against p105 inhibit ER-Golgi transport by preventing vesicle budding.[292] p105 is designated Sec24p. Sec23p (in either the monomeric form or the Sec23p-Sec24p complex) functions as a GAP (GTPase activating protein) for Sar1p to stimulate its GTP hydrolysis.[864] Sec23p alone stimulates GTP hydrolysis about 10- to 15-fold. When Sec12p (which stimulates the exchange of GDP for GTP) and Sec23p are both present, GTP hydrolysis by Sar1p can be enhanced more than 50-fold. Similar to Sec13p-Sec31p complex, Sec23p-sec24p complex is also a component of COPII coat for yeast ER derived vesicles.[47] Antibodies against Sec23p detected an 85 kDa counterpart in mammalian cells that is enriched in structures involved in ER-Golgi transport,[556] suggesting that a Sec23p homologue in mammalian cells may also be involved in ER-Golgi transport. A mouse cDNA clone encoding a Sec23p-like protein (Msec23p) has been sequenced.[819] The Msec23p has an estimated size of about 65 kDa and is composed of 573 amino acids. Its derived sequence is about 40% identical to the Sec23p. The N-terminal portion of Msec23p contains three successive immunoglobulin-like domains followed by α-helical regions, indicating that the N-terminal part of Msec23p may interact with other proteins. Whether the mammalian 85 kDa protein detected by antibodies against the yeast protein corresponds to the Msec23p remains to be established.

Bet1p/Sly12p

Cloning of the BET1 gene revealed that it is identical to SLY12, a suppresser of loss Ypt1p function.[160,558] Bet1p/Sly12p is an 18 kDa integral membrane protein composed of 142 amino acids. It is anchored to the membrane by a C-terminal 18-residue (residue 125-142) hydrophobic tail. The region between residue 61 to 124 is structurally related to synapto-

brevin and may form a coiled-coil domain. Bet1p/Sly12p is required for ER-Golgi transport. As a membrane component of ER derived transport vesicles, Bet1p/Sly12p may function as a SNARE. Functionally, Bet1p/Sly12, Sec22p/Sly2p and Bos1p are members of a group of interacting proteins involved in ER-Golgi transport.

Bet2p

BET2 was originally identified as a gene required for ER-Golgi transport. The cloned BET2 gene encodes a 36 kDa protein of 322 amino acids that is related (34% amino acid sequence identity) to Dpr1p (Ram1p), a component of the protein prenyltransferase that catalyses the attachment of prenyl group to the Ras protein. Bet2p is an essential component of the protein geranylgeranyltransferase that catalyses attachment of the geranylgeranyl group to Ypt1p, Sec4p and other members of the Ypt1p/Sec4p/Rab family. The attachment of the geranylgeranyl group to these proteins is essential for their membrane association and function. Therefore Bet2p is also required for other transport events that are regulated by members of the Ypt1p/Sec4p/Rab family.[645]

Bos1p

As a multicopy suppresser of the bet1 ts mutation, BOS1 encodes a 27 kDa membrane protein composed of 244 amino acid residues.[521,700] Anchored by its C-terminal 22-residue (residue 223-244) tail anchor, the rest (residue 1 to 222) of the polypeptide is oriented to the cytoplasmic face of the membrane. Like Sec22p/Sly2 and Bet1p/Sly12p, Bos1p is structurally related to synaptobrevin. Being essential for ER-Golgi transport, depletion of cellular Bos1p led to the accumulation of 50 nm vesicles and ER membrane. In perforated cells depleted of Bos1p, formation of ER derived vesicles is not affected. However, the resulting vesicles are incompetent for fusion with the Golgi. An antibody specific for Bos1p does not affect vesicle budding but blocks subsequent fusion of the budded vesicles.[413]

Bos1p is also a component of ER derived vesicles and functionally interacts with Sec22p/Sly2 and Bet1p/Sly12. Together with Sec22p/Sly2p, Bet1p/Sly12p and/or other components of the vesicle,[522] Bos1p may be a SNARE of ER derived vesicles and play a fundamental role in vesicle docking and fusion.

Sed5p

SED5, a multicopy suppresser for ERD2 disruption, encodes a 39 kDa integral membrane protein of 340 amino acids.[269] Sed5p is anchored by its C-terminal 16-residue hydrophobic tail. Structurally, Sed5p is related to syntaxins and syntaxin-like proteins such as Sso proteins and Pep12p. Depletion of cellular Sed5p leads to an inhibition of ER-Golgi transport and accumulation of vesicles and ER membrane. Sed5p is predominantly associated with cytoplasmic punctate structures, most likely the yeast Golgi apparatus. Sed5p can interact with Sly1p, in a way similar to the interaction between syntaxins and mammalian Sec1 proteins or between Sso proteins and Sec1p. When fusion of ER derived vesicles is blocked, antibodies against Sed5p immunoprecipitate a protein complex that contains Sed5p, Sly1p, Sec17p, Bos1p, Sec22p/Sly2p, Bet1p/Sly12p and three other proteins of about 28 (p28), 26 (p26) and 14 (p14) kDa, suggesting that Sed5p is an integral component of the SNARE complex formed between ER derived vesicles and the Golgi membrane.[720] Sed5p thus plays a critical role in ER derived vesicle docking and fusion.[36,269,720] p26 is a novel protein that was designated Ykt6p during the sequencing of the yeast chromosome XI. p26/Ykt6p is composed of 200 amino acids and is structurally related to Sec22p/Sly2p, Snc proteins and synaptobrevins, suggesting that it may function as a component of the SNARE complex. In contrast to Sec22p/Sly2p, Snc proteins and synaptobrevins, there is no hydrophobic tail at the C-terminal end of p26/Ykt6p. Instead it has a C-terminal CAAX (CIIM) consensus sequence for farnesyla-

tion, suggesting that p26/Ykt6p may be associated with the membrane via a lipid anchor. p14 is identical to Sft1p.[35]

Sft1p

SFT1 was isolated as a multicopy suppresser of sed5-1, a ts allele of SED5 created by a single Arg to Gly change at position 255 of Sed5p. It encodes an 11 kDa membrane protein of 97 amino acids characteristic of a v-SNARE.[35] Sft1p is associated with the Golgi apparatus and essential for normal growth. A ts allele (sft1-1) of SFT1 can be suppressed by SED5, providing further support for the interaction between Sft1p and Sed5p. Sft1p is involved in traffic from an early Golgi to a later Golgi compartment and it was proposed that Sft1p may function as a v-SNARE for retrograde transport from a later Golgi to the Sed5p-containing early Golgi compartment. Sft1p is identical to the p14 that is coimmunoprecipitated with Sed5p in fusion-defective cells and thus has an apparent size of about 14 kDa.[720]

Sly1p

The mutant SLY1 (Sly1-20) gene is a suppresser for the functional loss of Ypt1p.[160,558] The mutant protein results from a Glu to Lys mutation at position 532 of the wild-type Sly1p. SLY1 is an essential gene encoding a 75 kDa hydrophilic protein of 666 amino acids. Cells depleted of Sly1p are defective in protein transport from the ER to the Golgi. Since Sly1p is associated with Sed5p and is structurally related to Sec1p, Vps33p/Slp1 and Vps45p, it may be involved in docking and fusion of ER derived vesicles.

Mammalian cells defective in the exocytotic pathway have also been isolated via various approaches and have provided additional insight into intracellular vesicular transport.[256,348,381,599] As discussed previously, CHO idlF cells are defective in ER-Golgi transport at the nonpermissive temperature (39.5°C). Furthermore, the Golgi apparatus becomes dissociated into vesicles and tubules at 39.5°C. By transfecting a cDNA expression library into idlF cells for complementing growth at 39.5°C, a cDNA encoding the 36 kDa ε-COP was revealed to be able to rescue the ildF phenotype.[256] This provided strong in vivo evidence that ε-COP, and thus the coatomer, plays an essential role in establishing and maintaining the Golgi structure and in mediating ER-Golgi transport. CHO ildC cells are defective in multiple medial and trans Golgi-associated processes. By complementing the mutant phenotypes of ildC cells at the nonpermissive temperature, a human cDNA (LDLC) was isolated that could rescue the ildC phenotypes.[599] Unlike the parental CHO and another mutant CHO (ildB) cells, ildC cells have no detectable endogenous LDLC mRNA. LDLC encodes a 83 kDa hydrophilic protein (ldlCp) of 738 amino acids. ldlCp is evolutionally conserved and a C. elegans ldlCp-like protein is about 26% identical and 53% similar to the human protein. Immunofluorescence localization suggests that ldlCp is a peripheral protein associated with the Golgi apparatus. Furthermore, Golgi association of ldlCp is abolished in ildB CHO cells, although it is expressed at normal levels. ildB cells have mutant phenotypes similar to ildC cells, suggesting that Golgi association of ldlCp is essential for its normal function and the protein (ildBp) mutated in ildB CHO cells is necessary for Golgi association of ildCp. Identifying the mutant genes of other mammalian mutant cells and isolating additional mutant cells will certainly provide a better understanding of the secretory pathway.

CHARACTERIZATION OF PROTEINS ASSOCIATED WITH THE EXOCYTOTIC PATHWAY

Characterization of proteins associated with organelles of the exocytotic pathway has also provided significant knowledge about the molecular aspects of the secretory pathway. Some examples of proteins identified in this way are discussed later.

β-COP/110K protein

Monoclonal antibody M3A5, raised against microtubule-binding proteins, detects specifically a 110 kDa protein (110K protein) present in both cytosolic and Golgi-associated form.[8,189] cDNA cloning revealed that the 110K protein is β-COP of the coatomer.

TGN38/41

Using a polyclonal antiserum raised against membrane proteins of a Golgi-enriched fraction to screen a cDNA expression library, the cDNA encoding TGN38 was isolated.[446] TGN38 cDNA encodes a protein of 357 amino acids. The N-terminal 17 residues function as a signal peptide that is subsequently cleaved. The mature protein is a type I integral membrane protein of 340 residues; the N-terminal 286-residue sequence forms the luminal/extracellular domain followed by a 22-residue transmembrane domain (residue 287-308) and a 32-residue C-terminal cytoplasmic domain (residue 309-340). TGN41 was later identified to be an isoform of TGN38 generated by alternative splicing with TGN41 having a cytoplasmic domain of 55 residues, the first 30 residues of which are identical to TGN38.[622] TGN38 and TGN41 form a heterodimer (TGN38/41) that is predominantly localized to the trans Golgi network (TGN) and cycles between the TGN and the cell surface.[739] TGN38/41 may be involved in the formation of exocytotic vesicles from the TGN.[739] The cytoplasmic domain of TGN38 has been shown to interact with a cytosolic protein complex composed of a 62 kDa protein (p62) and Rab6, which is a member of the Ypt1p/Sec4p/Rab family of GTPases. p62 can be phosphorylated and its phosphorylation status may be coupled to its association and/or dissociation from the cytoplasmic tails of TGN38/41. The p62-Rab6 complex is involved in vesicle formation from the TGN because depleting this complex from cytosol abolishes in vitro vesicle budding from the TGN.

Peptides corresponding to the cytoplasmic tail of TGN38 or 41 could inhibit vesicle budding, possibly by their competition with TGN38/41 for the cytosolic p62-Rab6 complex.

p28

Using a fraction of rat liver Golgi integral membrane proteins as the antigen source, a hybridoma was obtained that secretes a monoclonal antibody (HFD9) against a 28 kDa integral membrane protein (p28).[751] p28 is widely-expressed and highly conserved. Immunolocalization suggests that p28 is enriched in the cis Golgi and its associated structures. The epitope recognized by the antibody is cytoplasmically oriented. cDNA cloning established that p28 is a membrane protein of 250 amino acids. The major portion of the polypeptide (residue 1-230) is anchored by its 20-residue C-terminal hydrophobic tail (residue 231-250). Preceding the hydrophobic tail, there are several regions (residue 67-94; 101-128; 128-152; and 163-226) that can potentially form coiled-coil domains. This topology of p28 is very similar to those proteins involved in vesicle docking and fusion. HFD9 monoclonal antibody inhibits in vitro ER-Golgi transport at a late stage and this inhibition can be neutralized by a recombinant cytoplasmic fragment (residue 1-229) of p28. Furthermore, this cytoplasmic fragment itself is inhibitory at a late stage in an in vitro ER-Golgi transport assay, suggesting that p28 may be involved in docking and fusion of ER derived vesicles with the cis Golgi in mammalian cells.[752] This interpretation is further supported by the demonstration that p28 may function as an integral part of the Golgi SNARE complex.

p200

A monoclonal antibody AD7 raised against canine liver Golgi membranes recognizes specifically a 200 kDa protein (p200).[512] p200 exists in cytosolic and membrane-bound forms and may cycle between the cytosol and membranes. p200 is preferentially associated with the Golgi membrane, behaving as a peripheral protein of the Golgi membrane. The Golgi association of p200 can be dissociated very rapidly by the fungal metabolite brefeldin A and enhanced by the activation of GTPases by GTP-γ-S or aluminum fluoride. In the Golgi, p200 is predominantly confined to the trans face of the Golgi, most likely the TGN. More interestingly, p200 is largely associated with coated vesicles on the TGN and these vesicles are different from coatomer-coated or clathrin-coated vesicles, suggesting that p200 may represent a novel class of coated vesicles that bud from the TGN.[513] Whether p200 is associated with vesicles that contain TGN38/41-p62-Rab6 remains to be confirmed. The functional and structural aspects of p200 have not been established.

ERGIC-53/p58

A 53 kDa human protein was identified by the monoclonal antibody G1/93. This protein was originally referred to as p53 and is now designated ERGIC-53.[679] ERGIC-53 is an integral membrane protein associated with the early secretory pathway, including the ER, the cis Golgi and an intermediate compartment between ER and the cis Golgi. The intermediate compartment is also termed as ER Golgi intermediate compartment (ERGIC). At 15°C, ER-Golgi transport as revealed by VSV G-protein and other cargo molecules is blocked at the ERGIC, suggesting that protein transport from the ER to the Golgi is mediated via the ERGIC.[74,303,431,598,661,680,681,770] The exact nature of ERGIC is unknown. It could represent either a specialized domain of the ER that is actively involved in vesicle budding, an independent membrane structure between the ER and the Golgi, or a specialized region of the cis Golgi (Fig. 10). ERGIC-53 has been shown to dynamically cycle between these early structures of the exocytotic pathway. cDNA cloning revealed that ERGIC-53 is a type I membrane protein synthesized as a precursor of 510 amino acids.[671] The first 30-residue sequence represents the signal peptide. A 450-residue sequence (residues 31-480) forms the luminal domain followed by a 18-residue transmembrane domain (residue 481-498) and a 12-residue cytoplasmic tail (residue 499-510). Interestingly, the cytoplasmic tail of ERGIC-53 ends with a KKFF sequence, which is a retrieval signal for ER localization. The cytoplasmic and luminal domains of ERGIC-53 have been shown to participate in the cycling between the ER, ERGIC and the cis Golgi.[323] Within the cytoplasmic tail, the RSQQE sequence adjacent to the transmembrane domain and the C-terminal KKFF are most critical for its cycling property. ERGIC-53 and a protein named VIP-36 are structurally related and both bear sequence homology with leguminous plant lectins.[212] In this respect, ERGIC-53 was also identified and cloned independently as a mannose-specific lectin referred to as MR60.[16] Although the cellular function of ERGIC-53 has not been established, the dynamic cycling between structures of the early exocytotic pathway and its ability to function as a mannose-specific lectin suggests that it may play a yet unknown role in protein transport, such as glycoprotein concentration during vesicle budding in the ER. Independently, polyclonal antibodies against a rat protein of about 58 kDa (p58) shows that it behaves much the same way as does ERGIC-53. p58 may be the rat counterpart of human ERGIC-53.[387,661]

Emp24p

Proteins present in a partially-purified yeast endosomal fraction were analyzed by microsequencing, leading to the cloning of a gene (EMP24) encoding a 203-residue precursor of a type I membrane protein.[670] The first 20-residue sequence is the signal peptide. Residues 21 to 172 form the luminal domain followed by a transmembrane domain (residue 173-193) and a 10-residue cytoplasmic tail (residue 194-203). Detailed studies establish that Emp24p is not an endosomal protein but rather a protein confined mainly to the endoplasmic reticulum. Emp24p is a membrane component of ER derived COPII-coated transport vesicles. When the

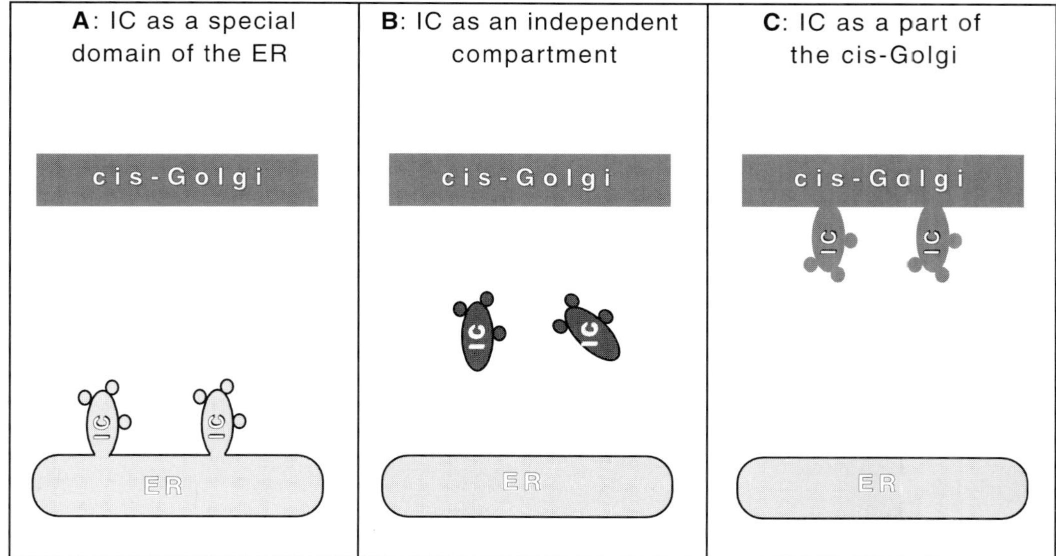

Fig. 10. Three possible models for the identity of the intermediate compartment (ERGIC) that is marked by ERGIC-53, p58 and KDEL receptor at 15 °C. The ERGIC could be a specialized region of the ER that is involved in vesicular budding (A), an independent structure between the ER and the cis Golgi (B), or a specialized region of the cis Golgi (C) that is the entry site for traffic derived from the ER.

EMP24 gene is disrupted, transport of invertase and the GPI-anchored surface protein Gas1p from the ER to the Golgi is delayed significantly. In wild-type cells, 90% of the invertase is transported from the ER to the Golgi and about 50% is secreted within 5 min at 37°C, while the half-time of secretion for invertase in mutant cells is about four times longer (20 min). Gap1p is normally transported from the ER to the Golgi with a half-time of 10 min, which is delayed to 30 min in EMP24-disrupted cells. In contrast, transport of α-factor precursor, acid phosphatase, two vacuolar proteins (carboxypeptidase Y and alkaline phosphatase) from the ER to the Golgi is not affected and occurs with normal kinetics in EMP24-disrupted cells. Since folding and oligomerization of invertase is not affected by EMP24 disruption, it is proposed that Emp24p may function as an integral component involved in sorting and/or concentration of a subset of cargo proteins during the budding process for COPII-coated vesicles, indicating that incorporation of cargo molecules in the budding vesicles may involve an active sorting/concentration step. Several Emp24p-like proteins exist in both yeast and mammalian cells, suggesting that they could play a similar role in vesicle budding along the exocytotic pathway.

p24

When the integral membrane proteins associated with in vitro Golgi-derived coatomer-coated vesicles were analyzed, a major 24 kDa protein (p24) was identified. cDNA cloning based on partial amino acid sequences of purified p24 shows that p24 is composed of 196 amino acids and is about 32% identical to Emp24p.[737] It is proposed that p24 plays a role in vesicle budding.

Many other proteins have also been identified by this approach, including VIP21/caveolin, VIP36, M6P receptors, coat proteins associated with the clathrin-coated vesicles, proteins associated with synaptic vesicles such as synaptobrevin, synaptophysin, synaptotagmin and proteins associated with the presynaptic membrane such as syntaxin 1A, syntaxin 1B and SNAP-25. They will be described in detail in relevant sections.

REVERSE-GENETICS

Cloning novel genes based on sequences of known proteins via cross-species hybridization, low stringency hybridization, gene complementation, polymerase chain reaction (PCR) or other genetic tools have expanded our list of proteins involved in vesicular transport along the exocytotic pathway. As discussed previously, Snc2p and the brain specific and ubiquitously-expressed mammalian homologues of Sec1p were identified by this approach. Other examples are discussed below.

Rab proteins

The importance of ras-like GTPases in diverse cellular processes[259,260,854] and the specific involvement of Ypt1p and Sec4p in ER Golgi and Golgi-surface transport, respectively, have triggered extensive efforts in cloning their mammalian homologues and related proteins that are involved in protein trafficking. This has led to the identification of over 30 distinct members of this Ypt1p/Sec4p/Rab small GTPase protein family, including Rab1a, Rab1b, Rab2, Rab3a, Rab3b, Rab3c, Rab3d, Rab4a, Rab4b, Rab5a, Rab5b, Rab5c, Rab6, Rab7, Rab8, Rab9, Rab10, Rab11a, Rab11b, Rab12, Rab13, Rab14, Rab15, Rab16, Rab17, Rab18, Rab19, Rab20, Rab22, Rab23, Rab24, Rab25 and Rab26 that may participate in the regulation of membrane trafficking along the exocytotic or endocytotic pathway.[123,129,201,237,278,312,372,388,443,444,486,539,547,548,715,764,800,820,865,866, 867]

ARFs and related Arls

Originally identified as a cofactor required for ADP-ribosylation of the Gs subunit of the trimeric G protein catalyzed by cholera toxin,[341,496] the ADP-ribosylation factor protein family contains six distinct but highly related members in mammals that fall into three classes.[69,73,342,496,604,693,795] Using an oligonucleotide probe based on the partial amino acid sequence of purified ARF, the bovine cDNA encoding ARF1 is cloned. cDNA encoding bovine ARF2 was cloned in a similar way. Using bovine ARF2 cDNA as a hybridization probe, cDNAs encoding human ARF1, ARF3, ARF5 and ARF6 were cloned. Similarly, library screening with bovine ARF1 also led to the cloning of human ARF1. Human ARF1 and bovine ARF1 are 100% identical in their primary amino acid sequence. Human ARF4 cDNA was serendipitously cloned during an attempt to isolate human cDNAs that could confer anchorage-independent growth to SW13 cells. ARF1, 2 and 3 form class I of ARFs with 181 amino acids. Class II includes ARF4 and 5 with 180 amino acids. ARF6 is the only member for class III with 175 amino acids. Members of class I (ARF1, 2 and 3) are over 95% identical to each other and about 80% and 68% identical to members of class II (ARF4 and 5) and III (ARF6), respectively. ARF4 and 5 are about 90% identical to each other and about 64% identical to ARF6. The yeast *S. cerevisiae* ARF1 gene was cloned by cross-species hybridization using the bovine ARF1 cDNA and encodes a polypeptide of 181 amino acid that is 77% identical to the bovine and human ARF1.[340,747] Screening a yeast genomic library with yeast ARF1 gene led to the isolation of yeast ARF2 gene encoding a polypeptide of 181 residues that is about 96% identical to yeast ARF1.[746,73] A third member (ARF3) was cloned from yeast by polymerase chain reaction.[399] Yeast ARF3, composed of 183 amino acids, is about 60% identical to human ARF6 and about 52-56% identical to the rest of mammalian ARFs. ARFs have also been cloned from other species and are evolutionarily well conserved. Recently, a number of ARF-like proteins (Arls) were identified,[131,441,496,675,768] including Drosophila Arl1 and 2, rat Arl1, 3, 4 and human Arl2. Arls are about 40-60% identical to

ARFs and about 40-45% identical to each other. Rat Arl1 is about 79% identical to the Drosophila Arl1 and Arl2 is about 76% identical between the human and Drosophila forms.

Syntaxin 2, 3, 4 and 5

Using cDNAs of brain specifc syntaxin 1A and 1B (which are 84% identical) as probes to screen various rat cDNA libraries at low stringency, cDNAs encoding syntaxin 2, 3, 4 were cloned.[59] Rat syntaxin 5 cDNA was cloned using a Drosophila homologue of yeast SED5 as a probe.[36,59] Mouse syntaxin 2 was also identified as epimorphin.[573] Syntaxin 2, 3, 4 and 5 are about 63%, 64%, 46%, 23% identical to syntaxin 1A, respectively. The identity between syntaxin 5 and Sed5p is about 35%, suggesting that syntaxin 5 could be the mammalian counterpart of Sed5p. In contrast to syntaxin 1A and 1B, syntaxin 2, 3, 4 and 5 are widely-expressed in rat tissues. Syntaxin 5 is primarily localized to the cis Golgi while other syntaxins may be targeted to the plasma membrane or endosomal structures.[36,59,159]

Cellubrevin

A genomic fragment encoding a synaptobrevin-related protein was isolated during a screen for genes encoding synaptobrevins, leading to the isolation of the cDNA encoding cellubrevin with 103 amino acids, which is about 59% identical to mammalian synaptobrevin I and II.[474] Most importantly, cellubrevin is ubiquitously-expressed and localized to early-endosomal structures. Cleavage of cellubrevin by tetanus toxins in permeabilized CHO cells impairs exocytosis of transferrin receptor-containing vesicles, suggesting that cellubrevin is involved in recycling of surface receptors from the endosomal system back to the plasma membrane.[219]

DIFFERENT COATS FOR TRANSPORT VESICLES

Thus far, three different types of coated vesicles have been characterized in terms of their coat components, the clathrin-coated vesicles, coatomer (COPI)-coated vesicles and COPII-coated vesicles. The clathrin-coated vesicles can be further divided into two different classes, the plasma membrane AP2-associated vesicles and the TGN AP1-associated vesicles. The protein components for these coats are compiled in Table III.

CLATHRIN-COAT

The characterization of the coat components for clathrin-coated vesicles was facilitated by the early purification of coated vesicles. This coat consists of clathrin triskelion and adapter proteins.[570,634] The coated vesicles originating from the TGN contain AP1 adapter proteins while those from the plasma membrane contain AP2 adapter proteins. The clathrin triskelion is composed of three molecules of clathrin heavy chain of about 180 kDa[360,400,685] and three molecules of clathrin light chain of about 30 kDa.[88,324,325,709] Two different forms of light chains have been identified. The AP1 Golgi adapter proteins contain γ-adaptin, β'-adaptin, a 47 kDa protein (AP47) and a 19 kDa protein (AP19).[220,358,359,456,507,511,591,601,612,633] The AP2 plasma membrane adapter proteins contain α-adaptin, β-adaptin, a 50 kDa protein (AP50) and a 17 kDa protein (AP17).[220,358,359,361,456,511,601,612,632,785] The AP1-associated Golgi clathrin-coated vesicles may mediate selective transport from the TGN to the late-endosome, while the AP2-associated surface clathrin-coated vesicles play a major role in receptor-mediated endocytosis from the plasma membrane. AP50 and AP47 have been shown to interact directly with cytoplasmic Tyr-based sorting signals.[543]

COATOMER (COPI)-COAT

As discussed earlier, the coatomer-coated vesicles (also referred to as COPI-coated vesicles) were purified from the in vitro intra-Golgi transport assay, in which the uncoating process is inhibited by GTP-γ-S. During the purification of the general transport factor p115/TAP, the components of the coat were fortuitously discovered to exist in a cytosolic protein complex (coatomer). This coat contains eight distinct proteins: the ARF and seven distinct subunits of the coatomer, including α- (160 kDa), β- (110 kDa), β'- (102 kDa), γ- (98 kDa), δ- (61 kDa), ε- (36 kDa) and ξ-COP (20 kDa).[253,256,609,647,649,766] Coatomer-coated vesicles have been implicated in mediating transport from the ER to the cis Golgi, from the cis to the medial cisternae of the Golgi apparatus and in retrograde transport from the cis Golgi back to the ER (to be discussed later).

COPII-COAT

The yeast in vitro assay for vesicle budding from the ER has uncovered a novel type of coat structure, the coat proteins II (COPII).[47,205a,626] The COPII coat consists of Sar1p, Sec16p, the Sec13p-Sec31p complex and the Sec23p-Sec24p complex. COPII-coated vesicles mediate transport from the ER to the early Golgi compartment.

In addition to these well established coats, there is evidence that other types of coats may be involved in other transport events.[513,523,588]

SNARE HYPOTHESIS AND SPECIFICITY OF VESICULAR TRANSPORT

The identification of synaptic vesicle-associated synaptobrevin II and presynaptic membrane-associated syntaxin 1A/B and SNAP-25 as the major brain SNAREs that can form the 20S NSF-SNAP-SNAREs fusion complex, the established functions of synaptobrevins, syntaxins and SNAP-25 in synaptic vesicle docking/fusion and the identification of proteins that are structurally related to these proteins but are associated with other cellular structures have led to the formulation of the SNARE hypothesis.[207,649,650,666,682,722,723,753,766,838] The SNARE hypothesis (Fig. 11) suggests that the specificity of vesicular transport is primarily determined by unique SNAREs associated with each type of transport vesicle (v-SNAREs) that specifically

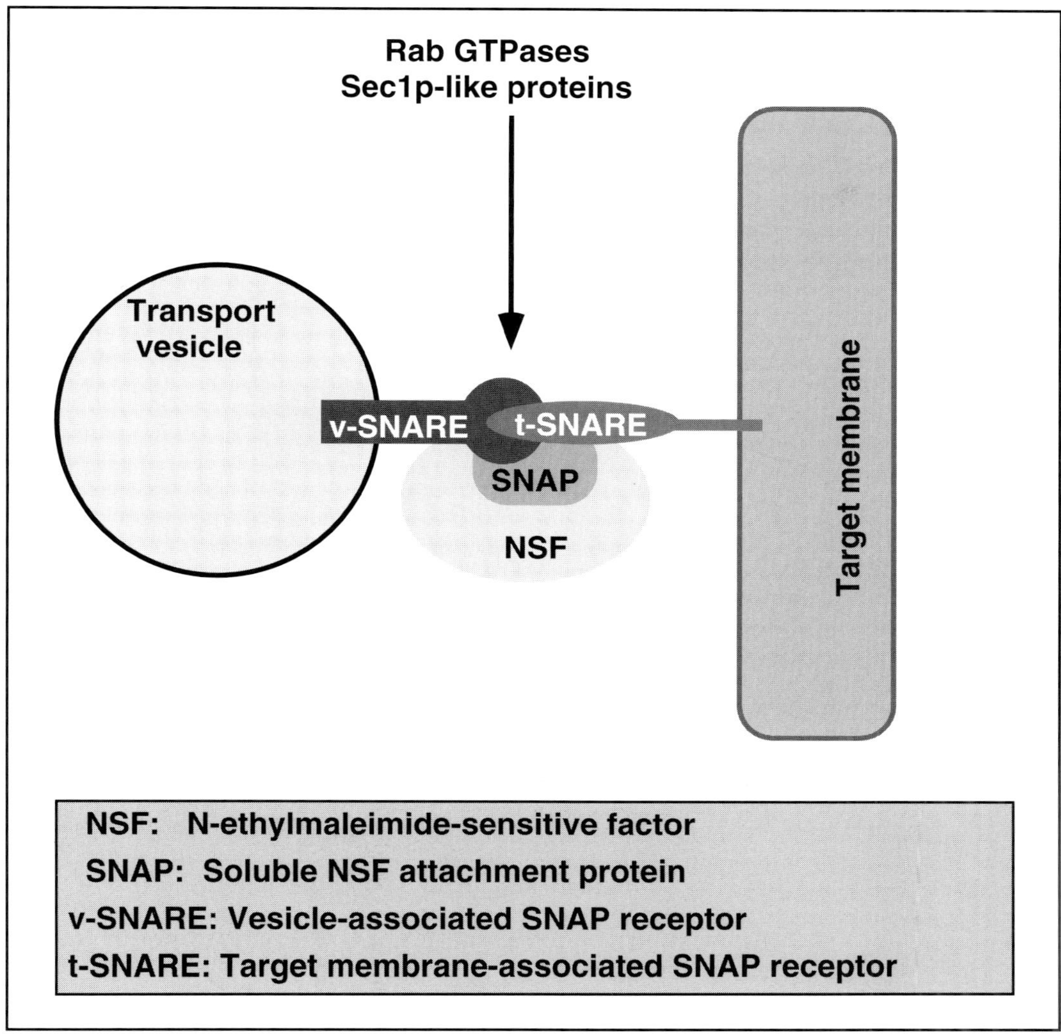

Fig. 11. Model for the SNARE hypothesis. This hypothesis states that the specificity of vesicle transport is determined by interaction between vesicle-associated proteins (v-SNAREs) with the cognate proteins of the target membrane (t-SNAREs). This interaction triggers the formation of a 20S fusion complex composed of v-SNAREs, t-SNAREs, SNAPs and NSF. This 20S fusion complex is also intimately involved in fusion between the vesicle membrane with the target membrane.

interact with the cognate SNAREs associated with the respective target membrane (t-SNAREs). The specific association of the v-SNAREs with the cognate t-SNAREs serves not only to dock the transport vesicles on the correct target membrane but also to recruit SNAPs and subsequently NSF on to the v-SNAREs-t-SNAREs complex, leading to the formation of the 20S v-SNAREs-t-SNAREs-SNAPs-NSF fusion complex. This 20S complex is intimately associated with the fusion of the docked vesicles with the target membrane. Activation of the ATPase activity of NSF upon the formation of this 20S fusion complex may play an essential role in the fusion of docked vesicles with the target membrane as well as in subsequent disassembly of this complex. According to this model, v-SNAREs should be selectively incorporated into the vesicles during the budding process and recycled from the target membrane back to the original donor membrane after fusion. The t-SNAREs of each target membrane should be excluded from entering into any budding vesicles originating from it. The formation of the specific SNAREs complex during the docking and subsequent formation of the 20S fusion complex may be regulated by additional cytosolic and membrane proteins to further ensure the specificity and fidelity of each intracellular vesicular transport step.

According to the SNARE hypothesis,[207,649,650,666,682,722] Sec22p/Sly2p, Bos1p, Bet1p/Sly12p and Ykt6p/p26 could be considered as v-SNAREs associated with ER derived vesicles, while Sed5p could function as a t-SNARE on the early Golgi membrane for the docking/fusion of ER derived vesicles. Sft1p/p14 has been suggested to function as a v-SNARE for retrograde transport between a later Golgi and the Sed5p-containing early Golgi compartment. Snc1p and Snc2p may function as the v-SNAREs for Golgi-derived vesicles destined for the plasma membrane, while the corresponding t-SNAREs may be Sso1p, Sso2p and Sec9p (structurally related to SNAP-25) localized to the plasma membrane. Synaptobrevin I and II represent v-SNAREs of the synaptic vesicles that will pair with the presynaptic t-SNAREs (syntaxin 1A, 1B and SNAP-25). VAP-33 (see chapter 8) has been suggested to be a novel t-SNARE for neurons. Cellubrevin could function as a v-SNARE for recycling vesicles derived from the endosomal system, although the corresponding t-SNAREs on the cell surface have not been defined. Syntaxin 2, 3 and 4 could function as t-SNAREs on the plasma membrane for vesicles derived from either the TGN or the endosomal system, although the relevant v-SNAREs for each of them remain to be established. Pep12p could be a t-SNARE for vesicles destined for the yeast vacuoles, although the corresponding v-SNAREs are not known. Syntaxin 5 on the cis Golgi of mammalian cells could act as a t-SNARE for fusion of ER derived vesicles, although no candidate v-SNARES operating in the mammalian early secretory pathway has been identified. The mammalian cis Golgi protein p28 is apparently a component of the SNARE complex for docking/fusion of ER derived vesicles with the cis Golgi. Whether p28 acts as a t- or v-SNARE remains to be established, although its cis Golgi localization indicates that it may be a t-SNARE. The known SNAREs are summarized in Table IV.

The functional state of SNAREs and/or the formation of the SNARE complex between vesicles and their target membrane are expected to be tightly-regulated. The function of Sed5p requires Sly1p because Sed5p and Sly1p physically interact with each other. Sec1p may be involved in the regulation of the SNARE complex assembly between the Golgi-derived vesicles and the plasma membrane because its mutations could be overcome by overexpression of syntaxin homologues Sso1p and Sso2p localized to the plasma membrane. Mammalian brain-specific Sec1p may also regulate SNARE complex formation because it could interact directly with syntaxin 1A and 1B and this interaction could prevent association of syntaxins with synaptobrevins. Synaptobrevins have been shown to associate with synaptophysin, which is another synaptic vesicle membrane protein, on the synaptic vesicles.[108,197] The interaction of synaptophysin with synaptobrevins may inactivate synaptobrevins on the vesicles because synaptobrevins bound with synaptophysin are no longer capable of interacting with syntaxins and SNAP25. The major regulatory proteins for the formation of the SNARE complex could be members of the Ypt1p/Sec4p/Rab family of small GTPases.[414,535,539,715,720,753,867] Ypt1p is required for the formation of the SNARE complex between the ER derived vesicles and the early yeast Golgi and its functional inactivation could be suppressed by overexpression of components involved in the regulation (Sly1p) or assembly (Sly2/Sec22p and Sly12/Bet1p) of the SNARE complex. Ypt1p may regulate Sly1p in such a way that it may control the functional activity of Sed5p. Furthermore, it has been shown that Bos1p and Sec22p/Sly2p pair

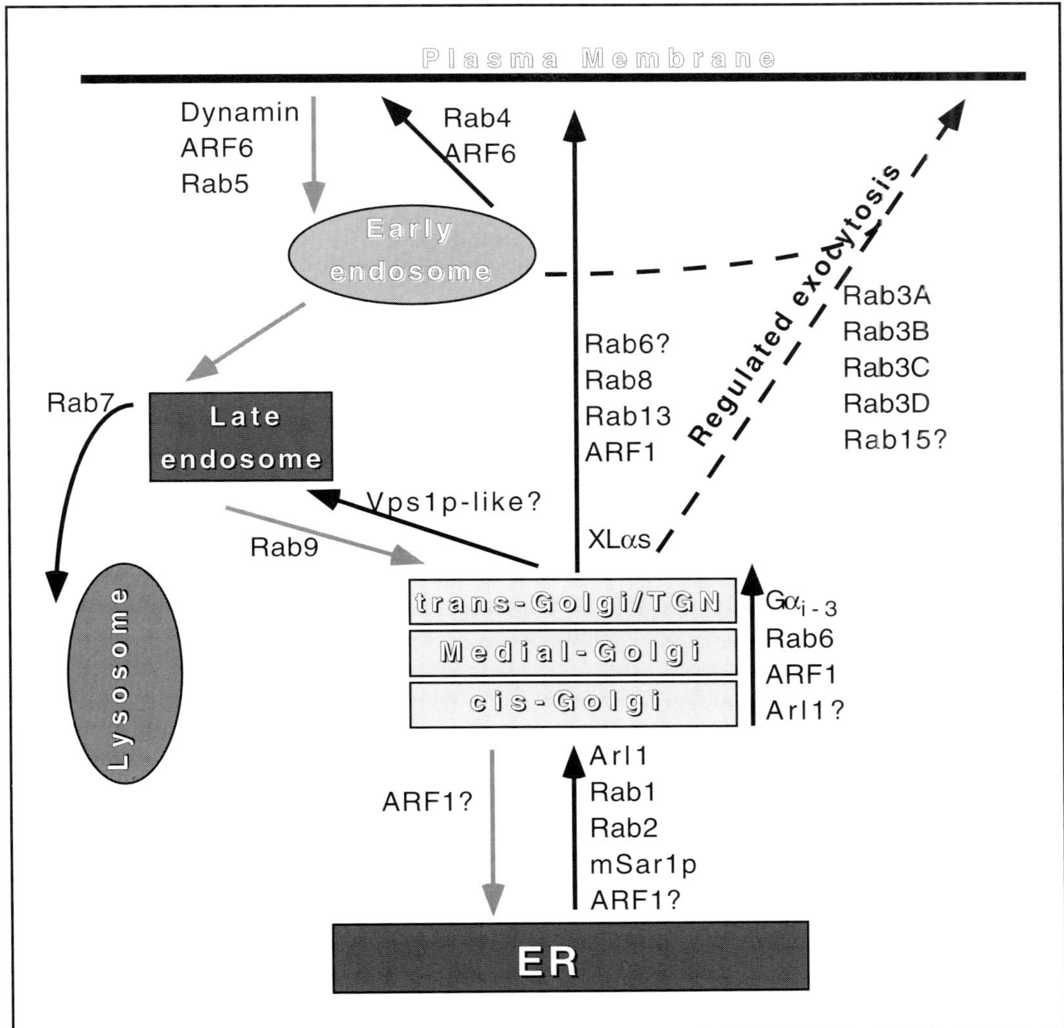

Fig. 12. *GTPases that are associated with membrane trafficking are indicated at the respective transport steps.*

with each other on ER derived vesicles and that this association requires functional Ypt1p, suggesting that Ypt1p may regulate the activation of SNAREs on the carrier vesicles, although it may act on the target membrane as well. Sec4p has been shown to regulate the SNARE complex fomation between the Golgi-derived vesicles with the plasma membrane. Sec9p, which is a component of the SNARE complex, may be the direct effector protein acted on by Sec4p. Rab3a is primarily involved in synaptic vesicle docking/ fusion and it may regulate the assembly of SNARE complex between the synaptic vesicles and the presynaptic membrane because it is well established that Rab3a is involved in the docking/fusion of synaptic vesicles with the presynaptic membrane and a protein complex containing Rab3a, SNAP-25, synaptobrevin 2 and syntaxins has been observed.

GTP-BINDING PROTEINS IN VESICULAR TRANSPORT

EXAMPLES OF GTPASES INVOLVED IN TRAFFIC

Since the discovery that Sec4p is a member of the small GTPase superfamily,

Table IV.

Protein	Type of SNARE	Step of Transport	Membrane attachment by
VAMP1	v-SNARE	Synaptic vesicle-Surface	C-terminal anchor
VAMP2	v-SNARE	Synaptic vesicle-Surface	C-terminal anchor
Cellubrevin	v-SNARE	Early endosome-surface	C-terminal anchor
Snc1p	v-SNARE	Late Golgi-surface	C-terminal anchor
Snc2p	v-SNARE	Late Golgi-surface	C-terminal anchor
Snf1p	v-SNARE	Intra-Golgi	C-terminal anchor
Ykt6p	v-SNARE	ER-early Golgi?	Lipid (farnesyl?)
Bos1p	v-SNARE	ER-early Golgi	C-terminal anchor
Sec22p/Sly2p	v-SNARE	ER-early Golgi	C-terminal anchor
Bet1p/Sly12p	v-SNARE	ER-early Golgi	C-terminal anchor
Syntaxin1A	t-SNARE	Synaptic vesicle-Surface	C-terminal anchor
SNAP-25	t-SNARE	Synaptic vesicle-Surface	Lipid (palmitate)
Syntaxin1B	t-SNARE	Synaptic vesicle-Surface	C-terminal anchor
VAP-33	t-SNARE	Synaptic vesicle-Surface	C-terminal anchor
Syntaxin2	t-SNARE	TGN or endosome-Surface?	C-terminal anchor
Syntaxin3	t-SNARE	TGN-late endosome?	C-terminal anchor
Syntaxin4	t-SNARE	TGN or endosome-Surface?	C-terminal anchor
Syntaxin5	t-SNARE	ER-early Golgi	C-terminal anchor
p28	t-SNARE?	ER-early Golgi	C-terminal anchor
Sso1p	t-SNARE	Late Golgi-surface	C-terminal anchor
Sso2p	t-SNARE	Late Golgi-surface	C-terminal anchor
Sed5p	t-SNARE	ER-early Golgi	C-terminal anchor
Pep12p	t-SNARE	Late Golgi-PVC	C-terminal anchor

Table IV. Summary of known v-SNAREs and t-SNAREs involved in vesicle docking/fusion

many other small GTPases have been established to play an important role in intracellular vesicular transport, including Sar1p, other members of the Ypt1p/Sec4p/ Rab family, some members of the ARF family and one member of the Arl subfamily.[29,167,208,535,539,715,740,789,867] Furthermore, trimeric GTPases have also been shown to be involved in protein trafficking.[49,137,184,374,390,539,848] A number of these proteins have been described in previous sections such as Sar1p and Sec4p, while some others will be discussed below. For GTPases, there are four conserved motifs (G1-4) in their primary sequence which are involved in GTP binding and hydrolysis. The first motif (G1) consists of GxxxxGK(S/T) that is involved in the interaction with the α- and β-phosphates of GDP and GTP. The second motif (G2) corresponding to the putative effector domain of Ras and Rab proteins contains a highly conserved Thr residue that interacts with a Mg^{2+} ion coordinated to the oxygen atoms of the β- and γ-phosphates of GTP. The Mg^{2+} is essential for GTP hydrolysis and also interacts with Ser/Thr residue in the G1 motif and, through an intervening H_2O molecule, a conserved Asp residue in the third motif (G3) consisting of DxxGQ sequence. The forth motif consisting of NKxD sequence interacts with the guanine nucleotide ring.[79,539,764] Examples of GTPases involved in trafficking are shown in Figure 12 and described below in detail.

Ypt1p

The YPT1 gene, originally isolated as a gene (YP2) located at the 5' region of the yeast (*S. cerevisiae*) actin gene, encodes a polypeptide of 206 amino acids of about 38% identical to the mammalian ras oncoprotein over a region of 168 amino acids.[221,672,673] Ypt1p also shares about 46% identity with Sec4p. The mammalian Rab1A is about 75% identical to Ypt1p and could functionally complement YPT1 in the yeast,[278] suggesting that Rab1A could be the mammalian functional homologue of Ypt1p. YPT1 isolated from *Schiz. pombe* encodes a protein (203 amino acids) that is 73% identical to the *S. cerevisiae* Ypt1p and could also substitute for Ypt1p function in *S. cerevisiae*. Ypt1p is associated with Golgi-like structures in the yeast and is required for ER-Golgi transport.[53,339,414,687,688] More specifically, Ypt1p is involved in the late docking/fusion stage of ER derived transport vesicles with the early Golgi.[25,28] Ypt1p may function to regulate the functional state of SNAREs associated with the vesicles (such as Sec22p/Sly2p and Bos1p)[414] and/or the target membrane (such as Sed5p) in such a way so as to control the SNARE complex assembly between the ER derived vesicles and the early Golgi membrane.

Rab1

There are two forms of Rab1, Rab1A and Rab1B, which are over 92% identi-

cal in their primary sequences. Antibodies against Rab1 reveal that Rab1 is associated with the ER, the intermediate compartment and the cis Golgi. Since Rab1 is highly related to Ypt1p and can functionally substitute for Ypt1p in yeast,[278] Rab1 may be involved in ER-Golgi transport in mammalian cells.[540,595,597,678,786,849] When the Asn residue at position 124 in the consensus G4 domain NKxD of Rab1A is mutated to Ile, the resulting mutant ($Rab1A_{N124I}$), is a potent inhibitor of ER-Golgi transport of VSV G-protein in vivo. $Rab1A_{N124I}$ is unable to bind GDP or GTP in vitro, suggesting that it is defective in guanine nucleotide binding or is in a conformation that has a high rate of guanine nucleotide exchange. $Rab1A_{N124I}$ may be equivalent to H-Ras/N116I, which is defective in guanine nucleotide binding and is transforming, indicating it is most likely in a constitutively active state. $Rab1A_{N124I}$ blocks a late (most likely docking/fusion) stage of VSV G-protein transport from the ER to the cis Golgi. When the Ser residue in the G1 motif (GxxxxGKS) of Rab1A is mutated into an Asn, which perturbs Mg^{2+} coordination and reduces affinity for GTP, the resulting mutant ($Rab1A_{S25N}$) is mostly restricted to the GDP-bound form and is also inhibitory to ER-Golgi transport. $Rab1A_{S25N}$ may be equivalent to H-Ras/S17N, a dominant negative mutant that inhibits cell proliferation and appears to be restricted to the GDP-bound form. Similar mutations in Rab1B have the same effects on ER-Golgi transport, Rab1 thus may play an essential role in transport from the ER to the Golgi apparatus.[540,595]

Rab2

Rab2 is about 40% identical to Rab1 and is preferentially associated with the intermediate compartment (ERGIC) involved in ER-Golgi transport.[122,678,786] Mutant Rab2 proteins also inhibit ER-Golgi transport, indicating that it may play an important role in trafficking in the early secretory pathway.

Rab3

Rab3 proteins are preferentially expressed in cells that possess regulated secretion, such as endocrine cells, exocrine cells and neuronal cells involved in synaptic transmission.[34,424,467,524,541,562] Four highly related isoforms of Rab3 have been described. Rab3A is mainly associated with neuronal cells that have regulated synaptic transmission. Rab3B and 3C are predominantly associated with the pituitary and insulin-secreting cells, respectively. Rab3D is expressed in adipocytes and is proposed to play a role in regulated delivery of glucose transport 4 (GLUT4) to the cell surface in response to insulin.[34] Rab3 proteins are primarily involved in the vesicle docking/fusion stage of regulated secretion.

Rab4

Rab 4 protein is associated mainly with the early endosomes and is phosphorylated in a regulated manner.[122,805,806,807] Overexpression of wild-type Rab4 has no effect on the early stage of endocytosis but causes a shift of endosomal transferrin receptor to the surface. Rab4 may play a critical role in positively-regulating the recycling rate from the early endosomes back to the surface. Two highly related Rab4 isoforms (Rab4A and Rab4B) are known and they play a similar role in the endocytotic pathway.

Rab5

Rab5 is associated with the plasma membrane, clathrin-coated vesicles, early endosomes and synaptic vesicles in neuronal cells.[44,100,122,243,411,744,745] Rab5 overexpression increases the rate of endocytosis from the cell surface and induces the appearance of large early endosomes, suggesting that it may be primarily involved in regulating the rate of the initial stage of endocytosis. Membrane recruitment of Rab5 is accompanied by an exchange of bound GDP for GTP,[798] suggesting that the putative membrane guanine nucleotide exchange factor may play a major role in

specifc membrane recruitment of Rab proteins. When the Gln residue at position 79 of the G3 motif (DxxGQ) is mutated into a Leu residue, the resulting mutant (Rab5$_{Q79L}$) exhibits markedly reduced intrinsic GTPase activity and is maintained mainly in the GTP-bound state.[745] Overexpression of Rab5$_{Q79L}$ led to a dramatic change in cell morphology with the appearance of unusually large endosomal structures, considerably larger than those formed by overexpression of the wild-type Rab5. Consistent with this, cytosol prepared from cells overexpressing Rab5$_{Q79L}$ stimulated homotypic early endosome fusion in vitro. Furthermore, an increased rate of transferrin endocytosis is observed, whereas recycling back to the surface is inhibited. Rab5$_{Q79L}$ may be equivalent to the activating H-Ras/Q61L mutant form. When the Ser residue at position 35 in the G1 motif (GxxxxGKS) is replaced with an Asn residue, the resulting mutant (Rab5$_{S35N}$), like the inhibitory H-Ras/S17N mutant, has a preferential affinity for GDP and remains predominantly in the GDP-bound form. Overexpression of Rab5$_{S35N}$ induces the accumulation of very small endocytotic profiles and inhibits transferrin endocytosis. This mutant also inhibits fusion between the early endosomes. The opposite effects of Rab5$_{Q79L}$ and Rab5$_{S35N}$ suggest that Rab5:GTP is required prior to membrane fusion, whereas GTP hydrolysis by Rab5 occurs after membrane fusion and functions to inactivate the protein. Rab5 proteins exist in three forms: Rab5A, Rab5B and Rab5C that are highly related and may be functionally equivalent. Three Rab5-like proteins have been identified in *S. cerevisiae* and referred to as Ypt51p, Ypt52p and Ypt53p that are about 54%, 55% and 52% identical to Rab5, respectively.[716] Endocytotic delivery to the vacuole of two markers, lucifer yellow and α-factor was inhibited in YPT51 disrupted cells and aggravated in the double YPT51-YPT52 and triple YPT51-YPT52-YPT53 gene-disruption mutants. Furthermore,

YPT51 is identical to VPS21, whose mutation is associated with mis-sorting of vacuolar proteins,[305] suggesting that these Rab5-like proteins are required for the endocytotic pathway and vacuolar sorting in the budding yeast. A Rab5-like protein (Ypt5p) was similarly identified in *Schiz. pombe* and is 62%, 55%, 57% and 47% identical to Rab5, Ypt51p, Ypt52p and Ypt53p, respectively.[17] Since Rab5 is associated with synaptic vesicles, it has been suggested that this protein is involved in the internalization of exocytosed components of synaptic vesicles.

Rab6

Rab6 is distributed from medial Golgi to the TGN.[13] Overexpression of wild-type Rab6 and the GTP-bound form (Rab6Q72L) greatly reduces transport between the cis/medial and the trans Golgi/TGN, without affecting transport from the ER to the cis/medial Golgi or from the TGN to the plasma membrane, suggesting that Rab6 may be primarily involved in intra-Golgi transport,[459] although studies on TGN38/41 has suggested a role of Rab6 in vesicle budding from the TGN.[739] The Ryh1p of *Schizo-Schiz. pombe* is about 73% identical to Rab6[287] and Rab6 could functionally substitute for Ryh1p in the fission yeast.

Rab7

Rab7 is associated preferentially with late endosomes.[122,714] A protein Ypt7p of 63% identical to Rab7 has been identified in budding yeast.[669,840] In YPT7 disrupted cells, uptake/endocytosis of α-factor is not affected, whereas degradation of internalized pheromone is severely inhibited. Cells lacking Ypt7p are also characterized by highly fragmented vacuoles and defects of vacuolar protein transport and maturation, suggesting that Ypt7p may play an important role in traffic from the late endosome to the vacuole. Similarly, Rab7 could regulate traffic from the late endosomes to the lysosomes.

Rab8

Rab8 is about 60% identical to Sec4p and is associated with the TGN, vesicular structures and the plasma membrane (the basolateral domain in epithelial cells).[129] A peptide corresponding to the hyper-variable C-terminal region of Rab8 inhibits transport from the TGN to the basolateral surface, while transport from the TGN to the apical surface or from the ER to the Golgi is not affected; suggesting that Rab8 may selectively regulate trafficking from the TGN to the basolateral plasma membrane.[311] Rab8 is preferentially found in the dendrites of fully polarized hippocampal neurons in culture and in both axons and dendrites of immature neurons. Depletion of Rab8 from mature polarized neurons with antisense oligonucleotides impairs the delivery of membrane protein to the dendritic membrane.[310] In developing neurons, Rab8 plays a key role in membrane traffic involved in neuronal process outgrowth. In nonpolarized cells, Rab8 may have a general role in transport between the TGN and the plasma membrane. A protein Ypt2p identified in the fission yeast is 55% identical to Sec4p and 63% identical to Rab8. Replacement of endogenous YPT2 with a mutant one in which a conserved Val residue at position 20 in the G1 motif (GDSGVGKS) is mutated into an Asn results in cells that are temperature-sensitive for growth. At the restrictive temperature, the mutant cells accumulate a population of vesicles and the secretory protein acid phosphatase in a form that appears to be fully glycosylated, indicating that Ypt2 may be involved in docking/fusion of Golgi derived vesicles with the plasma membrane in the fission yeast.[146,277]

Rab9

Rab9 associates predominantly with the late endosomes.[389,628,695] The in vitro system which reconstitutes the traffic of M6P receptor from the late endosomes to the TGN revealed an important role for Rab9 in this transport. Functional Rab9 stimulates and antibody against Rab9 inhibits this transport step.[389,628] Overexpression of a dominant negative GDP-bound mutant of Rab9, Rab9$_{S21N}$, strongly inhibits recycling of M6P receptor from the late endosomes to the TGN in the cell. Furthermore, overexpression of Rab9$_{S21N}$ was accompanied by a decrease in the efficiency of lysosomal enzyme sorting, consistent with reduced levels of the M6P receptor in the TGN. Like Rab5, membrane recruitment of Rab9 is accompanied by GDP exchange for GTP.[727]

Rab13

Rab13 cDNA was cloned from human intestinal epithelial (Caco2) cells.[866] Rab13 is about 61% identical to Rab8 and 56% identical to Sec4p. Most intriguingly, Rab13 is localized to tight junctions in polarized epithelial cells in culture and in tissues and this localization is dependent on the integrity of the tight junction. When tight junctions are disassembled and in nonpolarized cells, Rab13 is distributed to cytoplasmic vesicles. Rab13 may be involved in post-Golgi transport that is associated with the biogenesis of the plasma membrane and its specific structures such as the tight junction.

Rab15

Northern blot analysis of rat tissues shows that Rab15 is specifically expressed in the brain, suggesting that Rab15 may be involved in traffic associated with the brain, such as synaptic vesicle docking/fusion, or internalization of exocytosed proteins of the synaptic vesicles.[201]

ARF1

Among the six distinct mammalian ARFs, ARF1 has been studied most extensively.[32,73,158,180,200,279,285,343,401,496,526,584,690,693,738,747,774,775] It is the major form of ARF that is associated with the cytoplasmic surface of the Golgi apparatus and its associated vesicles. Furthermore, ARF1 is associated with coatomer-coated vesicles

derived from the in vitro intra-Golgi transport assay. Binding of ARF1 to the Golgi membrane is dependent on a membrane component[285] and is accompanied by bound GDP exchange for GTP, indicating that this membrane receptor may be part of the guanine nucleotide exchange factor that stimulates GDP exchange for GTP. Brefeldin A, a fungal metabolite, has diverse effects on the exocytotic and endocytotic pathways.[363,415,416] Brefeldin A causes a rapid dissociation of coatomer components (such as β-COP) and ARF from the Golgi membrane.[182,183] It is now known that brefeldin A actually inhibits Golgi membrane-catalyzed guanine nucleotide exchange onto ARF1, suggesting that exchange of ARF1 bound GDP for GTP is essential for Golgi recruitment of ARF and possibly for coatomer components.[181,286,614] Using an in vitro binding assay, it was established that ARF1 recruitment onto the Golgi membrane is indeed required for membrane binding of components of the coatomer, indicating that ARF1 may be involved in the budding process of coatomer-coated vesicles.[180,564] Together, ARF1 and the coatomer are sufficient for the assembly of coatomer-coated vesicles from the Golgi membrane in vitro.[647,649,690] A mutant ARF1, in which the Gln at position 71 in the G3 motif (DxxGQ) is mutated into a Leu, is expected to have a reduced rate of GTP hydrolysis, leading to its preferential existence in the GTP-bound active form. Overexpression of ARF1$_{Q71L}$ inhibits membrane traffic from the ER to the Golgi and between successive Golgi subcompartments.[158,868] Furthermore, ARF1$_{Q71L}$ also causes vesiculation of the Golgi apparatus. Overexpression of ARF1$_{Q71L}$ also renders Golgi association of coatomer components resistant to the effects of brefeldin A.[777] In the in vitro assay, ARF1$_{Q71L}$ inhibited the intra-Golgi transport but does not affect the assembly and budding of coatomer-coated vesicles. However, coatomer-coated vesicles assembled with ARF1$_{Q71L}$ was unable to uncoat and therefore were not functional, similar to those vesicles assembled in the presence of GTP-γ-S, suggesting that GTP hydrolysis by ARF1 plays a pivotal role in the uncoating process of the coatomer-coated vesicles. Another mutant of ARF1, in which the Thr at position 31 in the conserved G1 motif (GxxxxGKT) is mutated into a Asn, is expected to have a preferential affinity for GDP. ARF1$_{T31N}$ inhibits protein export from the ER and triggers a brefeldin A-like phenotype, resulting in the redistribution of coatomer components from the Golgi membrane to the cytosol and collapse of the Golgi into the ER, suggesting that ARF1 is indeed a key element involved in trafficking events that use coatomer-coated vesicles. Similar studies on yeast ARFs have come to the same conclusion. Recently, it is shown that ARFs may regulate/stimulate phospholipase D activity.[90,134,380,496]

ARF6

ARF6 is about 68% identical to ARF1 and has been shown to associate preferentially with the plasma membrane and the endocytotic pathway.[187,496,584] Overexpression of wild-type ARF6 and ARF6$_{Q67L}$ mutant leads to a shift of transferrin receptor to the cell surface and a decrease in the rate of receptor-mediated endocytosis (as measured by the uptake of transferrin). However, overexpression of ARF6$_{T27N}$ mutant results in a decrease of the surface pool of transferrin receptor due to an inhibition of the recycling of the receptor from the endosomal structures back to the surface. ARF6 thus plays a critical role in receptor-mediated endocytosis and recycling of internalized receptor back to the plasma membrane.[187] Morphological studies in cells overexpressing wild-type ARF6, ARF6$_{Q67L}$ and ARF6$_{T27N}$ similarly suggest that ARF6 is a key regulator of endocytosis and its recycling pathway.[584]

Arl1

Arl1 was originally identified in Drosophila and it plays an unknown but essential function because its loss is lethal.[768]

The rat Arl1 is about 79% identical to the Drosophila counterpart and is predominantly associated with the Golgi apparatus.[441] Antibodies against the C-terminal portion of Arl1 inhibit a late stage of in vitro ER-Golgi transport of VSV G-protein. Recombinant Arl1 fused to glutathione-S-transferase also inhibits the in vitro ER-Golgi transport. The inhibitory effect of the Arl1 antibodies and the recombinant fusion protein could be neutralized by the presence of an equal molar amount of the other. Arl1 may thus play a role in the docking/fusion stage of ER-Golgi transport.

Dynamin

Dynamin was initially identified as a microtubule-stimulated GTPase and is now revealed to play a vital role in the fission of endocytotic vesicles.[636,802] Several forms of dynamin have been thus far identified, with similar sequences and domain structures.[175a 636,802] The 851 amino acid sequence of rat brain dynamin can be divided into several domains. The N-terminal 300-residue sequence is the GTPase domain that is characterized by the consensus motifs involved in guanine nucleotide binding and GTP hydrolysis. The 100-residue C-terminal tail is extremely basic (pI=12.5) and rich in Pro (33%), within which there is an SH3-binding motif. There is a pleckstrin homology (PH) domain at residues 516-629. Between the PH domain and the C-terminal basic domain, there exist several regions that could potentially form coiled-coil structures. The involvement of dynamin in endocytosis was revealed when it was found that the Drosophila shibire gene encodes a homologue of dynamin.[125] At the restrictive temperature, flies with temperature sensitive alleles of shibire exhibit rapid paralysis associated with a depletion of synaptic vesicles from nerve terminals and accumulation of membrane invaginations and coated pits on the cell surface due to a specific defect in re-formation of synaptic vesicles at the nerve terminals and a general defect in endocytosis. In cells overexpressing dominant-negative mutants of dynamin, coated pits fail to become constricted and coated vesicles fail to bud, while coated pit assembly, invagination and the recruitment of receptors such as the transferrin receptor into the coated pit are unaffected.[155] It was recently revealed that dynamin can spontaneously self-assemble into rings and stacks of interconnected rings, comparable in dimension to the 'collar' observed at the necks of invaginated coated pits that accumulate at synaptic terminals in flies with mutant shibire genes at the restricted temperature.[298] When permeabilized nerve terminals were incubated in the presence of GTP-γ-S, long invaginations were formed which are coated with dynamin rings or stacks of interconnected rings.[765] Clathrin-coated bulbs are seen at the tips of these invaginations. Dynamin may thus play a key role in the fission step for clathrin-coated vesicles during endocytosis and reformation of synaptic vesicles. Dynamin may constrict the coated pit by the assembly of rings around their necks. A conformational change of dynamin triggered by GTP hydrolysis may close the rings and pinch off the budding coated vesicles. Recently, an 85 kDa yeast protein Dnm1p that is 41% identical to dynamin has been shown to participate in the internalization step of endocytosis.[222] Dnm1p is not involved in vacuolar sorting. Vps1p is another yeast protein structurally related (45% identity) to dynamin and is involved in vacuolar sorting of vacuolar enzymes and Golgi retention of late Golgi components.[651,808,845]

Trimeric GTPases

Heterotrimeric GTPases are composed of the catalytic α-subunit that binds and hydrolyzes GTP, a β- and a γ-subunit. A diverse number of α-subunits and several distinct β- and γ-subunits have been identified. Trimeric GTPases have been implicated in many processes associated with the exocytotic and endocytotic pathway.[49,137,184,390,848] The Golgi recruitment of coatomer components has been impli-

cated to be under the regulation of trimeric G proteins and vesicle formation may be the step where trimeric G proteins may execute a regulation.[184] The pertussis toxin sensitive $G\alpha_{i-3}$ trimeric GTPase is preferentially associated with the Golgi apparatus.[203] When $G\alpha_{i-3}$ was inducibly overexpressed in LLC-PK1 cells, constitutive secretion of heparan sulfate proteoglycans (HSPG) was retarded due to its accumulation in the Golgi.[749] This effect is reversed by treatment of cells with pertussin toxin which ADP-ribosylates and functionally uncouples $G\alpha_{i-3}$. A cholera toxin sensitive "extra large" G protein (XLαs) of about 92 kDa has been identified that consists of a 367 amino acid extra portion linked to a 348 amino acid Gαs truncated at the N-terminus. XLαs is specifically associated with the TGN and occurs selectively in cells possessing both the regulated and the constitutive pathway. XLαs may play an important role in the vesicle formation from the TGN.[351] Other uncharacterized trimeric G proteins may also be involved in the exocytotic pathway.

ENZYMES INVOLVED IN LIPIDATION OF GTPASES

CAAX Farnesyltransferase

For proteins such as Ras, yeast α-factor, lamins and γ-subunits of trimeric GTPases with a C-terminal CAAX motif (where C is Cys, A is aliphatic residue and X is usually Met or Ser), a 15-carbon farnesyl group is attached to the Cys residue via a thioether bond, followed by proteolytic removal of the AAX moiety and carboxy-methylation of the farnesylated Cys.[333,490,624] The farnesylation is catalyzed by CAAX farnesyltransferase, which is a heterodimer composed of a 49 kDa α-subunit and a 46 kDa β-subunit (Fig. 13). The β-subunit binds Zn^{2+} and the CAAX sequence, but requires α-subunit for the catalytic activity. The cloned rat cDNA for the α-subunit encodes a polypeptide of 377 amino acids, while cDNA for rat β-subunit encodes 437 amino

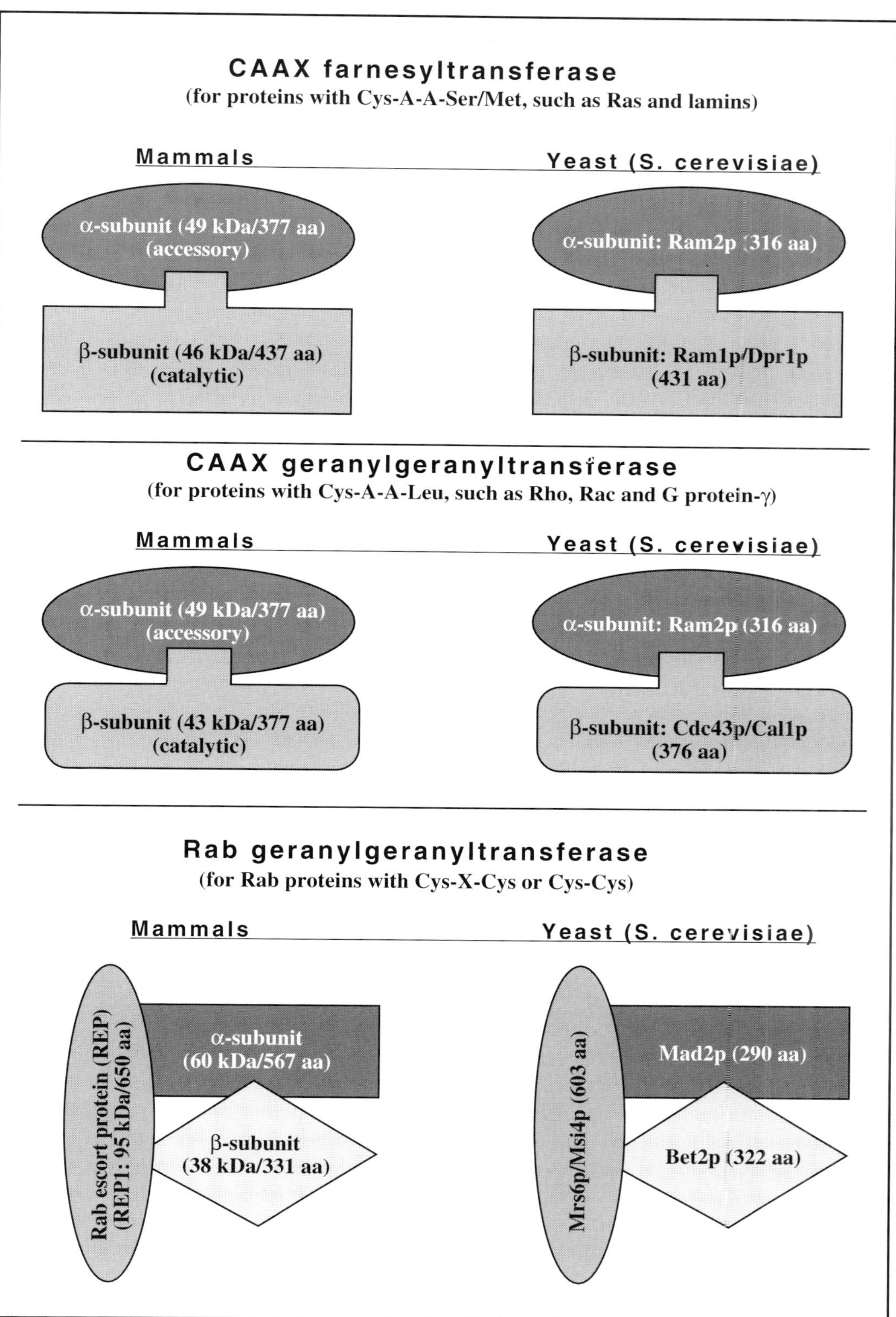

Fig. 13. The subunit compositions of three distinct prenyltransferases.

acids.[11,127,128] The α-subunit and β-subunit of yeast CAAX farnesyltransferase are encoded by RAM2 and RAM1/DPR1, respectively.[238,282,373,412] Ram2p (316 amino acids) and Ram1p/Dpr1p (431 amino acids) proteins are about 30% and 37% identical to the mammalian α-subunit and β-subunit, respectively.

CAAX Geranylgeranyltransferase

Some Ras-like small GTPases such as Rho, Rac, Cdc42p and the γ-subunit of trimeric GTPases terminate with a similar C-terminal sequence CAAX (where C is Cys, X is an aliphatic residue and X is usually Leu).[331,490] The Cys residue of this motif is lipidated by the addition of a 20-carbon geranylgeranyl group, a reaction catalysed by CAAX geranylgeranyltransferase (also referred to as type I protein geranylgeranyltransferase), followed by proteolysis of AAX and methylation of the lipidated Cys. The CAAX geranylgeranyltransferase is a heterodimer that shares the same α-subunit with the CAAX farnesyltransferase[684] but has a distinct β-subunit of about 43 kDa (Fig. 13). The primary sequence of the β-subunit derived from cDNA cloning consists of 377 amino acids that are over 97% identical between rat and human. Furthermore, the β-subunit is about 30% identical to Cdc43p/Cal1p (composed of 376 amino acids), which has been shown to be a component of yeast CAAX geranylgeranyltransferase.[869] The yeast CAAX geranylgeranyltransferase is therefore composed of Ram2p (α-subunit)[412] and Cdc43p/Cal1p (β-subunit).[213,333] Recently, it was revealed that CAAX geranylgeranyltransferase can transfer farnesyl as efficiently as geranylgeranyl to RhoB, suggesting that this enzyme is able to transfer both farnesyl and geranylgeranyl groups and the choice (ratio) of the prenyl group attached may be dictated by the nature of the acceptor proteins.[20]

Rab Geranylgeranyltransferase

For members of the Ypt1p/Sec4p/Rab protein family, their C-terminal Cys-X-Cys or Cys-Cys sequences are lipidated by the addition of one or two 20-carbon geranylgeranyl groups to the Cys residues.[357,490,683,846] This process is dependent on the C-terminal Cys-X-Cys or Cys-Cys motif as well as upstream structures. The enzyme that catalyzes this reaction is designated Rab geranylgeranyltransferase (also referred to as type II protein geranylgeranyltransferase) and is composed of two components (Fig. 13). Component A is a 95 kDa protein referred to as the Rab escort protein (REP),[7] while the component B is a heterodimer consisting of a 60 kDa α-subunit (with 567 amino acids) and 38 kDa β-subunit (with 331 amino acids).[21] The β-subunits of CAAX farnesyltransferase, CAAX geranylgeranyltransferase and Rab geranylgeranyltransferase are all structurally related, suggesting a similar mechanism during the lipidation catalysed by these three distinct transferases. Component B has the catalytic activity that is stimulated by the REP. REP functions to present newly made Rab proteins to the catalytic component B for lipidation. The α-subunit and β-subunit of component B are encoded by MAD2 and BET2, respectively, in the budding yeast.[92,329] Two mammalian REPs have been identified. REP-1 is encoded by the choroideremia gene in humans, whose mutations are associated with retinal degeneration. Rat REP-1 is composed of 650 amino acids and has two regions (residues 1-50 and residues 356-401) showing sequence similarity with RabGDI (residues 1-46 and residues 210-255, respectively).[12] A second form of REP (REP-2) has been identified that is highly related to REP-1 and possesses the escort activity.[147] Full activity of Rab geranylgeranyltransferase can be reconstituted by recombinant 60 kDa α-subunit, 38 kDa β-subunit and REP-1 or REP-2. The budding yeast MRS6/MSI4 encodes a protein with 603 amino acids that is about 30% identical to the mammalian REPs.[216,330] When recombinant Mrs6p/Msi4p is mixed with Mad2p (the yeast α-subunit with 290 amino acids) and Bet2p (the yeast β-subunit with 322 amino acids), Rab geranylgeranyltransferase activity is reconstituted, establishing that Mrs6p/Msi4p is the yeast REP. The present model for Rab geranylgeranyltransferase is that newly-synthesized Rab proteins are first bound by REPs,[7] which then present them to the catalytic component B, which in turn catalyzes the addition of geranylgeranyl groups to the Rab proteins at the C-terminal Cys-X-Cys or Cys-Cys sequences. The geranylgeranylated Rab proteins are then presented by the REPs to their respective target membranes. The geranylgeranylation is essential for membrane association and functional activity of members of the Ypt1p/Sec4p/Rab protein family. Mutations of Rab proteins (or their yeast counterparts such as Ypt1p and Sec4p) that prevent geranylgeranylation or mutations in the components (such as mutations in BET2 or human REP-1) of Rab geranylgeranyltransferase abolish their membrane attachment and cellular functions.[487,645,846]

Protein myristoyltransferase

Proteins such as ARFs, the oncoprotein Src and the α-subunit (Gα) of trimeric GTPases are lipidated with myristate, a 14-carbon saturated fatty acid via an amide linkage to the N-terminal Gly.[501] This modification is required for cellular functions of the lipidated proteins and is catalyzed by myristoyl-CoA:protein N-myristoyltransferase.[239,334,637,638] The cloned human myristoyltransferase predicts a protein of 416 amino acids that is 44% identical to its counterpart in budding yeast.[193] For proteins destined for N-myristoylation, the second residue in the primary translation product must be Gly. Furthermore, the N-terminal Met must be removed by methionylaminopeptidase, which recognizes the NH_2-MetGly- in order to expose the Gly as the N-terminus for myristoylation. In addition to the removal of the initiation Met and the penultimate Gly, other structural features downstream of the MetGly motif are involved in the determination of myristoylation. If we define the Gly as position number 1, introduction of bulky hydrophobic residues at position 2 inhibits myristoylation, suggesting that bulky hydrophobic residues are not allowed at position 2. Ser at position 5 promotes high affinity interaction with the myristoryltransferase and therefore high efficiency of myristoylation. Basic residues at position 7 and 8 result in better substrates. For some myristoylated proteins such as Src and Gα subunit, sequences even beyond the N-terminal eight amino acids appear to play a role in their recognition by the myristoryltransferase.

PROTEINS THAT FUNCTIONALLY INTERACT WITH GTPASES

Since guanine nucleotide exchange and GTP hydrolysis play a fundamental role in mediating the biological function of GTPases, proteins that can regulate these two processes are intimate partners for GTPases. Several types of proteins are known that regulate the function of GTPases, including those that enhance the exchange of bound GDP for GTP (guanine nucleotide exchange factors, GEF), inhibit this exchange (GDP dissociation inhibitor, GDI), or stimulate the hydrolysis of GTP (GTPase activating proteins, GAP) or inhibit GTP hydrolysis (GTPase inhibiting proteins, GIP). In addition to these GTPase regulatory proteins, the other class of proteins that interact with GTPases are effector molecules that are coupled to the activation of GTPases by the nucleotide exchange for GTP or to the GTP hydrolysis. As discussed previously, Sec12p is a GEF for Sar1p, while Sec23p is a GAP for Sar1p. Sec9p could be an effector for Sec4p. Some other examples are discussed below.

RabGDI

RabGDI was originally identified by its ability to inhibit GDP dissociation from Rab3A (termed Rab3AGDI). It is now established that it is a general GDI that has a similar effect on a broad spectrum of different Rab proteins (renamed RabGDI).[62,178,199,224,465,579,590,720] Three properties of RabGDI are intimately associated with its function. First, RabGDI interacts with Rab proteins that are geranylgeranylated and the geranylgeranyl group is part of the structure that is directly recognized by RabGDI. Associated with this property, RabGDI is able to extract GDP-bound Rabs from the membrane to form the RabGDI-Rab:GDP heterodimer that behaves like a cytosolic complex. Furthermore, RabGDI-Rab:GDP is the protein complex that will deliver the bound Rab to the target membrane, accompanied by the dissociation of RabGDI and subsequent nucleotide exchange for GTP of the transiently membrane bound Rab:GDP. RabGDI therefore functions as both a solubilizing as well as a membrane targeting factor for Rabs. In fact, most soluble Rab proteins exist as a heterodimer with RabGDI and this heterodimer is the functional form of Rab proteins. What triggers the membrane targeting of RabGDI-Rab:GDP complex is unknown, although regulated phosphorylation of RabGDI could be involved.[741] Functionally, excess free RabGDI has been shown to inhibit in vitro ER Golgi and intra-Golgi transport events.[199,580] These inhibitory effects of RabGDI are associated with the removal of Rab proteins from the membrane. Several structurally related forms of RabGDI have been identified. The original RabGDI has been referred to as RabGDI-1 and the bovine and rat cDNA for RabGDI-1 both predict a protein with 447 amino acids and an estimated size of about 51 kDa. The bovine and rat RabGDI-1 molecules are 99% identical. A novel form of RabGDI (designated RabGDI-2 or RabGDI-β) is composed of 445 amino acids in mouse, rat and human and is over 80% identical to RabGDI-1.[533,705,706,862] A yeast gene (GDI1) has been cloned that encodes a 451-residue protein (Gdi1p) of about 50% identical to mammalian RabGDIs.[224,225] GDI1 is essential for cell viability and cells depleted of Gdi1p show multiple defects in protein transport and accumulate ER, Golgi and transport vesicles. Consistent with its role in extracting membrane Rab:GDP, depleting Gdi1p also causes the loss of a soluble pool of Sec4p. In addition, sec19-1 is a mutant allele of GDI1.

Gyp6p

A GAP for Ypt6p is encoded by yeast GYP6 and is composed of 456 amino acids.[750] Ypt6p is most similar to fission yeast Yyh1p and mammalian Rab6 and is implicated in vacuolar sorting/transport in the budding yeast. Gyp6p is the first GAP thus far identified for members of the Ypt1p/Sec4p/Rab family of GTPases and it is highly specific for Ypt6p, whose GTP hydrolysis could be stimulated over 100-fold by Gyp6p.

Dss4p and Mss4

A spontaneous dominant suppresser of the temperature-sensitive sec4-8 mutant was isolated and named DSS4-1.[500] Cloned DSS4 encodes a protein of about 17 kDa with 143 amino acids. Recombinant Dss4p stimulates GDP dissociation from Sec4p. A mammalian homologue of Dss4p was identified by its ability to suppress the sec4-8 mutation and was designated as Mss4.[106] Mss4 consists of 123 amino acids with about 27% identity to Dss4p and stimulates GDP dissociation from Sec4p, consistent with its identification as a cross-species suppresser. Mss4 is primarily a soluble protein with wide tissue distribution.[105] It can bind directly to a selective subset of Rab proteins, including Rab1, Rab3, Rab8, Sec4p and Ypt1p but not others such as Rab2, Rab4, Rab5, Rab6, Rab9 and Rab11. Furthermore, Mss4 could be coimmunoprecipitated with Rab3A in a brain extract by antibodies against Rab3A. Under identical conditions, Mss4 is not coimmunoprecipitated with Rab5. Recombinant Mss4 and Rab3A could form a stable complex in the absence of guanine nucleotide. Mss4 could stimulate GDP dissociation from and promote association of GTP-γ-S onto those interacting Rab proteins. Functionally, Mss4 facilitates neurotransmitter release when microinjected into squid giant nerve terminals.

Rabphilin-3A

Using a crosslinking approach, an 85 kDa protein was identified that interacts selectively with GTP bound Rab3A and has been referred to as Rabphilin-3A.[703] cDNA cloning for bovine Rabphilin-3A predicts that it is composed of 704 amino acids, which can be divided into two domains.[702] The N-terminal region (residues 1 to 390) binds Rab3A and contains a cluster of nine Cys residues, eight of which are arranged in four pairs of CxxC sequences that are typical for Cys-mediated metal-binding motifs, while the C-terminal region (residue 391 to the C-terminus) is structurally related to synaptotagmin (with 33% identity). Like synaptotagmin (see chapter 8), this C-terminal region of Rabphilin-3A contains two copies (residue 396-530 and 548-669) of an internal repeat that are homologous to the C_2 domains of protein kinase C and these two copies of C_2 domains are about 35% identical. Like other C_2 domains, the C2-domain-containing region of Rabphilin-3A binds both phospholipid and Ca^{2+}. Furthermore, Rabphilin-3A and its N-terminal region can inhibit Rab3A GAP-stimulated GTPase activity, indicating that Rabphilin-3A could be an effector for Rab3A:GTP and functions to maintain Rab3A in the GTP-bound active form.[362] Rabphilin-3A also interacts with Rab3C and it was suggested that Rab3A and Rab3C may function in recruiting Rabphilin-3A to the synaptic membrane in a GTP-dependent manner.[409] Alternatively, it has also been suggested that Rabphilin-3A may bind to a membrane protein of the synaptic vesicle in a Ran3A-independent manner.[704]

Rabin3

Rabin3 was identified by an interaction screen using the yeast two hybrid system and is a 50 kDa protein of 460 amino acids.[96] Rabin3 protein and mRNA are distributed widely. Rabin3 interacts with Rab3A and Rab3D but not with others such as Rab3C, Rab2, Ran and Ras. There exists a 59-residue region (residue 154-212) in Rabin3 that is about 44% identical to Sec2p (residue 69-120). How Rabin3 is involved in the function of Rab3A remains to be investigated.

MODELS FOR VESICLE-MEDIATED TRANSPORT

COATOMER-COATED VESICLES

Transport mediated by coatomer-coated vesicles can be divided into seven highly coupled steps.[200,552,553,559,589,614,647,649,774] The first step is the membrane recruitment of ARF1 and its activation by exchange of bound GDP for GTP, this process is mediated by a putative membrane receptor for ARF and a putative GEF. The ARF receptor and the GEF may be integrated into a single protein or a protein complex, distinct subunits of which may function as the receptor and the GEF, respectively. What triggers the membrane recruitment and nucleotide exchange of ARF1 is unknown, although the involvement of heterotrimeric G proteins has been suggested. Whether cargo proteins are involved in this process remains to be explored. The second step is membrane recruitment of the coatomer, formation of coated pits and incorporation of cargo molecules and v-SNAREs into the coated pits. The seven distinct COPs are recruited as a protein complex, the coatomer. How membrane recruitment and GTP-loading of ARF1 results in coatomer recruitment is not clear, but it seems possible that upon membrane recruitment and GTP loading, a conformational change of ARF1 may expose a high affinity binding site for coatomer. Which (if any) COP subunit(s) of the coatomer interacts directly with ARF1 has not been determined. Recruitment of coatomer to the membrane via ARF1 may also trigger an interaction of coatomer with other proteins in the budding sites. Association of coatomer with the membrane results in membrane deformation and formation of coated pits and buds, into which cargo molecules and v-SNAREs are incorporated. The next step is the fission process where budding vesicles pinch off into coatomer-coated free vesicles. This fission process requires acyl-CoA. What proteins generate the force to scissor the neck of the budding vesicles into free coated vesicles is unknown, but the coatomer components or some membrane components may be involved because no other cytosolic proteins are required for the formation of coatomer-coated vesicles in vitro. The budded coated vesicles are then translocated into the vicinity of the acceptor membrane. This translocation process could be governed by components of the cytoskeleton and some motor molecules associated with the vesicles. The fifth step is the uncoating step in which GTP hydrolysis by ARF1 leads to a conformational change of ARF1, resulting in dissociation of coatomer from ARF1:GDP and the vesicles. What triggers GTP hydrolysis by ARF1 and the uncoating process is not known. Alternatively, the uncoating process may occur before or concomitantly with the translocation process. The next step is the docking of the uncoated vesicle with the target membrane. This process involves the specific interaction of v-SNAREs on the vesicle with the t-SNAREs on the target membrane to form the v-SNARE-t-SNARE complex, which is regulated by members of the Ypt1p/Sec4p/Rab GTPase family. This docking process is intimately coupled to the fusion step in that the formation of the v-SNARE-t-SNARE complex triggers the recruitment of α-SNAP and γ-SNAP, resulting in the exposure of a high-affinity binding site for NSF on the SNAPs due to a conformational change of SNAPs. NSF is then recruited to the docking site by binding with the SNAPs, leading to the formation of the v-SNARE-t-SNARE-SNAPs-NSF fusion complex at the interface between the docked vesicles and the target membrane and the activation of the ATPase activity of NSF. ATP hydrolysis by NSF may be coupled directly to the fusion process and the disassembly of the fusion particle. The v-SNAREs are then recycled back to the original membrane.

COPII-Coated Vesicles

COPII-coated vesicles have an essential role in transport from the ER to the early Golgi of yeast cells and may play a similar role in mammalian cells.[47,205a,231a,554,556,626,655] Transport mediated by COPII-coated vesicles is fundamentally similar to that mediated by coatomer-coated vesicles. The first step involves membrane recruitment of Sar1p and its activation by exchange of bound GDP for GTP, which is mediated by Sec12p on the ER membrane. Recent studies also suggest that Sed4p, in conjunction with the new COPII component Sec16p, may also play a critical role in Sar1p recruitment. Sed4p and Sec12p are type II integral membrane proteins whose cytoplasmic domains interact specifically with Sar1p and function as GEFs for Sar1p. Membrane association of Sar1p and its GTP loading results in the membrane recruitment of other components of COPII coat, which is composed of the Sec13p-Sec31p complex and the Sec23p-Sec24p complex. Which subunit(s) interacts directly with Sar1p has not been defined. Similarly, it is not known whether the Sec13p-Sec31p complex and the Sec23p-Sec24p complex are recruited sequentially, simultaneously or as an integrated unit. Membrane association of COPII coat components results in the formation of coated pits and budding vesicles and incorporation of cargo molecules and v-SNAREs into these structures. A fission step is necessary for the excision of the budding vesicles into free COPII-coated vesicles, but the protein(s) is involved in this fission process are not known, although it seems no additional cytosolic proteins are involved because Sar1p, Sec16p, the Sec13p-Sec31p complex and the Sec23p-Sec24p complex are sufficient for the formation of COPII-coated vesicles in vitro. The free COPII-coated vesicles are then translocated to the vicinity of the early Golgi membrane. GTP hydrolysis by Sar1p may be involved in the uncoating process and Sec23p is the GAP for Sar1p. How Sec23p GAP activity and Sar1p GTPase activity are specifically activated during the uncoating process is not clear, although this could be coupled to the fission reaction. The uncoating process may occur before or concomitantly with the translocation step. The uncoated vesicle are then docked onto the membrane of the early Golgi by the interaction of v-SNAREs with the t-SNAREs to form a docking complex. This docking SNARE complex results in the recruitment of Sec17p, a conformational change of Sec17p and exposure of a highly affinity binding site for Sec18p. These events lead to the recruitment of Sec18p to form a fusion complex composed of v-SNAREs (Sec22p/Sly2p, Bet1p/Sec12p, Bos1p and Ykt6p/p2), t-SNARE (Sec5p), Sec17p and Sec18p as well as the activation of the ATPase activity of Sec18p. The formation of the docking complex and subsequent fusion complex require Ypt1p. ATP hydrolysis by Sec18p may be coupled to the fusion process and the disassembly of the fusion complex. The v-SNAREs are then recycled back to the ER.

CLATHRIN-COATED VESICLES

Transport mediated by clathrin-coated vesicles is mechanistically similar to that described for coatomer- and COPII-coated vesicles.[39,56,118,570,634,635,649,728] The role played by clathrin-coated vesicles in receptor-mediated endocytosis and lysosomal targeting of lysosomal hydrolases is well established. The formation of clathrin-coated vesicles on the plasma membrane may be triggered by the interaction between the cytoplasmic domain of cell surface receptors with the AP2 adaptin complex composed of α-adaptin, β-adaptin, AP50 and AP17. This interaction could be initiated, facilitated, or enhanced by ligand binding with the extracellular domain of the receptors, leading to a conformational change of the cytoplasmic domain and/or alteration in the oligomeric state. Membrane association of the AP2 complex then triggers membrane association of clathrin triskelion through interaction between the AP2 complex and the clathrin triskelion, resulting in the formation of coated pits/

buds and incorporation and concentration of cell surface receptors and their ligands. Membrane proteins such as v-SNAREs are also incorporated. The fission process is catalysed by the GTPase, dynamin, which forms ring or interconnected stack of rings on the neck of the budding vesicles. GTP hydrolysis by dynamin may cause a conformational change of dynamin which pinches off the clathrin-coated vesicles. The next step is the uncoating process, which results in the dissociation of clathrin triskelion and the AP2 adaptin complex. The uncoated vesicles may fuse with each other and/or with the pre-existing early endosomes, this process may be governed by pairing of SNAREs on the vesicles with the early endosome. Several steps, such as the recruitment of AP2 adaptin complex, uncoating process and docking/fusion could be under the regulation of small GTPases of the Ypt1p/Sec4p/Rab family (for example, Rab5) and/or ARF family (for example ARF6), details of which remain to be established. Formation of clathrin-coated vesicles in the TGN could be initiated by binding of lysosomal enzymes (through the M6P signal on the N-linked glycans) with the M6P receptors on the TGN. This luminal interaction may trigger recruitment of the AP1 protein complex, composed of γ-adaptin, β'-adaptin, AP47 and AP19, onto the cytoplasmic domain of the M6P receptors followed by the association of clathrin triskelion and the formation of coated pits and budding vesicles. Cargo molecules (the lysosomal enzymes, M6P receptors and others) are concentrated and v-SNAREs are incorporated in these coated structures. The fission process results in free coated vesicles. Whether dynamin or a dynamin-like protein is involved in the fission of clathrin-coated vesicles in the TGN is unknown. The coated vesicles then undergo an uncoating process and the uncoated vesicles are specifically docked onto late endosomes via interactions between the v-SNAREs on the vesicles and t-SNAREs on the endosomes, followed by the fusion step.

ACTIVE VS DEFAULT TRANSPORT

Protein transport along the exocytotic pathway has been proposed to occur by default.[590a,842a] This model suggests that proteins are transported nonselectively from the ER to the plasma membrane and also that proteins that are retained intracellularly at each compartment or diverted to other pathways (such as the lysosomal pathway) are mediated by a signal-dependent process. Many signals have been identified that serve to retain proteins at various compartments or to sort them into a diverted pathway. Strictly speaking, identification of signals required for retention or sorting does not exclude the possibility that incorporation of proteins into transport vesicles during each budding process may be an active process that may involve some kind of interaction of the cargo molecules with components of the budding machinery such as a sorting/concentration receptor. During this interaction, the putative sorting/concentration receptor may recognize some kinds of structures, such as conformational information that are coupled to the folding/assembly of the proteins, that are fundamentally different from the linear signals involved in intracellular retention or pathway divergence. The demonstration that vesicle budding processes are regulated by various GTPases suggest strongly that vesicle budding is tightly regulated and thus an active event. Some other evidence supportive of this is discussed below. It seems that active and passive processes during each vesicular transport step may not be mutually exclusive and could occur simultaneously.

CONCENTRATION OF CARGO MOLECULES DURING THE BUDDING PROCESS

Recently, it was revealed that several cargo proteins are concentrated many-fold during the budding process.[32a,47,486a] One such example is the soluble secretory protein serum albumin. Using immuno-electron microscopic localization, it was observed that albumin is first concentrated at isolated sites within the ER before it is transferred to the cis face of the Golgi apparatus in Hep-G2 human hepatoma cells. Similarly, the VSV G-protein is observed to be concentrated 5- to 10-fold during vesicle budding from the ER in vitro. During the budding step, G-protein is efficiently sorted from the resident ER proteins, suggesting that the concentration is a selective process. Another example is that the ER form of pro-α-factor (pαf/ER) is concentrated many-fold and efficiently sorted from resident ER proteins during its in vitro incorporation into COPII-coated vesicles. Early observations that different proteins exported from ER at markedly different rates are consistent with the existence of a sorting/concentration process during export from the ER.[426a] Cargo molecules involved in this concentration/sorting process may not share any apparent primary structures but may bear a common biophysical property to interact with one of a few sorting/concentration receptors. This phenomenon could be compared to the diverse antigenic structures recognized by one of only two major antigen-presenting molecules, the class I and class II major histocompatibility complex antigens. This kind of interaction between the cargo molecules and the sorting/concentration receptor may render it impossible to define an obvious linear structure/motif that is shared by different cargo molecules.

GOLGI RETENTION OF MUTANT PROTEINS

Another line of evidence indicating that some types of structure are involved during budding comes from expression and localization of mutant and chimeric proteins. Many mutant and chimeric proteins are rendered incompetent for export from the ER mainly due to a failure to fold/assemble properly. However, a number of mutant and chimeric proteins are efficiently exported from the ER and transported to the Golgi apparatus.[435] However, further transport within the Golgi and/or from the TGN to the surface is abolished or delayed. One example is a chimeric protein (DAD) derived from two surface proteins, dipeptidylpeptidase IV (DPPIV) and aminopeptidase N (ApN). DAD is identical to DPPIV with the exception that its transmembrane domain has been replaced by that of ApN. Both DPPIV and ApN are efficiently transported along the exocytotic pathway up to the plasma membrane (apical surface domain in polarized epithelial cells). However, DAD is selectively retained in the trans Golgi/TGN, while transport from the ER to the Golgi and dimerization of the molecule occur as efficiently as for DPPIV.[435] Although this could be explained by the fortuitous creation of a Golgi-retention signal in DAD, it could also be explained by proposing that the transmembrane domain of DPPIV is necessary for the formation of a structure that is involved in its efficient export from the TGN. Furthermore, when the extracellular domain of DPPIV is replaced by the α-subunit of human chorionic gonadotrophin (αhcg), which can be efficiently secreted without an accompanying β-subunit, a significant amount of the resulting chimeric protein (D4αhcg) is accumulated intracellularly in the Golgi. Further studies show that transport between the cis Golgi compartment and subsequent structures is significantly-delayed for D4αhcg, although the chimeirc protein is eventually transported to the cell surface, indicating that the extracellular domain of DPPIV is involved in the formation of a structure that may facilitate transport from the cis Golgi to the downstream compartment.[436] Two temperature-sensitive mutants (ts482 and ts651) of Fowl Plague virus hemagglutinin are selectively retained in the post-medial compartments of the Golgi apparatus at the restrictive temperature, although trimerization and ER-Golgi transport are not affected. At permissive temperatures, these two mutant proteins are transported to the cell surface. These results suggest that some structural information of hemagglu-

tinin is required for efficient transport from the Golgi to the surface, and this information is abolished at the restrictive temperature for these two mutant proteins. The H1 subunit of human asialoglycoprotein receptor is transported to the cell surface. However, when the cytoplasmic domain of H1 subunit is truncated, the mutant molecules are retained in the Golgi, suggesting that its cytoplasmic domain may be involved in transport from the Golgi to the cell surface.[821] Taken together, it seems that transport events within the Golgi and from the TGN to the surface may involve an active process.

EMP24P AS A CONCENTRATOR

As mentioned earlier, the yeast Emp24p is the first membrane protein that is implicated in the sorting/concentration of a subset of cargo molecules during the budding of COPII-coated vesicles from the ER.[670] Emp24p is a type 1 ER membrane protein that is incorporated into COPII coated vesicles. EMP24 disruption is not lethal but delays significantly the export rate of invertase and a GPI-anchored surface protein Gas1p. This effect is selective since export from the ER of several other proteins is not affected. Emp24 may thus function as a sorting receptor for concentrating a set of cargo molecules such as invertase and Gas1p during their export from the ER. Other sorting receptors may facilitate the concentration of other proteins during their budding process from the ER. Recently, it was revealed that there are at least five Emp24p-like proteins both in yeast and mammalian cells.[737] A 24 kDa type I membrane protein with 32% identity with Emp24p has been cloned from CHO cells and is one of the major integral proteins of coatomer-coated vesicles. The existence of several Emp24p-like proteins suggest that they may play a similar role in sorting/concentrating cargo molecules in diverse compartments of the secretory pathway. The localization and function of these Emp24-like proteins remains to be established.

BASOLATERAL SIGNALS

The identification of cytoplasmic signals (to be discussed in chapter 7) required for efficient transport from the TGN to the basolateral surface of polarized epithelial cells is consistent with the idea that this vesicular transport may be an active process involving the interaction of cargo molecules with the budding machinery of vesicles destined for the basolateral surface.

SIGNAL-MEDIATED ER LOCALIZATION

LUMINAL C-TERMINAL KDEL MEDIATES ER LOCALIZATION

As discussed in the first chapter, the lumen of the ER contains several abundant soluble proteins involved in folding/assembly of nascent proteins of the exocytotic pathway, including Bip/GRP78, GRP94, PDI and others. Examination of the primary sequence of Bip/GRP78, GRP94 and PDI reveals a common motif, -Lys-Asp-Glu-Leu-COOH (KDEL), at the C-terminal end of these proteins.[572] This finding prompted a thorough investigation of the KDEL motif as a possible signal for ER location of these proteins.[503,572] When overexpressed in transfected cells, wild-type Bip/GRP78 is efficiently retained in the cells and no secretion into the culture media is observed. However, several Bip/GRP78 mutants with altered C-terminal ends are found to be secreted into the media, suggesting that the intact C-terminal KDEL sequence is necessary for intracellular ER accumulation and for preventing its secretion. Particularly, when the C-terminal 60-residue sequence of BiP/GRP74 is replaced with an 11-residue epitope of Myc protein, the mutant (D60Myc) is secreted. However, attachment of the last six residues (SEKDEL) of BiP/GRP74 to the C-terminus of the D60Myc mutant can restore its intracellular accumulation and prevent its secretion, suggesting that secretion of D60Myc is not due to a global change of its conformation but rather due to the lack of the KDEL sequence. Similarly, when the C-terminal 15-residue sequence is replaced either with the Myc epitope (D15Myc) or with the Leu-Asp sequence (D15LD), the mutants are secreted. Furthermore, addition of the Gly-Leu sequence to the C-terminus of wild-type BiP/GRP74 results in the secretion of the modified protein, suggesting that the KDEL sequence must be located at the C-terminus. The C-terminal KDEL sequence is therefore necessary for its intracellular ER retention. To examine whether the C-terminal KDEL is sufficient for ER localization, experiments were performed to determine if the C-terminal sequence of BiP/GRP74 would prevent secretion of a reporter molecule which is normally secreted. For this purpose, chicken lysozyme, which is a small, highly basic protein that is constitutively-secreted, was used as a reporter. Addition of the Myc epitope to the C-terminus of lysozyme resulted in a modified protein (LyMyc) that was secreted as efficiently as lysozyme. However, addition of the last six-residue sequence of BiP/GRP74 (SEKDEL) to LyMyc (generating LyMycSEKDEL) prevented its secretion. To rule out the possibility that the non-secretion of LyMycSEKDEL is due to a conformational change rather than the C-terminal SEKDEL sequence, the last two residues of LyMycSEKDEL were changed into Ala-Ser and the resulting protein (LyMycSEKDAS) was again secreted efficiently. Furthermore, immunofluorescence microscopy reveals that LyMycSEKDEL is retained intracellularly in the ER. These experiments firmly establish that the C-terminal SEKDEL sequence of BiP/GRP74 is both necessary and sufficient for its accumulation in the lumen of the ER. Since only the KDEL sequence is conserved, it was proposed that this C-terminal tetrapeptide sequence is the signal for ER localization. Subsequently, it was revealed that the C-terminal KDEL sequence is also shared by other ER luminal proteins (such as CaBP1, PDI and ERp72) and involved in their ER localization. Extensive mutations revealed that some KDEL-related sequences such as RDEL, KNEL, KEEL and DKEL can also confer ER localization in mammalian cells.[572]

Sequences related to KDEL (such as HDEL, DDEL, ADEL, or SDEL) are present in the ER luminal proteins of diverse species.[575] In yeast *S. cerevisiae*, the HDEL sequence is found in major ER proteins such as Bip, GRP94 and PDI. When the C-terminal 6-residue sequence (FEHDEL) of the yeast Bip is fused to the C-terminus of the Myc epitope-tagged lysozyme (LyMyc), the resulting protein is efficiently secreted in mammalian cells, suggesting that HDEL is not an efficient

ER retention signal in mammalian cells. However, the FEHDEL is a functional signal for ER localization in *S. cerevisiae*. This is demonstrated by expressing various modified versions of invertase, which is normally secreted from yeast cells and remains trapped under the cell wall. Sucrose is the substrate for invertase and can not pass through the plasma membrane and therefore can not be used as a substrate for intracellular invertase in intact yeast cells. Appending the FEHDEL sequence to the C-terminus of invertase has no effect on the secretion and is most likely due to the masking of the attached sequence. When invertase is fused to the mature BiP molecule, the resulting chimeric protein (invertase-BipFEHDEL) is efficiently retained in the cells when expressed as an integrated chimeric gene under the control of a weak promoter, and only 3-5% of the total invertase activity of the fusion protein is secreted. However, when the FHHDEL sequence is replaced by SEKDEL, the resulting variant (invertase-BipSEKDEL) is poorly retained and more than 50% of the fusion enzyme is secreted, demonstrating that the FEHDEL of yeast Bip but not the SEKDEL of mammalian Bip is able to mediate intracellular accumulation in the yeast.[575] Furthermore, retention of invertase-BipFEHDEL is saturable as more than 15% of it can be secreted when it is expressed from a stronger promoter that produces 6- to 7-fold more protein.

When the KDEL sequence is appended to the C-terminus of a type II membrane protein (DPPIV), the majority of the resulting protein is localized to the ER, suggesting that KDEL may also function as an ER localization signal for membrane proteins whose C-termini are luminally localized. Two yeast type II membrane proteins (Sec20p and Sed4p) contain the C-terminal HDEL motif. These results suggest that the KDEL signal can also be used for ER localization of membrane proteins.

KDEL MEDIATES RETRIEVAL FROM POST-ER COMPARTMENTS

Several lines of evidence suggest that the C-terminal KDEL (HDEL in yeast) sequence functions as a retrieval signal for escaped ER proteins at post-ER compartments. When the SEKDEL sequence is attached to a modified cathepsin D that has a myc-epitope at the C-terminus (CDM), the resulting protein (CDM*KDEL*) is efficiently retained in the ER as assessed by immunofluorescence microscopy. Like other lysosomal hydrolases, cathepsin D and CDM are modified by two Golgi enzymes (N-acetylglucosaminyl-1-phosphotransferase followed by N-acetylglucosaminyl phosphoglycosidase) to generate the M6P signal for lysosomal targeting (see chapter 6). Interestingly, ER localized CDM*KDEL* is efficiently modified by N-acetylglucosaminyl-1-phosphotransferase, suggesting that it can reach the cis Golgi compartment which harbors the enzyme. However, the N-acetylglucosaminyl group is not removed in the majority of CDM*KDEL*, suggesting that the most of the CDM*KDEL* does not reach the later Golgi which contains N-acetylglucosaminyl phosphoglycosidase. Furthermore, it is seen that modification of the ER localized CDM*KDEL* by N-acetylglucosaminyl-1-phosphotransferase is greatly inhibited at 20°C and essentially abolished at 15°C, conditions known to reduce and block ER-Golgi transport, respectively, suggesting that modification of CDM*KDEL* requires its transport from the ER to the cis Golgi. These observations suggest that CDM*KDEL* is initially transported from the ER to the cis Golgi but is efficiently retrieved to the ER.[571] Similarly, the C-terminal HDEL sequence is shown to function as a retrieval signal from the early Golgi of yeast, which is marked by α1,6-mannosyltransferase, when the signal is attached to normally secreted proteins such as invertase and pro-α-factor.[168] Furthermore, modification of the ER localized invertase-BipFEHDEL

and pro-α-factorFEHDEL by α1,6-mannosyltransferase is dependent on the function of Sec18p, suggesting that transport from the ER to the early Golgi is necessary for this modification and that the major site of retrieval is the early Golgi.

In most cases, attachment of the KDEL sequence to the C-terminus of secretory or lysosomal proteins results in efficient ER localization achieved solely by the KDEL-mediated retrieval process. However, when the KDEL motif of luminal ER proteins (such as BiP/GRP74) is mutated, they are exported from the ER very poorly as compared to secretory proteins and lysosomal enzymes and a steady state accumulation of the mutant proteins in the ER can still be observed, although a significant portion of the mutants are detected in the culture medium. These results suggest that ER luminal proteins may have another primary mechanism that prevents their export from the ER and this process may be independent of the KDEL sequence.[572,725] The KDEL motif may therefore function as a safe-guard system for luminal proteins that have escaped from the primary retention mechanism.

THE KDEL RECEPTORS: STRUCTURE AND DYNAMIC LOCALIZATION

The identification of the KDEL receptor was achieved through a genetic approach. Since invertase-BipFEHDEL is efficiently retained intracellularly when expressed in yeast, a yeast strain expressing invertase-BipFEHDEL was mutagenized and screened for mutants that failed to retain invertase-BipFEHDEL intracellularly, resulting in the secretion of invertase-BipFEHDEL and high external invertase activity.[268] Several mutants fell into a complementation group (erd2). erd2 mutants not only secreted invertase-BipFEHDEL but also endogenous BiP. The ERD2 gene encodes a protein (Erd2p) of 219 amino acids.[689] Cloning and sequencing of four independent erd2 alleles shows that they all have mutations in the coding region of the ERD2 gene. Two of the erd2 alleles result from a single base change that introduces a termination codon 12 amino acids from the C-terminus, resulting in a mutant Erd2p with a truncation of the C-terminal 12 amino acids. This demonstrates that the last 12 residues of the C-terminus of Erd2p are essential for its function. Biochemical characterization have established that Erd2p is an integral membrane protein.

The identity of Erd2p as the HDEL receptor is suggested by two major observations. Firstly, it was noted that the levels of Erd2p regulate the capacity of the HDEL retrieval system.[689] Since HDEL-mediated retrieval is saturable, high levels of expression of pro-α-factorFEHDEL resulted in secretion of not only itself but also endogenous Bip. However, when Erd2p is overexpressed in this yeast strain, secretion of pro-α-factorFEHDEL and Bip is greatly reduced. Furthermore, overexpression of Erd2p also reduces the extents of modification of pro-α-factorFEHDEL by α1,6-mannosyltransferase, probably due to faster and more efficient retrieval. When the extents of retention of a HDEL-tagged invertase fusion protein are measured in strains engineered to contain different levels of Erd2p, it is found that the efficiency of retention is proportionally correlated with the levels of Erd2p. During the course of screening for ERD2, it was found that four (sec17, sec18, sec20 and sec22) sec mutants at permissive temperature and a yeast strain (erd1) with ERD1 disruption have some erd phenotype in that they secrete endogenous BiP to different extents. This is most likely due to an indirect effect of these mutations on the retrieval system. Importantly, overexpression of Erd2p in these cells can suppress their erd phenotype and secretion of endogenous Bip is prevented. The other line of evidence comes from the observation that the specificity of the retrieval system is determined by Erd2p.[408] In the budding yeast *K. lactis*, the ER luminal proteins terminate either with

HDEL or, in the case of Bip, with DDEL, suggesting that the putative receptor in *K. lactis* can recognize both HDEL and DDEL. The ERD2 of *K. lactis* was cloned by hybridization at low stringency with the *S. cerevisiae* ERD2 as the probe and encodes a protein which is 59% identical to *S. cerevisiae* Erd2p. *S. cerevisiae* does not recognize DDEL efficiently, because an invertase fusion protein ending with the YFDDEL motif of *H. lactis* Bip is poorly retained. However, when *H. lactis* ERD2 is introduced into *S. cerevisiae*, this YFDDEL-tagged invertase fusion protein is now retained efficiently, suggesting that Erd2p determines the specificity of retrieval mediated by KDEL and KDEL-related sequences. Residues between positions 51 and 57 of Erd2p are important in determining the ligand specificity. In addition to the HDEL-mediated retrieval, Erd2p also plays a vital role in the integrity of the Golgi and cell growth because ERD2 disruption leads to a defect in passage of proteins through the Golgi, accumulation of intracellular membranes and cessation of cell growth.[689,790,792]

Based on the conserved ERD2 sequences between *S. cerevisiae* and *H. lactis*, polymerase chain reaction (PCR) was used to isolate a human homologue of Erd2p, the human KDEL receptor or hERD2.[309,406,772] cDNAs or genes for the KDEL receptor were subsequently cloned from diverse species. A second form of the KDEL receptor (ELP-1) was cloned from humans and its 214 amino acid sequence is about 83% identical to hERD2.[309] When either hERD2 or ELP-1 is overexpressed in cells, brefeldin A-like effects, characterized by the disassembly of the Golgi apparatus and inhibition of ER-Golgi transport, are observed, suggesting that the levels of human Erd2p-like proteins are intimately associated with the integrity and function of the early secretory pathway.[309,415,416,772] When the C-terminus of human KDEL receptor is tagged with a Myc epitope and expressed in transfected cells, the tagged receptor is found primarily in the Golgi apparatus. Consistent with its nature as the KDEL receptor, increased levels of human KDEL receptor by transfection improve retention of lysozyme DDEL, which is weakly recognized by the endogenous retrieval system. Furthermore, the localization of the epitope-tagged receptor could be specifically shifted from the Golgi to the ER by overexpression of recognizable ligands such as lysozyme KDEL and lysozyme DDEL but not by lysozyme AARL (AARL is not a functional retrieval signal), while other Golgi markers are not affected. This provides in vivo evidence for direct ligand interaction with the KDEL receptor.[407] As discussed above, Erd2p determines the specificity of the retrieval. The specificity of the human receptor could be modified by site-directed mutagenesis. When the ligand binding region (DLFTNYI) of human KDEL receptor is replaced by the sequence NLFTKYT to resemble the *H. lactis* Erd2p binding region (NLFTKWT), the Golgi localization of the resulting variant of the human KDEL receptor is no longer shifted by overexpressed lysozyme KDEL but can still be shifted by lysozyme DDEL, consistent with the interpretation that KDEL-containing ligands interact directly with the human KDEL receptor and the dynamic localization of the KDEL receptor is modulated by the load of ligands. Further in vitro binding studies established that peptides ending with KDEL and related sequences can bind with membrane-associated human KDEL receptor and that this interaction is pH-dependent with optimal binding at pH 5.0-5.5.[849]

The primary sequence of Erd2p and its homologues in other species predicts the existence of seven hydrophobic regions in the protein. A seven-transmembrane-domain (7TM) model has been proposed for the KDEL receptor.[792] An alternative six-transmembrane-domain (6TM) model for the KDEL receptor has also been independently established (Fig. 14).[717] The major differences between these two topological models include: (1) the orientation of the N-terminus (cytoplasmic for

the 6TM and luminal for the 7TM model, respectively) and therefore the polarity of the first hydrophobic region (residue 4-21) as a transmembrane domain ($N_{out}C_{in}$ in the 6TM and $N_{in}C_{out}$ in the 7TM model, respectively); and (2) whether the second short hydrophobic region (residue 36-46) is used as a transmembrane domain (used in the 7TM model but associated only with the luminal surface of the membrane in the 6TM model). The hydrophilic sequence (residue 22-35) between the first and the second hydrophobic region is predicted to be on the cytosolic side for the 7TM and luminal side for the 6TM model, respectively. The major ligand specificity-determining sequence from residue 50 to 56 (DLFTNYI) is predicted as part of an exposed hydrophilic luminal region in the 6TM model, while this region is defined as the luminal portion of the second transmembrane domain. In both models, each of the remaining five hydrophobic domains are predicted to function as a transmembrane domain with the C-terminal 13-residue tail oriented in the cytoplasmic side of the membrane. The cytoplasmic orientation of the C-terminal 13-residue tail has been confirmed by specific antibody binding when the plasma membrane but not internal membranes are selectively permeabilized. As mentioned, truncation of the C-terminal last 12-residue sequence results in an erd phenotype. Recent evidence suggests that some cytosolic factor(s) may interact with the cytoplasmic side, including the C-terminal tail of the receptor.[773]

The localization and dynamic trafficking of endogenous mammalian KDEL receptor have been examined with antibodies specific for the C-terminal 13-residue cytoplasmic tail (which is conserved between two different human forms) of the KDEL receptor.[252,772] Immunofluorescence microscopy in diverse cell types shows that the receptor molecules are primarily localized to the Golgi apparatus and some cytoplasmic spotty structures, which correspond to the intermediate compartment (ERGIC) between the ER

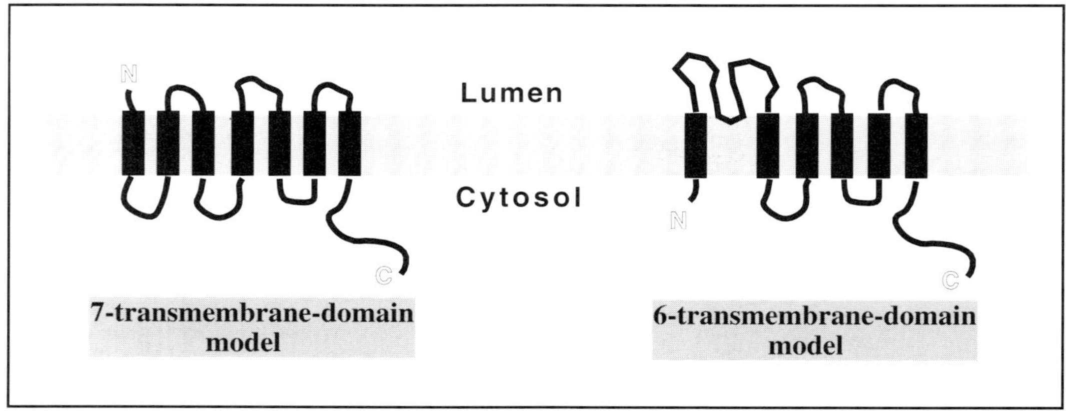

Fig. 14. Two different models of the KDEL receptor topology.

and the Golgi. The majority of the receptors could be selectively accumulated in the ERGIC at 15°C or by treatment with brefeldin A. The KDEL receptors then migrate from the ERGIC to the Golgi upon incubation at 37°C or removal of brefeldin A, suggesting that the receptors cycle between the Golgi and the ERGIC. Immunoperoxidase and immunogold labeling of the receptors under electron microscopy revealed that the majority of the receptors is present in the ERGIC and the cis side of the Golgi.[252] A small but significant amount of the receptors is found on the trans side of the Golgi under normal physiological conditions.[252] However, conditions which induce cell stress such as viral infection and heat-shock could induce a shift of the receptor from the cis to the trans side of the Golgi.[252] Based on these morphological observations, it was proposed that retrieval of KDEL-containing proteins occurs at multiple post-ER compartments up to the TGN along the exocytotic pathway, and that within this pathway, the amounts of the receptors in different compartments vary according to physiological conditions. In addition, the effects of several toxins ending with KDEL or related sequences are dependent on their C-terminal sequences and addition of KDEL to toxins lacking the signal can increase their toxicity, sug-

gesting that endocytosed toxins may use the retrieval pathway mediated by the KDEL receptor for their effects. Since the TGN is the site where endocytotic pathway meets with the biosynthetic pathway, these observations are consistent with the functional existence of KDEL-mediated retrieval from the TGN and the localization of the receptors in the TGN. Direct experimental evidence demonstrating that KDEL-mediated retrieval can indeed occur in the TGN has been obtained.[480] A radiolabled peptide (YHPNSTCSEKDEL) that contains the KDEL signal and a motif (NST) for N-linked glycosylation was introduced into the TGN by a TGN38/41 dependent cell surface to TGN recycling pathway. Significantly, this peptide is delivered and retained in the ER as revealed by its acquisition of N-linked glycan, which takes place only in the ER. This delivery from the TGN to the ER is dependent on the C-terminal KDEL because addition of two additional residues to mask the KDEL signal abolishes the delivery. The retrieval of escaped ER proteins bearing KDEL thus occurs at multiple post-ER compartments up to the TGN. Since the majority of the receptors is present in the cis Golgi and cis Golgi is the first compartment encountered by the escaped proteins, the cis Golgi represents the major site of this retrieval.

C-TERMINAL KKXX MEDIATES POST-ER RETRIEVAL OF ER MEMBRANE PROTEINS

Cytoplasmic C-terminal KKXX or related sequences (such as KXKXX and RXKXX) have been shown to function as ER localization signals of membrane proteins with C-terminal tails oriented on the cytoplasmic side of the membrane (Fig. 15). This ER localization signal was originally identified in the adenovirus E3/19K protein,[326,327,530] which is a type I membrane protein composed of a 104-residue luminal domain, a 23-residue transmembrane domain and a 15-residue cytoplasmic domain (KYKSRRSFIDEKKMP). In adenovirus-infected cells, E3/19K binds to the class I major histocompatibility complex (MHC) antigens to abrogate their cell surface expression, a means by which the virus evades host immune surveillance. When the cytoplasmic domain of E3/19K is truncated by deleting the last eight amino acids of the C-terminus, the mutant is transported to the surface, suggesting that this C-terminal sequence is necessary for its ER localization. To examine whether the cytoplasmic domain of E3/19K is sufficient for ER localization, this region was used to replace the cytoplasmic domain of the cell surface proteins CD4 and CD8. The resulting chimeric proteins (CD4/E19 and CD8/E19, respectively) were found to be specifically confined to the ER, suggesting that the cytoplasmic domain of E3/19K can indeed function as a transferable autonomous signal for ER localization. When the C-terminal Pro residue is deleted, the resulting mutant is only partially ER localized. When the penultimate Met as well as the terminal Pro are both deleted, the mutant protein is no longer ER retained but transported to the cell surface. Further analysis established that the C-terminal last six residues (DEKKMP) form the signal for ER localization. Site-directed mutagenesis of DEKKMP sequence as well as replacing the cytoplasmic domain of CD8 with that of other cellular ER proteins revealed that the cytoplasmic C-terminal

KKXX, KXKXX or RXKXX is the signal for ER localization of membrane proteins. When the C-terminal last 13-residue sequence of CD4 is truncated, the resulting mutant carries a terminal SEKKTC sequence and is confined to the ER.[701] For the KKXX sequence to mediate ER retrieval, it must be exposed and accessible to the retrieval machinery.[404]

The KKXX and related sequences are found in glycoproteins of several foamy viruses[236] and a significant numbers of endogenous ER membrane proteins with their C-terminal tail exposed to the cytosol (type I and some type III proteins) (Fig. 15).[326,530,701] Examples of these proteins include gp25H/TRAPβ,[818] gp25L,[818] ERGIC53,[671] calnexin,[818] Wbp1p,[143] Vma21p,[297] Gaa1p,[257] HMG CoA reductase,[326] the TRAM protein,[240] different forms of UDP-glucuronosyltransferase (UDPGTs),[326,701] TCRβ,[701] mouse CD3δ,[701] and 53K SER.[398] The cytoplasmic tail of several UDPGTs, 53K SER, HMG CoA reductase have each been shown to confer ER localization to the respective chimeric CD8 molecules.[326]

Biochemical analysis of ER retained CD4/E19 and CD8/E19 reveals that they can receive modifications by enzymes localized to post-ER compartments. Particularly, the addition of GalNac (the first sugar residue of O-linked glycans) and palmitate to CD8/E19 is readily detected. The conclusion is that, like the KDEL sequence, the KKXX and related sequences function as retrieval signals for ER localized membrane proteins and that the retrieval occurs at multiple post-ER compartments. Since yeast Wbp1p contains the KKXX sequence and is localized to the ER, it was experimentally established that the cytoplasmic C-terminal KKXX sequence can confer ER localization to reporter proteins in yeast. The resulting ER localized proteins are modified by early Golgi α1, 6-mannosyltransferase, suggesting that the KKXX motif is also a retrieval signal in yeast.[226,791] The yeast Vma21p is an ER membrane protein having two potential transmembrane domains with

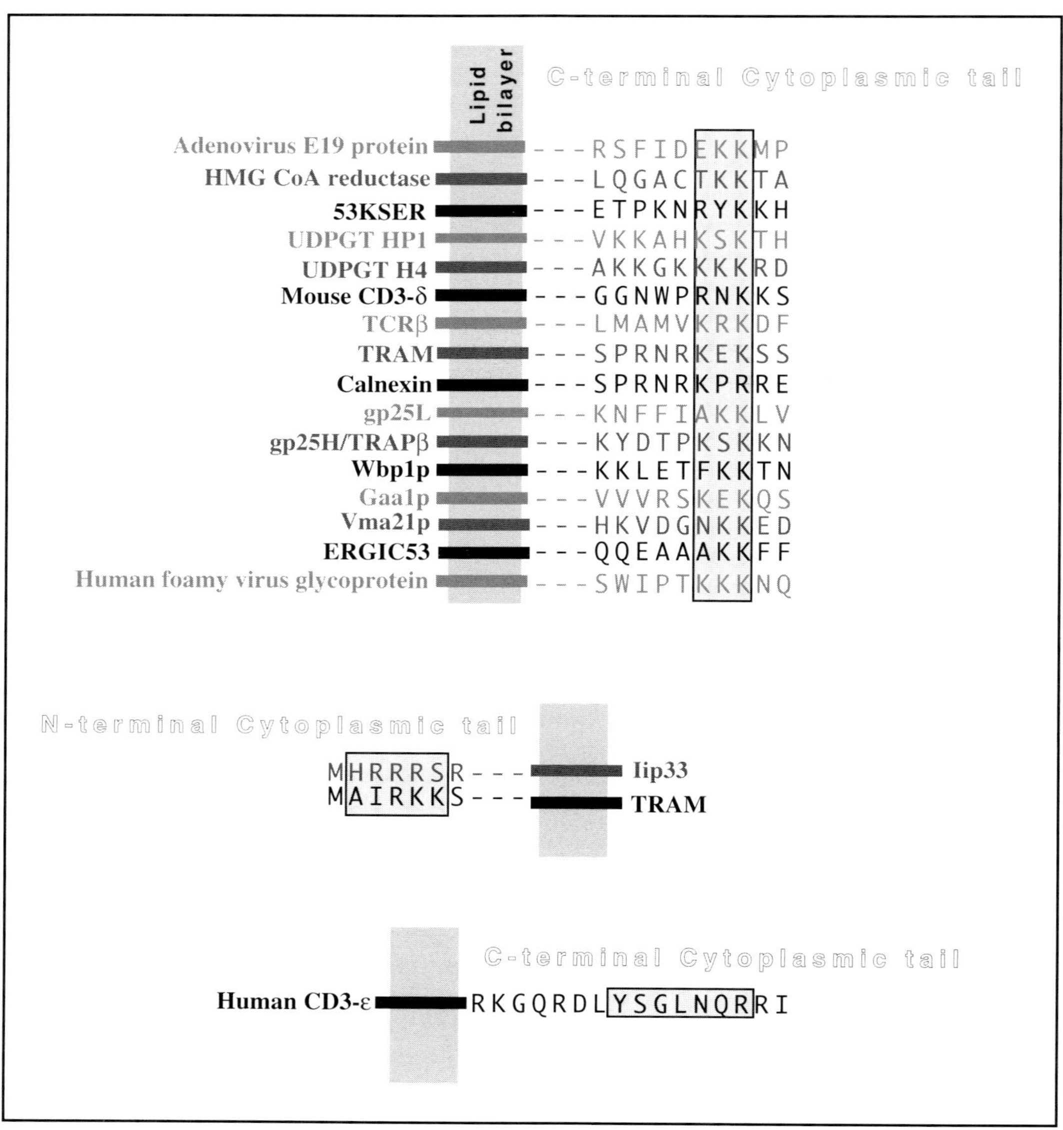

Fig. 15. Alignment of the C-terminal cytoplasmic KKXX and related ER retrieval signals. The N-terminal cytoplasmic ER retrieval sequences of Iip33 and TRAM and the Tyr-based ER localization signal of Human CD3-ε are also shown.

both N- and C-termini oriented on the cytoplasmic side of the membrane. Importantly, Vma21p terminates with a cytoplasmic KKED sequence. Mutating the Lys residues in this motif into Gln results in failure of the mutant protein to localize in the ER. Rather, the mutant is transported to the vacuole, demonstrating that the KKXX motif is necessary for its ER localization.[297]

ROLE OF COATOMER IN KKXX-MEDIATED RETRIEVAL

The mechanism of KKXX-mediated retrieval has been investigated by both biochemical as well as genetic approaches. The coatomer has been revealed to play a pivotal role in this retrieval.[143,574] When the cytoplasmic C-terminal sequence of Wbp1p is fused to the C-terminus of glutathione-S-tranferase (GST), the resulting fusion protein (GST-WBP1) is found to selectively interact with coatomer in yeast cell lysate. Furthermore, the interaction of GST-WBP1 with coatomer is abolished when the two Lys of the KKXX motif are mutated into Ser or Arg residues. Similarly, addition of four Ser residues to the C-terminus of GST-WBP1 (so that the KKXX sequence is now not at the extreme C-terminus) also abolishes this interaction. Interaction of GST-WBP1 with mammalian coatomer is also observed when COS-1 cell lysate is used. The cytoplasmic domain of E3/19K, when fused to GST, is also able to interact with coatomer in total cell lysates. A partial coatomer composed of α-COP, β'-COP and ε-COP is also able to interact specifically with the cytoplasmic domain of either Wbp1p or E3/19K presented as GST-fusion proteins, suggesting that one or more of these three COPs may directly contact the KKXX motif. These biochemical observations indicate that the coatomer may also be involved in the retrieval pathway mediated by the KKXX motif, in addition to its involvement in ER to Golgi and intra-Golgi transport.[143]

In order to identify components involved in KKXX-mediated retrieval in vivo, a genetic approach was used to identify and isolate yeast mutants defective in this retrieval process. Ste2p is expressed on the plasma membrane of MATα yeast cells and is essential for mating with MATα cells by functioning as the receptor for α-factor. Ste2p is a type III membrane protein with seven transmembrane domains and the C-terminal tail oriented on the cytoplasmic side of the membrane. When a chimeric Ste2p (Ste2-Wbp1p), in which most of the cytoplasmic domain of Ste2p is replaced by that of Wbp1p, was expressed in MATα yeast cells deleted for the STE2 gene, Ste2-Wbp1p was found in the ER but with modification attributable to early Golgi α1,6-mannosyltransferase. The resulting MATα cells cannot respond to α-factor and are defective in mating. However, when the KKXX motif in Ste2-Wbp1p is mutated or masked, these vari-

ants of Ste2-Wbp1p are expressed on the cell surface and can restore response to α-factor. These findings suggest that yeast mutants with defective KKXX retrieval would express Ste2-Wbp1p on the cell surface and be competent for mating.[403]

Screening available yeast ts mutants has identified sec21-1 and sec27-1 that are defective in KKXX-mediated retrieval. Since Sec21p and Sec27p are yeast γ-COP and β'-COP, respectively, it is possible that coatomer may be involved in the KKXX retrieval pathway.[403,574] New mutants defective in retrieval of Ste2-Wbp1p (ret mutants) have also been isolated by mutagenizing cells expressing Ste2-Wbp1p and screening for mating-competent mutants. One of the mutants is a new allele of sec21 and was referred to as sec21-2. Eight temperature sensitive mutants fall into one complementation group (ret1). A typical ret1 mutant allele, ret1-1, exhibits efficient mating, normal growth at 24°C and a conditional growth defect at 37°C.[403] The ret-1 mutation shows synthetic lethal interaction with sec21-1 and sec27-1 at 24°C, suggesting that the product of RET1 may interact with Sec21p and Sec27p. RET1 encodes α-COP of yeast coatomer, further strengthening the interpretation that coatomer plays an essential role in KKXX-mediated retrieval in vivo. Consistent with defective KKXX-mediated retrieval in vivo, coatomer in lysates from sec27-1 and ret1-1 cells fail to bind with peptide corresponding to the cytoplasmic domain of Wbp1p, although coatomer from wild-type cells shows specific binding under the same conditions. Since α-COP, β'-COP and ε-COP are part of the core structure of coatomer for binding with the KKXX sequence and Sec21p is γ-COP, coatomer in lysate from sec21-1 and sec21-2 cells is capable of binding with the cytoplasmic domain of Wbp1p. The defective retrieval in sec27-1 and ret1-1 cells is therefore due to a failure of the coatomer to bind the KKXX motif. In sec21-1 and sec21-2 cells, events following coatomer binding seem to be affected for the retrieval. The combined biochemical and genetic evidence demonstrate that coatomer plays a direct role in KKXX-mediated retrieval (Fig. 16).[143,403]

The demonstration of a direct role for coatomer in the KKXX-mediated retrieval pathway, the identification of COPII coat proteins and their established role in ER-Golgi transport have led to the argument that the coatomer (COPI) may be involved only in retrograde retrieval and that the suggested role of coatomer in ER-Golgi transport and intra-Golgi transport may be due to an indirect role of coatomer in these processes such as the retrieval of membrane proteins involved in budding and/or fusion (v-SNAREs) from post-ER compartments back to the ER.[574] It remains to be resolved whether coatomer is only directly involved in retrograde transport from the Golgi back to the ER or is directly involved in multiple transport events, including ER-Golgi transport, intra-Golgi transport and retrograde transport from the Golgi back to the ER.[467,469] Since coatomer can interact directly with the KKXX motif, coatomer-mediated retrieval may be independent of ARF1. Alternatively, ARF1 may play a regulatory role in the retrieval. In contrast to the retrieval model, direct binding of coatomer to the KKXX sequence could also imply that the coatomer may mediate ER retention of KKXX proteins by binding to the KKXX sequence and directly immobilizing them in the ER, because the majority of endogenous KKXX-containing ER proteins has not been demonstrated to export from the ER and then be retrieved from post-ER compartments.

In addition to the well studied KDEL and KKXX signals for ER retrieval, other types of structures have been identified that could confer ER localization. Using alternative initiator Met residues,

the human invariant (Ii) chain mRNA can give rise to two type II membrane proteins, Iip33 and Iip31. Iip31 is transported out of the ER, guiding MHC class II antigens to the endocytotic pathway. Iip33, which differs from Iip31 by a 16-residue extension at the N-terminus (MHRRRSRSCREDQKPV), is selectively retained in the ER, suggesting that this extra 16 residues contain an ER localization signal. Deleting four residues (HRRR) after the initiator Met abolishes the ER localization. Furthermore, the N-terminal five residues (MHRRR) are sufficient to confer ER localization when attached to the N-terminus of transferrin receptor. Detailed analysis shows that the minimum requirement for this targeting motif are two Arg residues located at positions 2 and 3, 3 and 4, or 4 and 5, or split by a residue at position 2 and 4 or 3 and 5. Lys can substitute for one of the Arg residues of the RR motifs but never for both. The RR motif is also found in the TRAM (MAIRKKS) (Fig. 15).[676] The RR motifs may therefore function as a retrieval signal from post-ER compartments. Human CD3ε is a type I membrane protein that is retained in the ER when expressed alone. Its cytoplasmic tail contains an ER localization signal that is different from the KKXX motif. Detailed mutagenesis revealed that a YxxLxxR motif (YSGLNQR) (Fig. 15) in the cytoplasmic tail is the major determinant for ER localization and this motif adopts a helix-turn structure.[454,455] Several C-terminal tail-anchored ER proteins, including microsomal aldehyde dehydrogenase, protein tyrosinephosphatase 1B and cytochrome b5, have also been studied and it was discovered that the hydrophobic tail and their respective flanking hydrophilic sequences are the determinants for ER localization.[461,676]

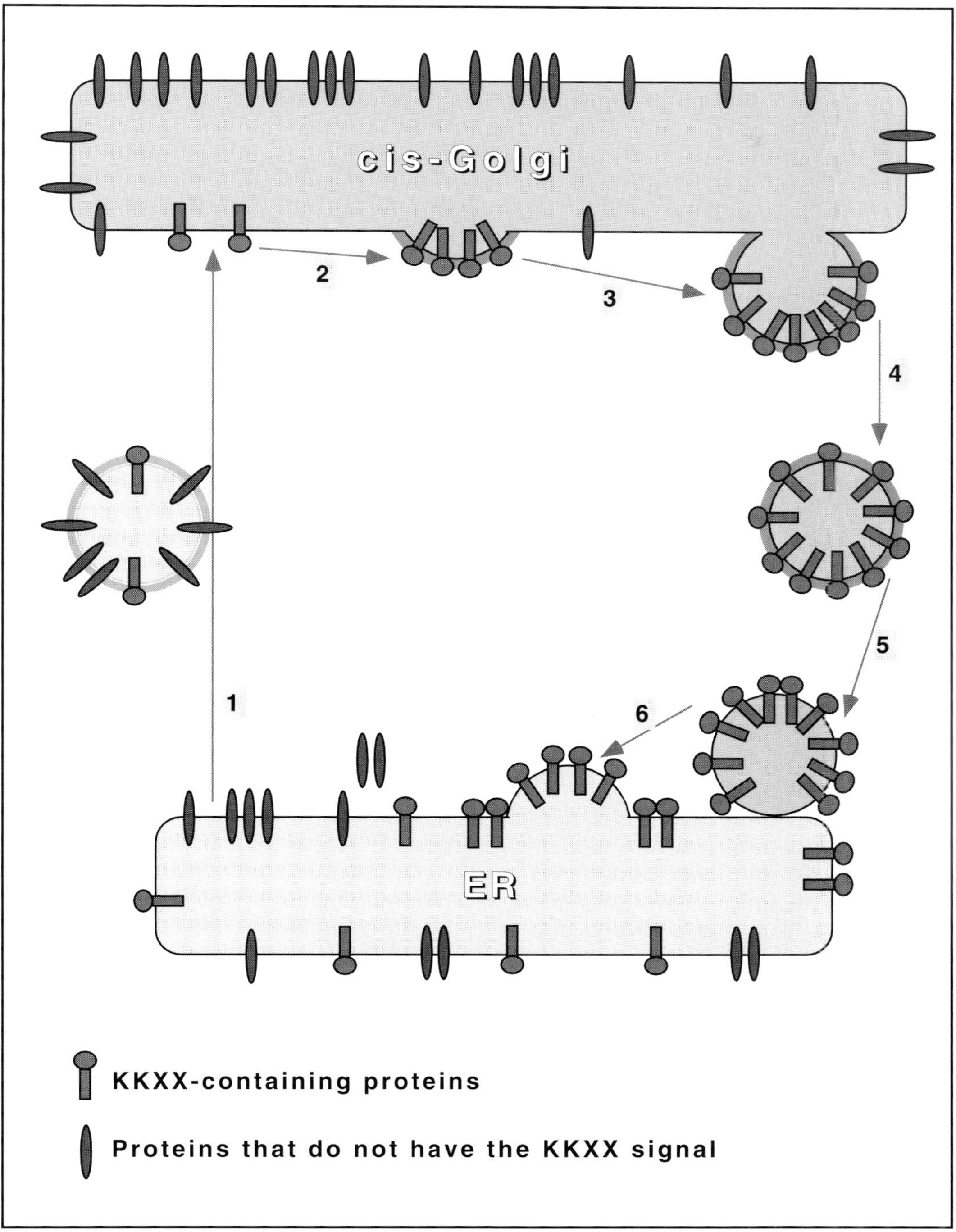

Fig. 16. Model for the KKXX-mediated retrieval pathway. Upon transport from the ER to the cis Golgi (step 1), the KKXX-containing proteins are recognized by COPI (coatomer) complex, resulting in the formation of coated buds (step 3) and coated budding vesicles (step 3). The coated vesicles pinch off (step 4) from the cis Golgi membrane and then become uncoated (step 5). The uncoated vesicles then dock and fuse with the ER membrane (step 6).

CHAPTER 5

Golgi Retention

GOLGI STRUCTURE AND FUNCTION

The Golgi apparatus plays a central role in the exocytotic pathway.[205b,253a,303,433,477] Its major functions include modification of N-linked glycans, initiation and subsequent modification of O-linked glycans, sorting and retrieval of proteins back to the ER via the retrograde transport pathway (such as those mediated by the KDEL signal and the cytoplasmic KKXX motif) and sorting of proteins to various post-Golgi structures of the anterograde pathway. The modification of glycans is achieved sequentially when glycan-containing molecules move along the Golgi apparatus from the cis to the trans face due to the polarized distribution of enzymes involved in the modifications. Although the sorting process for retrograde transport could potentially occur in the entire Golgi, the major site for retrograde sorting is probably the cis region of the Golgi as this is the first post-ER compartment encountered by proteins to be sorted back and the sorting machinery (such as the KDEL receptor and coatomer components) is enriched in the cis Golgi. Due to the decreasing gradient-like distribution of the sorting machinery from the cis to the trans Golgi and the vectorial movement of bulk traffic from the cis to the trans Golgi, the retrograde sorting in the Golgi can be pictured to occur sequentially and vectorially in a way similar to fractional distillation. The trans Golgi, particularly the TGN, is where proteins destined for various post-Golgi structures (such as the endosomal/lysosomal system, regulated exocytosis pathway and different domains of the plasma membrane), are sorted into different transport vesicles.[253a] In spite of these dynamic activities, the structure and architecture of the Golgi must be and is well maintained. The signals that determine the Golgi localization of several resident and cycling proteins have been defined and the mechanistic aspects are currently being investigated.

TRANSMEMBRANE DOMAIN-MEDIATED GOLGI RETENTION

Viral Membrane Proteins

Studies aimed at revealing structural determinants for Golgi localization have been performed using both viral as well as endogenous

Golgi proteins.[299,303,433,447,448,477] The M glycoprotein (formerly called E1 glycoprotein) of the avian coronavirus, infectious bronchitis virus (IBV) has been studied extensively. The M glycoprotein is a 225-residue type III integral membrane protein having three transmembrane domains with its N-terminal 20-residue hydrophilic region and the C-terminal 124-residue hydrophilic tail oriented in the luminal and cytoplasmic side of the membrane respectively. When expressed in transfected cells, M glycoprotein is specifically localized to the cis Golgi.[450] When its second and third transmembrane domains are deleted, the mutant protein (MGm1), containing only the first transmembrane domain, is also localized to the Golgi apparatus. However, when the first and second transmembrane domains are deleted, the mutant containing only the third transmembrane domain is transported to the cell surface, indicating that the first transmembrane domain of M glycoprotein may contain the information for cis Golgi localization.[451] When transferred to a surface protein such as the VSV G protein (resulting in Gm1), the first transmembrane domain is able to retain the reporter molecule in the cis Golgi, suggesting that the first transmembrane domain of M glycoprotein can function as a transferable autonomous signal for cis Golgi localization.[760] Under the same condition, the third transmembrane domain of M glycoprotein is unable to retain VSV G-protein in the cis Golgi. One important feature of the first transmembrane domain (composed of 22 residues) is the existence of uncharged polar residues which line one face of a predicted α helix.[449] Mutagenesis studies performed on Gm1 reveal that Asn, Thr, Thr and Gln residues at position 2, 6, 13 and 17 respectively, in this transmembrane domain play an important role in Golgi retention. When the first transmembrane domain of the homologous M glycoprotein from coronavirus mouse hepatitis virus (MHV) is used to replace the transmembrane domain of VSV G-protein, the resulting chimeric protein is transported to the cell surface, although the first transmembrane domain of MHV M-protein is about 50% identical to that of IBV M-protein and three of the four important residues (Asn2, Thr13, Gln17) are also conserved.[18,19] These results suggest that other structural features of the first transmembrane domain of IBV M-protein may also be involved in Golgi retention. When Golgi-associated Gm1 was analyzed by sucrose gradient centrifugation, it was found to exist as a large oligomeric form and a significant fraction of this oligomer is resistant to SDS.[832] Gm1 mutants with single amino acid substitutions in the transmembrane domain that are Golgi-retained are also in the oligomeric form, while other mutants that are not retained in the Golgi (transported to the cell surface) do not form this oligomer. The formation of this oligomer occurs gradually with a lag of about 10 min and may be coupled to vesicular transport to the cis Golgi. The cytoplasmic domain of Gm1 is not required for the formation of the oligomer but is involved in conferring the oligomer's resistance to SDS. In cells treated with cytochalasin D, the SDS-resistant oligomer is not observed, although Gm1 can still form the SDS-sensitive oligomer. It has been suggested that the transmembrane domain of Gm1 is involved in the formation of the oligomer and that the cytoplasmic domain may confer resistance to SDS by interactions with an actin-based cytoskeletal matrix.

The MHV M-protein is localized to the trans Golgi and TGN when expressed in transfected cells, in contrast to the cis Golgi localization of IBV M-protein.[18,19] Similar to IBV M-protein, MHV M-protein (of 228 residues) is composed of three transmembrane domains with N-terminal 25 hydrophilic residues in the lumen and C-terminal 122 hydrophilic residues oriented on the cytoplasmic side of the membrane. When the cytoplasmic C-terminal 18- or 22-residue sequence is deleted, the resulting mutants are delivered to the cell surface, suggesting that the cytoplasmic C-terminal sequence is necessary for trans

Golgi/TGN localization.[369] When the second and third transmembrane domains are deleted, the mutant is defective in ER export and retained in the ER. Mutants lacking the first or the first and second transmembrane domains are not efficiently retained in the trans Golgi/TGN, suggesting that the transmembrane domains are also involved in retention of HMV M-protein in the trans Golgi/TGN. Taken together, this indicates that the localization of HMV M-protein in the trans Golgi/TGN involves two different determinants, the transmembrane domain and the cytoplasmic C-terminal 22-residue sequence. When analyzed by sucrose gradients, it was found that the Golgi-localized HMV M-protein migrated as a large heterogeneous complex, while early intermediates of the M-protein in the ER and intermediate compartment migrated as a monomer.[370] The mutant lacking the C-terminal 22-residue sequence, which is transported to the cell surface, migrated similarly as heterogeneous oligomers, albeit smaller in size, and persists at the plasma membrane. It is suggested that Golgi localization of MHV M-protein is governed by two coupled mechanisms: oligomerization upon entry into the Golgi, possibly mediated by the transmembrane domains, and binding of its cytoplasmic tail to some cellular factors in the trans Golgi/TGN.

Resident Golgi Glycosyltransferases

The Golgi apparatus contains a variety of enzymes involved in the modification of glycans, such as glycosidases and glycosyltransferases. The most extensively studied Golgi enzymes include mammalian α2,6-sialyltransferase (ST),[135,136,150,214,502,769,856] β1,4-galactosyltransferase (GT),[214,462,531,653,778,860] and N-acetylglucosaminyltransferase I (NTI)[103,104,769,771] as well as yeast α1,2-mannosyltransferase (Mnt1p)[120] and α1,3-mannosyltransferase (Mnn1p).[247,258] Golgi α2,6-sialyltransferase is localized primarily in the trans Golgi and TGN. Rat ST is a type II membrane protein with N-terminal 9 residues (1-9) oriented on the cytoplasmic side of the membrane. Following this N-terminal region is a 17-residue hydrophobic transmembrane domain (residues 10-26) and a large luminal domain. The luminal domain has been divided into a stem region (residues 27-62) and the catalytic domain (residues 63 to the C-terminus). When the N-terminal 57-residue sequence of ST is replaced by a cleavable signal peptide, the resulting luminal domain is secreted,[135] suggesting that the N-terminal 57 residues, which contain a type II signal/anchor sequence are necessary for Golgi localization. When N-terminal 64-, 51-, 46-, 40- or 33- residue sequences were used to replace the N-terminal 32 residues of the surface protein DPPIV, which also has a type II signal/anchor sequence, the resulting chimeric proteins (S64D, S51D, S46D, S40D and S33D, respectively) are localized to the trans Golgi/TGN in stably-transfected MDCK cells, suggesting that the N-terminal 33-residue sequence of ST is necessary for Golgi localization in MDCK cells. Further studies suggest that the 17-residue transmembrane domain is sufficient for conferring Golgi localization in MDCK cells.[856] Similarly, when the N-terminal 44 residues of ST are fused to the mature polypeptide of chicken lysozyme, the resulting molecule (GSSS) is also retained in the trans Golgi/TGN when transiently expressed in COS cells. Detailed analysis of GSSS also suggests that the transmembrane domain is a key component for Golgi retention, while the hydrophilic sequences flanking the transmembrane domain increase the efficiency of Golgi localization. Furthermore, when the cytoplasmic and luminal flanking sequences are appropriately spaced by hydrophobic residues, they can confer Golgi localization independent of the transmembrane domain, suggesting that the flanking sequences may also contain a Golgi retention signal.[150] These results collectively suggest that both the transmembrane domain as well as the flanking sequences of ST are involved in its efficient Golgi retention. Although the transmembrane domain of ST alone is sufficient for Golgi retention in MDCK and COS cells, it is not able to confer Golgi localization in CHO cells. Furthermore, even S33D, which contains the N-terminal 33-residue sequence of ST, is not retained in the Golgi of CHO cells. The unretained chimeric proteins are instead transported to the lysosomal compartment. When the ST sequence is increased to include the N-terminal 40-residue or more sequence, the chimeric S40D, S46D, S51D and S64D are efficiently retained in the Golgi of CHO cells. These results suggest the that the N-terminal 40 residues of rat ST, containing the 9-residue cytoplasmic tail, the 17-residue transmembrane domain and the luminal 14-residue flanking sequences, function in general Golgi localization, regardless of the cell types.[769]

Studies performed on GT of diverse species, including human, bovine and mouse have similarly concluded that the transmembrane domain plays a major role in conferring retention in the trans Golgi, while the flanking sequences play an accessory role.[2462,531,653,778] When the human GT is tagged at the C-terminus with the C-terminal 59 residues of αhcg, the resulting tagged protein (GT/hcg) is effectively retained in the Golgi. Deletion of 19 residues (residues 5-23) of the 23-residue cytoplasmic tail has no effect on the Golgi localization of GT/hcg, nor does a deletion that removes 10 hydrophilic residues (residue 45-54) flanking the luminal side of the transmembrane domain (residue 24-43). Replacement mutagenesis of the transmembrane domain of GT/hcg reveals that the sequence Cys29-Ala30-Leu31-His32-Leu33 is necessary for Golgi localization.[14] Site-directed mutagenesis of this sequence shows that Cys29 and His32 are important for Golgi localization. A double-mutated GT/hcg in which Cys29 and His32 are mutated into Ser and Leu, respectively, (GT/hcg-SL) is delivered to the cell surface. Biochemical analysis of Golgi-localized GT derivatives suggest that GT can form homodimers as well as large oligomers in vivo, and the formation of the homodimer is dependent on Cys29 and His32 since GT/hcg-SL, which is not Golgi-associated, is defective in dimer formation. Furthermore, the GT oligomer is associated with α- and β-tubulin and other cellular proteins, suggesting that microtubule-based cytoskeleton may have a supportive role for Golgi localization of GT.[860]

NTI is predominantly confined to the medial Golgi. Human NTI contains a 6-residue (residues 1-6) cytoplasmic tail, a 23-residue (residues 7-29) transmembrane domain and a large luminal domain. The transmembrane domain of human NTI is sufficient to confer Golgi localization in MDCK and COS cells, but fails to do so in CHO cells.[769,771] The N-terminal 41-53- residue sequences of human NTI were found insufficient to specify Golgi localization in CHO cells. Only partial Golgi localization was observed in CHO cells for a fusion protein that contains the N-terminal 65-residue sequence of NTI, suggesting that general Golgi localization of NTI may be determined by the combined effects of the cytoplasmic tail, the transmembrane domain and the luminal domain.[769] Similar results were seen with rabbit NTI, suggesting that all the domains of the enzyme are involved in efficient Golgi localization.[103,104] The mechanism of Golgi retetion for medial Golgi enzymes has been studied. When the cytoplasmic tail of NTI, mannosidase II (MII) and GT is each replaced by that of the p33 invariant chain, which contains an ER retention signal, the resulting p33/NTI, p33/MII and p33/GT are retained in the ER. Expression of p33/NTI in Hela cells selectively causes relocation of endogenous MII to the ER and expression of p33/MII has similar effects on endogenous NTI. Neither endogenous NTI nor MII is relocated by the expression of p33/GT or p33. These results suggest that a kin recognition between medial Golgi enzymes contributes to their Golgi retention.[529,532] Furthermore, a Golgi matrix has been isolated that binds medial Golgi enzymes,[719] indicating that other structural proteins may be involved in Golgi retention of medial Golgi enzymes.

Molecular dissections of yeast Och1p (α1,6-mannosyltransferase for O-linked glycans) and Mnt1p (α1,2-mannosyltransferase involved in O-glycosylation) have also led to the conclusion that the transmembrane domain and flanking hydrophilic sequences play a dominant role in their retention in the yeast Golgi. Genetic screens have identified several genes that are involved in the organization of the early secretory pathway and are therefore involved in the Golgi retention of these transferases. One of the genes (ANP1) encodes a type II membrane protein localized to the ER.[120] Anp1p is a member of a new protein family with two other members, Van1p and Mnn9p that are also involved in the organization and proper functioning of the exocytotic pathway. The mechanism of Golgi retention for yeast Mnn1p (α1,3-mannosyltransferase) has also been investigated. A significant fraction of Mnn1p is mislocalized to the plasma membrane in clathrin heavy chain ts

mutants, suggesting that a clathrin-dependent process may participate in its Golgi localization.[247,248] Since the cytoplasmic tail of Mnn1p is not essential for its Golgi localization, clathrin may play an indirect role in its Golgi localization.

CYTOPLASMIC SIGNAL-MEDIATED ACCUMULATION IN THE TRANS GOLGI NETWORK

Proteins that are enriched in TGN and the yeast late Golgi compartment have been studied for their putative localization signals. TGN38 and furin represent two well studied mammalian proteins. The yeast Kex1p, Kex2p and dipeptidyl aminopeptidase A (DPAP A) are predominantly localized in a late Golgi compartment equivalent to the TGN of mammalian cells. The cytoplasmic domains of these proteins have been shown to play a major role in their late Golgi localization. The signals for these proteins are shown in Figure 17.

TGN38 is predominantly associated with the TGN[446] and is normally dimerized with TGN41. In contrast to resident

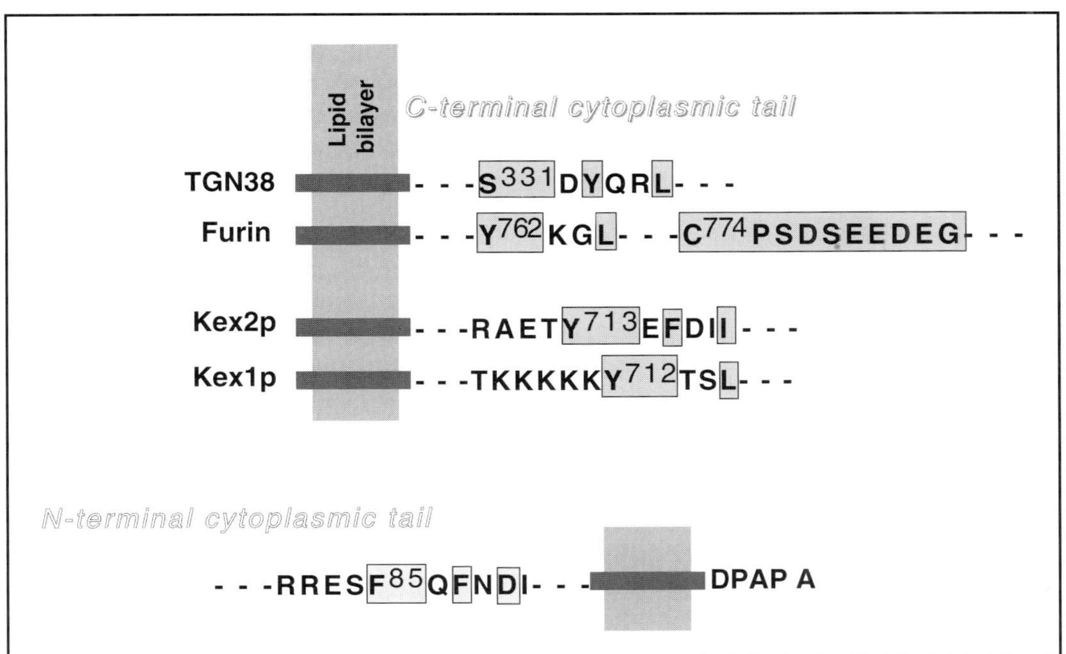

Fig. 17. Summary of cytoplasmic signals involved in TGN (mammalian cells) or late-Golgi (yeast) localization.

Golgi glycosyltransferases, TGN38 cycles between the TGN and the plasma membrane and is a type I membrane protein with a C-terminal cytoplasmic tail of 32 residues.[613,621] The cytoplasmic domain of TGN38 is sufficient for TGN localization when it is used to substitute for the cytoplasmic domain of cell surface type I membrane proteins such as glycophorin A, CD4, CD8, Tac (interleukin-2 receptor α-chain) and LDL receptor.[76,313,445,600,857] Detailed mutagenesis of this 32-residue cytoplasmic domain shows that Ser, Tyr and Leu residues in the context of SDYQRL are essential for TGN localization. Furthermore, the SDYQRL sequence is sufficient to confer significant TGN localization when transferred to the cytoplasmic tail of a reporter molecule.[857] Two-dimensional NMR analysis of peptides corresponding to the cytoplasmic domain suggests that this signal lies within a nascent helix.[841] This cytoplasmic signal is able to mediate both TGN localization and cycling between the TGN and the plasma membrane. In addition to this cytoplasmic signal, the transmembrane domain of TGN38 also contains an autonomous transferable signal for Golgi retention.[600] However, Golgi localized proteins mediated by the transmembrane domain of TGN38 are not able to cycle between the Golgi and the cell surface. Because of the cycling nature of TGN38, the cytoplasmic signal may play a dominating role, while the signal on the transmembrane domain may be normally masked or has an accessory function. In addition to its cycling between the cell surface and the TGN, TGN38 has two other properties that distinguish it from other resident Golgi proteins such as glycosyltransferases. First, upon treatment with brefeldin A, TGN38 accumulates in a collapsed membrane structure that is associated with the microtubule organizing center (MTOC) and early endosomal components are also redistributed into this TGN38-containing membrane structure, while Golgi glycosyltransferases and glycosidases are shifted into the ER. Furthermore, TGN associa-

tion of TGN38 is dependent on the acidic environment of the TGN and the endosomes.[119] Agents, such as chloroquine and monensin that neutralize acidic organelles abolish TGN association of TGN38. In contrast, Golgi association of resident proteins is not affected by chloroquine or monensin. The unique responses of TGN38 to brefeldin A and acid-neutralizing agents could result from its unique feature of cycling between the TGN and the plasma membrane. The SDYQRL sequence of TGN38 can interact directly with the AP50 subunit of the surface AP2 complex as well as the AP47 subunit of TGN AP1 complex, suggesting that AP2 and AP1 complex may, respectively, be involved in recycling and TGN association of TGN38.[543]

Furin, a type I membrane protein enriched in the TGN, also cycles between the TGN and the cell surface in a way similar to TGN38.[488,812] Its responses to brefeldin A and acid-neutralizing agents are therefore similar to TGN38.[119] Furin is a substilisin-like endoprotease responsible for proteolytic cleavage of cellular and viral proteins transported via the constitutive secretory pathway. The cytoplasmic tail of furin carries the signal for TGN association and for cycling between the TGN and the cell surface because the cytoplasmic tail is both necessary and sufficient for TGN association and cycling.[77] Molecular dissections of the cytoplasmic tail have identified two independent targeting signals, the acidic sequence CPSDSEEDEG and a tyrosine-based tetrapeptide YKGL, that determine TGN localization and internalization from the cell surface respectively.[664] The acidic signal is necessary and sufficient to localize reporter molecules to the TGN, whereas the YKGL sequence is involved in internalization from the cell surface for delivery to the endosome. Newly made furin molecules may be retained in the TGN by the acidic signal. Molecules that have escaped from the TGN are delivered to the plasma membrane, where the YKGL signal mediates its internalization and transport to the endosome. For retrieval from the endosome to the TGN, the acidic signal may also be involved. Therefore, the acidic signal is used both in the exocytotic pathway for TGN accumulation as well as in the endocytotic pathway for retrieval from the endosome back to the TGN.

CYTOPLASMIC SIGNAL-MEDIATED LOCALIZATION OF MEMBRANE PROTEINS IN THE YEAST LATE GOLGI COMPARTMENT

Three functionally distinct Golgi compartments have been suggested for the yeast Golgi complex.[215,246,568] The early compartment is marked by α1,6-mannosyltransferase and its activity has been used to follow protein transport from the ER to the early Golgi both in vivo and in vitro. Examples of proteins associated with the medial compartment include α1,2- and α1,3-mannosyltransferases. The late compartment (equivalent to mammalian TGN) has been proposed to sort proteins destined for post-Golgi structures, including the plasma membrane and the vacuole. Three widely studied membrane proteins of this late compartment are Kex2p, Kex1p and dipeptidyl peptidase A (DPAP A),[534,844] which are proteases intimately associated with processing of the precursor of the mating hormone α-factor. Kex2p and Kex1p are type I membrane proteins, while DPAP A is a type II membrane protein. Deletion of the cytoplasmic domains of these proteins leads to their delivery to the vacuole, which may be the default compartment for membrane proteins in the yeast, suggesting that the cytoplasmic domains of these proteins harbor signals for their selective localization in the late Golgi compartment.

Detailed mutagenesis of Kex2p has revealed that the Tyr residue at position 713 and its flanking sequences in the C-terminal cytoplasmic domain constitute the signal for late Golgi localization. The signal for DPAP A has been mapped to an 8-residue sequence (residues 85-92: FQFNDIEN) in the N-terminal cytoplas-mic domain. The cytoplasmic domain of DPAP A is sufficient to retain a vacuolar membrane protein alkaline phosphatase (ALP) in the late Golgi compartment. A similar sequence is present in the cytoplasmic domain of Kex1p (Fig. 17). Although the signals of Kex2p and DPAP A do not share an apparent primary sequence, the retention mechanism for them is similar and saturable, as overexpression (15-fold) of Kex2p (but not its retention-defective mutants) reduces the efficiency of late Golgi localization mediated by the cytoplasmic domain of DPAP A.

The mechanism of late Golgi localization has been investigated using several yeast mutants.[534,568,685,686] In cells with clathrin heavy chain (CHC1) gene disruption, 70% of Kex2p and 30% of DPAP A are mis-localized to the plasma membrane in a Sec1p-dependent way, suggesting that they are mis-sorted to vesicles destined for the plasma membrane in these cells. In cells carrying a ts CHC1 gene, these proteins are localized to the late Golgi at the permissive temperature. Upon shifting to the restrictive temperature, Kex2p and DPAP A are mis-sorted to the plasma membrane within 30 min. The relatively rapid onset of the mis-sorting effect in the CHC1 ts mutant suggests that clathrin may have a direct role in the late Golgi localization of these proteins. Another protein involved in late Golgi localization is Vps1p, a protein related to mammalian dynamin. VPS1 was isolated based on its participation in vacuolar sorting of soluble enzymes.[651,808,845] In cells with defective Vps1p, Kex2p is mistargeted to vacuoles and rapidly degraded. Cells carrying a ts VPS1 also mis-sort Kex2p rapidly at the nonpermissive temperature, suggesting a direct role for Vps1p in late Golgi localization of Kex2p. The vacuolar delivery of Kex2p in Vps1p-defective mutant cells is Sec1p-dependent, suggesting that Kex2p is first transported to the plasma membrane and then delivered to the vacuole by the endocytotic pathway.

A post-Golgi and prevacuolar compartment (PVC) equivalent to the mammalian late endosome/prelysosomal compartment is accumulated in class E vps mutants.[814] This PVC receives traffic from both the biosynthetic (via the late Golgi) pathway (Fig. 20) and the endocytotic pathway (possibly via an early endosome-like compartment) and can also be identified in wild-type cells. The late Golgi localization of Kex2p, Kex1p and DPAP A has been proposed to occur via a retrieval mechanism from this PVC. Clathrin and Vps1p have been suggested to participate in processes related to this retrieval, details of which remain to be experimentally established. One hypothesis is that Kex2p, Kex1p and DPAP A are sorted in the late Golgi to the PVC by a process that is dependent on clathrin and Vps1p and they are then selectively retrieved from the PVC back to the late Golgi by a process that is also dependent on Vps1p and clathrin. If either clathrin or Vps1p is mutated, transport from the late Golgi to the PVC is prevented, resulting in their incorporation into vesicles that fuse with the plasma membrane in a Sec1p-dependent manner. These late enzymes remain associated with the plasma membrane in cells with defective clathrin, possibly because clathrin plays a role in their entry into the endocytotic pathway. However, in cells with defective Vps1p, these enzymes can be endocytosed from the plasma membrane in a clathrin-dependent but Vps1p-independent manner and then delivered to the vacuole via the PVC, because retrieval from PVC to the late Golgi is abrogated due to its requirement for functional Vps1p. Consistent with this, another dynamin-like protein Dnm1p, but not Vps1p, has been shown to participate in endocytosis from the surface. Other models have also been proposed and future biochemical and genetic analysis should shed more light on this issue.

═══ CHAPTER 6 ═══

LYSOSOMAL/ENDOSOMAL TARGETING

MANNOSE-6-PHOSPHATE AS THE SORTING SIGNAL FOR LUMINAL LYSOSOMAL ENZYMES

The lysosomal/endosomal system plays a vital role in cellular degradation of various internalized and endogenous macromolecules by the diverse acid hydrolases present in the lumen of lysosomes. All lysosomal components are derived originally from the biosynthetic pathway and are then selectively accumulated in the lysosome by sorting events that occur at the TGN or in the endocytotic pathway.[153,171,376] The importance of lysosomal function is highlighted by the association of several clinical syndromes with various defects in lysosomal function.[518] Many different forms of lysosomal storage diseases are due to the deficiency of a single hydrolase, which results in the massive accumulation of materials that are normally degraded by the enzyme. Early studies with cultured fibroblasts from patients with genetic disorders of mucopolysaccharide catabolism have shown that culture media from normal fibroblasts can correct the phenotype exhibited by mutant fibroblasts, suggesting that lysosomal hydrolases are secreted by the normal cells and are then internalized and delivered to the lysosome by the mutant cells. Another widely-studied disorder is the I cell disease (mucolipidosis II), a rare autosomal recessive disorder characterized by the lysosomal deficiency of many different hydrolases and elevated levels of these hydrolases in the patients' plasma. Studies with I cells show that lysosomal hydrolases secreted by normal cells can be efficiently taken up and delivered to the lysosome of I cells, while the hydrolases secreted by the I cells cannot be internalized by either I cells or normal cells, suggesting that different hydrolases may possess a common signal for internalization and lysosomal delivery, and the I cell is defective in adding this signal to its hydrolases. Mannose-6-phosphate (M6P) on N-linked glycans of lysosomal hydrolases is now known to be the lysosomal targeting signal. This M6P signal targets newly-synthesized hydrolases to the lysosome via the TGN as well as secreted enzymes to the lysosome via the endocytotic pathway. The cellular receptor that recognizes this signal is the M6P receptor.

M6P signal is associated with high-mannose type or hybrid-type N-linked glycans of mature lysosomal hydrolases. It is constructed by a two-step process catalyzed by two different enzymes. First, α-N-acetylglucosamine-1-phosphate is added from UDP-N-acetylglucosamine to the hydroxyl group on C-6 of a selected mannose residue by the UDP-N-acetyl-glucosamine:lysosomal enzyme N-acetyl-glucosamine-1-phosphotransferase (NGPT), resulting in the formation of a phospho-diester. The N-acetylglucosamine group is then removed by α-N-acetylglucosaminyl phosphodiester glycosidase (NGPG), re-sulting in the formation of the M6P monoester. The first enzyme NGPT is defective in patients with I cell disease and a related disorder, pseudo-Hurler poly-dystrophy (mucolipidosis III), consistent with the original hypothesis that these patients' cells are defective in the construc-tion of the M6P signal. The first step of M6P synthesis occurs in the cis Golgi, while the second step may occur in the medial and/or trans Golgi. Detailed cel-lular localizations of these two enzymes remain to be determined.

Since M6P is added selectively to all lysosomal hydrolases but not to other gly-coproteins that travel together with them in the secretory pathway, there must be a common structure(s) that is shared by all lysosomal enzymes, which is recognized by NGPT in the cis Golgi. Therefore, there are two coupled recognizing processes for lysosomal enzyme sorting. The first is the recognition of common structural information in all newly made enzymes by NGPT, a critical step for the forma-tion of the M6P signal involved in the second recognition event mediated by the M6P receptor. This coupling of two rec-ognition events for lysosomal enzymes may serve to increase the fidelity of this sort-ing process. The structural information of lysosomal enzymes for recognition by NGPT has been investigated and it seems that a three-dimensional signal patch may be involved.[41,42,43,194] Two aspartyl pro-teases, cathepsin D, a lysosomal enzyme[171]

and a glycosylated form of pepsinogen, a secretory protein, which are 45% identi-cal in their primary sequences and have similar three dimensional structures, have been used to identify regions of cathep-sin D that conferred upon pepsinogen the property of being recognized by NGPT. Combined substitutions of two noncon-tinuous regions (amino acids 188-230, particularly Lys 203; and 265-292) from cathepsin D onto the corresponding re-gions of pepsinogen were sufficient to cause the resulting chimeric molecule to be recognized and modified by NGPT. These two regions, located in the carboxyl lobe of cathepsin D, are in direct appo-sition on the surface of the molecule, which, in addition, has an amino lobe and a pro-piece. Reciprocally, when these two regions in cathepsin D are replaced by the homologous sequences of pepsinogen, the resulting chimeric protein is poorly rec-ognized, suggesting that they are also nec-essary for NGPT recognition in the na-tive cathepsin D. Interestingly, when either one of these regions is substituted indi-vidually in cathepsin D, the resulting molecules are recognized in an amino lobe-dependent manner, indicating that there are other structural features in the amino lobe of cathepsin D that also contribute to this recognition. Further studies with chimeric proteins that contain either the amino lobe or carboxyl lobe of cathepsin D fused to the carboxyl or amino lobe of pepsinogen respectively, suggest that there are independent NGPT recognition signals associated with these two lobes and each is sufficient to confer recognition.[41,110] Further dissections have identified several regions in the amino lobe that will en-hance recognition of cathepsin D by NGPT.[194] Cathepsin D remains as the only lysosomal hydrolase whose signals for NGPT recognition have been investigated. It seems that other lysosomal hydrolases should be studied in a similar way or via independent approaches in order to re-veal the general structural feature for NGPT recognition that is shared by at least 40 different lysosomal hydrolases.

THE M6P RECEPTOR: STRUCTURE AND DYNAMIC LOCALIZATION

The concerted action of NGPT and NGPG in the cis and medial/trans Golgi respectively, results in the addition of M6P signal onto all newly made lysosomal enzymes, the majority of which is sorted in the TGN by interaction with the M6P receptor and subsequently incorporated into clathrin-coated vesicles destined for the prelysosomal compartment (late endosome) (Fig. 18).[153,376] A small fraction of newly made lysosomal hydrolases may escape this TGN sorting and is secreted; these proteins can be re-captured by the M6P receptor on the cell surface and endocytosed to the late endosome via the early endosome. In the late endosome, the acidic pH causes a dissociation of the hydrolases from the receptor and hydrolases are then delivered to the lysosome, while the M6P receptor is recycled back to the TGN with a small amount of the receptor possibly recycling back to the plasma membrane either via the TGN or the early endosome. Two distinct M6P receptors have been identified. The large (about 270 kDa) receptor binds M6P in a cation-independent way and has been referred to as CIMPR, while binding of M6P to the small receptor (46 kDa) (referred to as CDMPR) is dependent on divalent cations (such as Ca^{2+}). The CIMPR sorts lysosomal enzymes in both the biosynthetic pathway (via TGN) as well as in the endocytotic pathway (via endocytosis from the plasma membrane), while the CDMPR may play a role only in the biosynthetic TGN sorting. Furthermore, in normal cells, the majority (about 70%) of M6P sorting in the TGN is mediated by the CIMPR with the remaining sorting capacity contributed by the CDMPR.

Amino acid sequences derived from cDNA cloning have been established for both CIMPR and CDMPR in several species. Both receptors are type I integral membrane proteins with an N-terminal extracellular/luminal (E/L) domain followed by a transmembrane domain and a C-terminal cytoplasmic domain. Interestingly, CIMPR is identical to the receptor for insulin-like growth factor II. The E/L domain of CIMPR has a highly repetitive structure consisting of 15 contiguous repeating units that are about 147 amino acids in length and with amino acid identities ranging from 16 to 38%. The CDMPR is structurally related to CIMPR in that its E/L domain is characterized by the presence of one such domain that is similar to each of the repeating units of CIMPR. The E/L, transmembrane and cytoplasmic domain of mature bovine CIMPR are 2269-, 23- and 163 amino acids, respectively, in length.[425,557] The bovine CDMPR has 165, 25 and 67 amino acids in its E/L, transmembrane and cytoplasmic domain, respectively.[152]

Binding of the M6P signal with the receptors is restricted to the E/L domain of the receptors.[151,153] Although CDMPR is a homodimer, soluble monomeric E/L domain is capable of binding M6P, suggesting that dimerization is not necessary for its binding with the M6P signal. Although CIMPR has 15 repeating units, it has been shown that each receptor molecule has only two binding sites for M6P, one associated with repeating units 1-3 and the other with units 7-11.[835] Since the major cycling route of CIMPR is between the TGN and the late endosome (TGN sorting) with a minor route between cell surface and the endosomal system (surface internalization), identification of structural determinants governing its trafficking has been one important aspect (Fig. 19). To identify these structural determinants, normal CIMPR and various mutants with deletions in the 163-residue cytoplasmic domain were expressed in CIMPR-deficient mouse L cells.[426] The normal receptor mediates both TGN sorting as well as surface internalization. Mutant receptors with 40 and 89 residues deleted from the carboxyl terminus function normally in surface internalization but are partially impaired in TGN sorting, suggesting that the C-terminal 89-residue

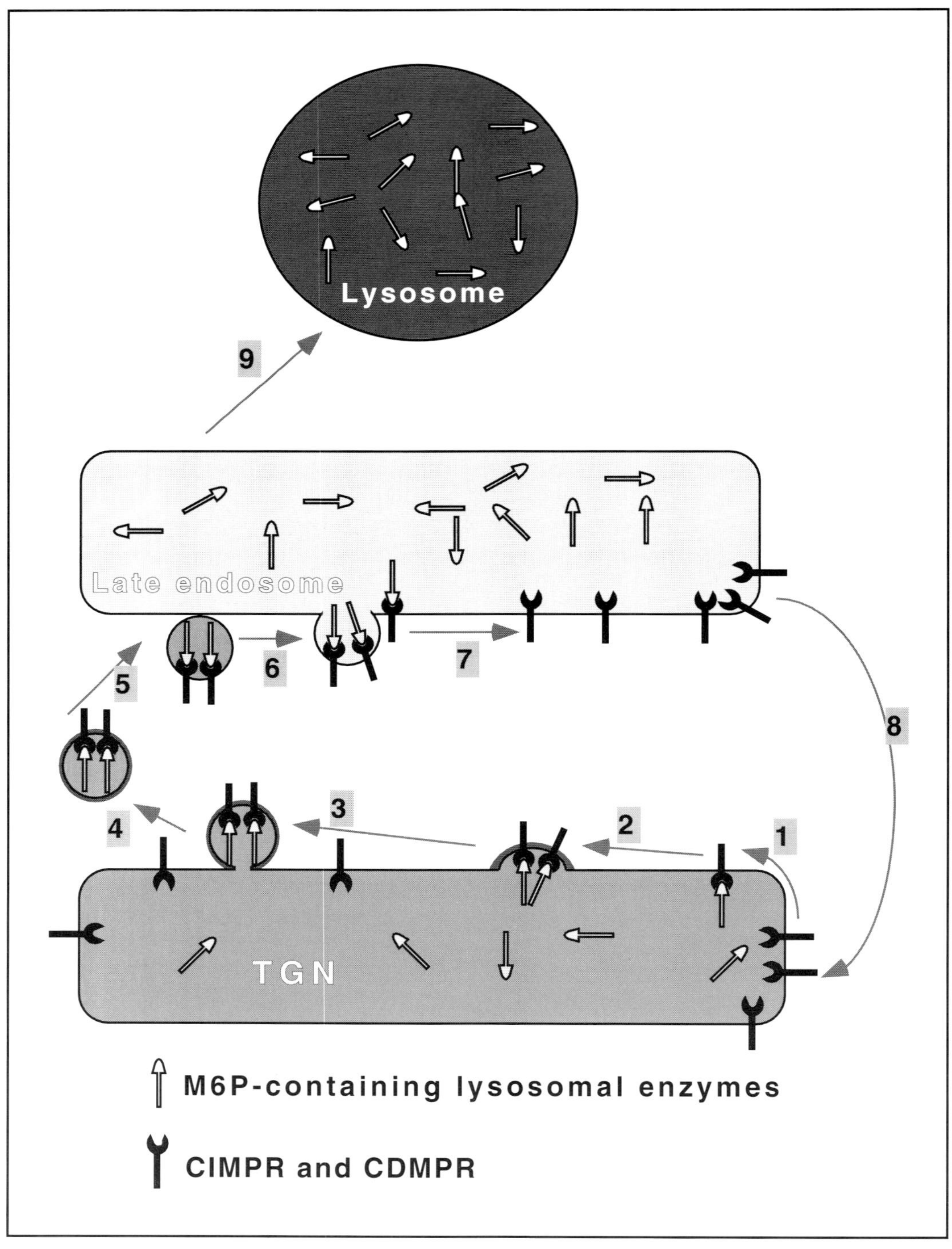

Fig. 18. Model for M6P receptors (CIMPR and CDMPR)-mediated TGN sorting of lysosomal enzymes. Upon arrival at the TGN, M6P-bearing lysosomal enzymes are recognized by the receptor (step 1) and this process initiates the formation of coated buds by the TGN AP1 complex (step 2) and progresses to coated budding vesicles (step 3). The coated vesicles pinch off (step 4) and become uncoated (step 5) and then dock with and fuse to the membrane of the late endosome (step 6). The acidic pH of the late endosome causes dissociation of lysosomal enzymes from the receptor (step 7). The receptors are recycled to the TGN (step 8), while lysosomal enzymes are delivered to the lysosome.

sequence of the receptor plays a critical role in TGN sorting but not in surface internalization. However, mutant receptors with larger deletions retaining only 7 and 20 residues of the cytoplasmic domain are defective in both TGN sorting and surface internalization. With regards to surface internalization, mutant receptors containing a cytoplasmic domain of 29 or more residues function normally, suggesting that the signal for surface internalization is located within the first 29 residues of the 163-residue cytoplasmic domain. Ala scanning mutagenesis of the region between amino acids 19 and 30 revealed that Tyr 26 and Val 29 are the most important residues for surface internalization. Tyr 24 and Lys 28 also contribute to the signal while others are not critical, establishing that the signal for surface internalization of CIMPR is contributed by the sequence Y^{24}-K-Y-S-K-V^{29}.[109] Deletion of the C-terminal LLHV residues impairs TGN sorting, resulting in the diversion to the surface of a portion of receptor-ligand complex formed in the TGN. The same phenotype is observed for a mutant with C-terminal 134 residues deleted. Taken together, these observations suggest that the C-terminal LLHV sequence is as important as the C-terminal 134-residue sequence in TGN sorting and that the first 29-residue sequence also contributed to TGN sorting to some extent.[124] Disruption of the surface internalization signal Y^{24}-K-Y-S-K-V^{29} alone has little effect on TGN sorting, indicating that the LLHV signal may be more critical for TGN sorting. Combined disruptions of the surface internalization signal Y^{24}-K-Y-S-K-V^{29} and the LLHV signal totally abolish TGN sorting These results suggest that the C-terminal LLHV sequence, in conjunction with the Y^{24}-K-Y-S-K-V^{29} signal, contributes fully to the signal for efficient TGN sorting of the normal receptor, while surface internalization requires only the Y^{24}-K-Y-S-K-V^{29} signal.[338] The Y^{24}-K-Y-S-K-V^{29} signal therefore has dual functions. Although necessary, the cytoplasmic domain alone of CIMPR is not sufficient to confer to a heterologous protein with these cycling properties.[176,177] When the entire cytoplasmic domain of CIMPR is fused to the E/L and transmembrane domain of the EGF receptor, the resulting chimeric protein binds EGF and is efficiently endocytosed and recycled back to the cell surface in the presence or absence of EGF. In contrast to CIMPR, which is predominantly localized in the TGN and the late endosome, the chimeric protein is primarily localized to the cell surface, suggesting that the chimeric protein has only gained the surface internalization but not the TGN sorting property. These studies suggest that the cytoplasmic signals of CIMPR function in the context of the entire receptor molecule.

The CDMPR has a role in TGN sorting of lysosomal hydrolases. Due to its extremely poor affinity for the M6P signal on the plasma membrane, CDMPR contributes little to the surface recapture of extracellular lysosomal hydrolases, even though it also cycles between the surface and the endosomal system. The 67-residue C-terminal cytoplasmic domain of CDMPR has also been studied in terms of the structural elements involved in its internalization from the plasma membrane (which, although, is not involved in surface recapture of lysosomal enzymes) as well as in TGN sorting. Two internalization signals have been identified in the cytoplasmic domain, one of which includes Tyr45 in the sequence context of $AAY^{45}RGVGDD$ and the other includes Phe 13, Phe18 and possibly Trp19 in the context of $MEF^{13}PHLAFWQ$.[336] The latter signal is more potent than the Tyr45-based one and both are required for maximal rates of internalization. With regards to TGN sorting mediated by CDMPR, the C-terminal HLLPM sequence has been identified to be essential.[337] Mutant receptors with this sequence deleted or replaced by Ala residues do not possess the TGN sorting function. When the outermost PM residues are replaced with Ala residues, the resulting receptor with a

HLLAA C-terminal sequence has a higher sorting capacity, suggesting that the HLL sequence is the key element for TGN sorting, which is similar to the LLHV signal for TGN sorting of CIMPR. In addition, the cytoplasmic domain of CDMPR has a determinant that prevents its trafficking to lysosomes with residues 34-39 being critical for this property.[640] The transmembrane domain of CDMPR also contributes to this avoidance of lysosomes. CDMPR has been shown to play a physiological role in lysosomal sorting using mice with disrupted CDMPR

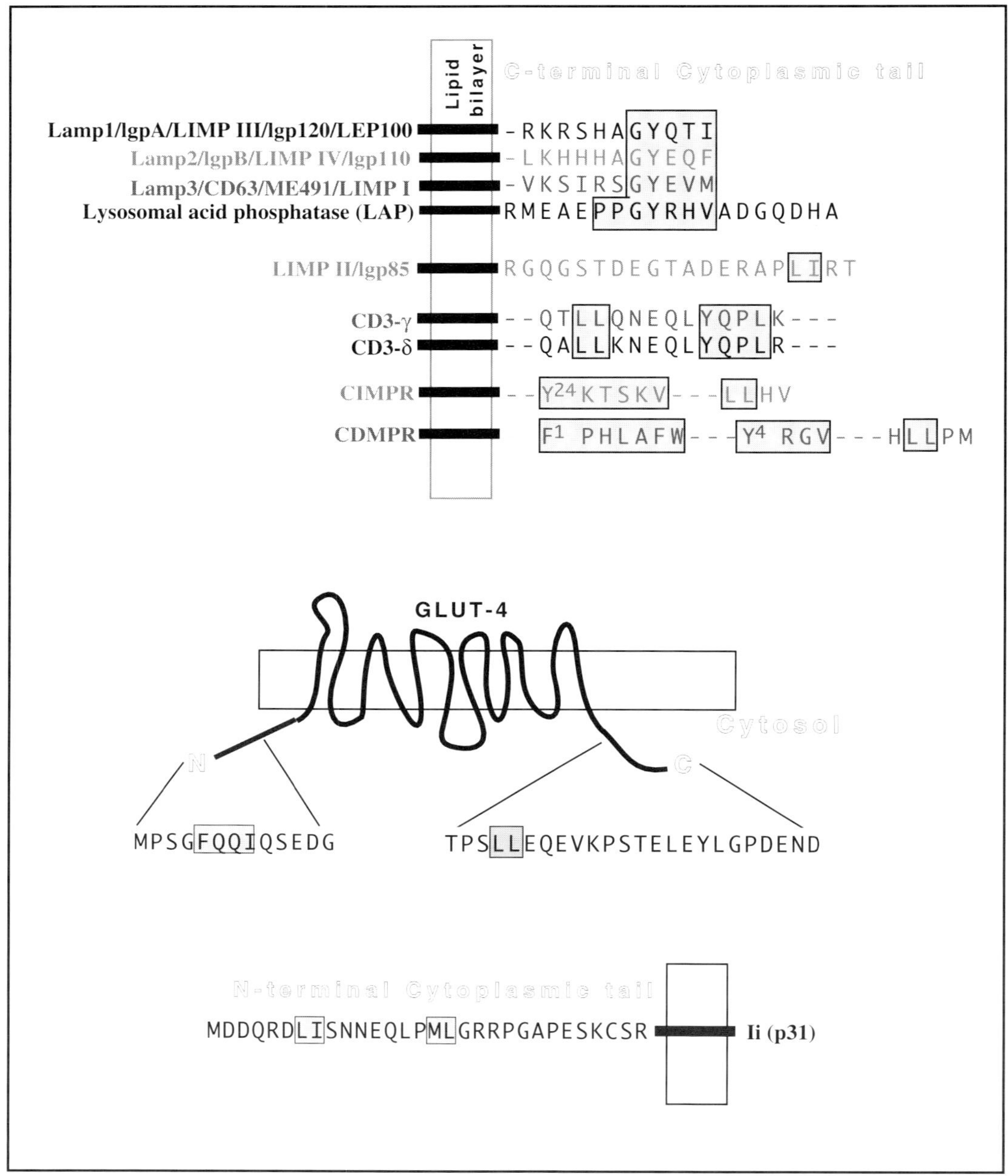

Fig. 19. A compilation of cytoplasmic sorting signals for the endosomal/lysosomal system.

gene.[377,442] Although phenotypically normal, these mutant animals and cells derived from them have an enhanced secretion of lysosomal hydrolases and reduced intracellular levels of lysosomal enzymes. This is consistent with a role for CDMPR in lysosomal sorting and with the fact that CIMPR may contribute the major sorting capacity for lysosomal enzymes in the TGN and almost all the recapturing capacity on the cell surface.

Since both CIMPR and CDMPR have two distinct sorting signals, one based on the double-leucine (LL) critical for TGN sorting and the other based on Tyr and/or Phe primarily involved in surface internalization, the mechanisms underlying their sorting processes have been molecularly investigated. It seems that LL-based and Tyr/Phe-based signals mediate selective packaging into clathrin-coated vesicles at the TGN and the plasma membrane, respectively. Recombinant cytoplasmic domain of CIMPR has been shown to interact in vitro with both the surface AP2 adaptin complex as well as the TGN AP1 adaptin complex.[118,234,569,570] Furthermore, binding of these two adaptin complexes is not mutually exclusive, suggesting that each of the adaptin complexes binds to distinct nonoverlapping sites in the cytoplasmic domain of CIMPR. When the Tyr at positions 24 and 26 are changed into Ala and Val, respectively, the mutant cytoplasmic domain binds normally to the AP1 complex but not the AP2 complex, demonstrating that the Tyr-based signal of CIMPR is required for interaction with AP2 complex in vitro and may be recognized by the surface AP2 complex in vivo for surface internalization.[234,728] Although not experimentally established, the LL-based motif may be the one that mediates an interaction with AP1 complex in vitro and in vivo for TGN sorting. The LL-based motif in the cytoplasmic domain of CDMPR may also be recognized by the AP1 complex for TGN sorting. The interaction of the LL-based signals with AP1 complex may initiate the formation of clathrin-coated vesicles on the TGN

membrane for targeting to the prelysosomal compartment (the late endosome), while the recognition of the Tyr/Phe-based motif by the AP2 complex may trigger the formation of clathrin-coated vesicles in the plasma membrane. Recent results show that AP47 of AP1 complex and AP50 of AP2 complex can both interact directly with Tyr-based sorting signals.[543] Future experiments are required to establish if AP1 adaptin complex in the TGN is directly involved in sorting lysosomal hydrolases to the late endosome by demonstrating that inactivation of AP1 complex will impair the sorting and lead to enhanced secretion of lysosomal hydrolases.

Once delivered to the late endosome, the luminal acidic pH causes a dissociation of lysosomal enzymes from the M6P receptors because the affinity of the receptors for the M6P signal is reduced dramatically at this acidic pH. From the late endosome, lysosomal hydrolases are delivered to the lysosome by a poorly understood mechanism,[472] while the receptors are recycled back to the TGN (Fig. 18). A signal has been determined in CDMPR that prevents its transport to the lysosome.[640] The recycling of CIMPR from the late endosome to the TGN has been reconstituted in vitro and this revealed that Rab9 plays a key role in this trafficking event.[389,628]

CYTOPLASMIC SIGNALS MEDIATE LYSOSOMAL TARGETING OF MEMBRANE PROTEINS

The lysosome contains a number of heavily glycosylated integral membrane proteins whose functions are unknown.[217,376,659] In contrast to lysosomal hydrolases, these integral membrane proteins do not contain the M6P signal, suggesting that they are sorted to the lysosome via a mechanism that is not mediated by the M6P receptors.[659] The major membrane proteins targeted to the lysosome include lamp1 (also termed lgpA, LIMP III, lgp120 or LEP100), lamp2 (also referred to as lgpB, LIMP IV or lgp110), lamp3 (also named

CD63, ME491 and LIMP I), LimpII (also designated lgp85) and the precursor for lysosomal acid phosphatase (LAP precursor), which, upon delivery to the lysosome, is processed into soluble LAP. Two major routes have been shown to contribute to lysosomal sorting of these proteins and other nonresident proteins that are transported to the lysosome. The first route (direct route) is intracellular sorting at the TGN to the late endosome en route to the lysosome while the other route (surface route) involves initial appearance on the plasma membrane followed by selective internalization to the endosomal system for final lysosomal delivery. The exact route for lysosomal sorting depends on the particular protein, its expression level and the cell type. Furthermore, these two pathways are not mutually exclusive and both can be used by the same protein. Examples of direct lysosomal sorting in the TGN are lamp1/lgp120 in normal NRK cells and in transfected CHO and MDCK cells,[271,304] Limp II and CD3 γ or δ chain when expressed alone. However, the same rat lamp1/lgp120 uses the surface route in rat hepatocytes, the chicken form of lamp1/lgp120 (LEP100) in chicken fibroblasts and the human lamp1 in human leukemia cells also use the surface route.[463] In addition, the canine lamp2 (AC17) is transported to the lysosome by the surface route in MDCK cells.[505] LAP precursor has been clearly shown to be sorted to the lysosome by the surface route.[85]

Whatever route is used for lysosomal targeting, the cytoplasmic tails of these proteins have been shown to possess the sorting information for lysosomal targeting (Fig. 19).[659,843] The C-terminal cytoplasmic sequences for the major lysosomal integral membrane proteins are shown in Figure 19. Two types of sorting signals have been uncovered, one based on a critical Tyr residue and the other based on double Leu or LL-like motifs. The C-terminal cytoplasmic domain of lamp1/lgp120 contains only 11 amino acids (LKRSHAGYQTI in rat and

RKRSHAGYQTI in human, mouse and chicken). This cytoplasmic domain has been shown to contain the sorting signal for several targeting processes, including direct lysosomal sorting, surface endocytosis and basolateral surface delivery in polarized cells. The importance of this short cytoplasmic tail of lamp1 has been investigated using human, rat, mouse as well as the chicken proteins. When expressed in COS-1 cells, human lamp1 is targeted correctly to the lysosome. A mutant with the last eight residues deleted is not targeted to the lysosome but rather transported to the cell surface. Further dissections suggest that the cytoplasmic tail is sufficient to sort a chimeric protein to the lysosome and that the sole Tyr residue in the cytoplasmic tail plays a key role in lysosomal targeting. A more recent study using rat lamp1 has clearly established that G^7-Y^8-X-X-I^{11} constitutes the sorting signal for direct intracellular sorting to the lysosome.[304] Within this defined sequence, G^7 is critical only for direct sorting in the TGN, while Y^8 and I^{11} are equally important for lysosomal sorting, endocytosis and basolateral targeting. The sorting information in this short cytoplasmic domain of lamp1 can thus be deciphered at several cellular sites.[304] LAP precursor is sorted to the lysosome via the surface route and the PPGYRHV sequence in its 19-residue cytoplasmic tail (RMQAQPPGYRHVADGEDHA) represents the sorting signal.[582] Structural examination suggests that the PPGY sequence in this sorting signal is part of a β-turn.[196] Similarly, Tyr-containing lysosomal sorting signals have also been identified in the cytoplasmic domains of lamp2/lgp110/LIMP IV/lgpB (GYEQF) and LIMP I/ME491/CD63 (GYEVM). For these Tyr-based sorting signals, there is a Gly preceding the Tyr (defined as position 1) and a bulky hydrophobic residue (I, F, M, V) in the forth position, giving a consensus motif of **GYXXB** (B stands for a bulky hydrophobic residue).

When expressed in COS cells, rat LIMP II is targeted directly from the TGN to the lysosome. Since the 20-residue cytoplasmic tail (RGQGSTDEGTADERAPLIRT) of LIMP II does not contain any Tyr residue, the sorting process must involve a novel type of signal. Substitution of its cytoplasmic tail for the native cytoplasmic domains of surface protein CD36 and CD8 result in the direct transport of the chimeric proteins to the lysosome, suggesting that its Tyr-lacking cytoplasmic tail contains an autonomous signal for lysosomal sorting in the TGN.[809] Similarly, a chimeric protein (hcg/G/L) composed of hcg as the E/L domain, the transmembrane domain of VSV G-protein and the cytoplasmic tail of rat LIMP II, is sorted to the lysosome. A variant of hcg/G/L with three residues (IRT) deleted from the C-terminus is no longer transported to the lysosome, suggesting that the last three residues are part of the sorting signal. Further dissections reveal that Leu and Ile residues in LIRT sequence represent the key element for lysosomal targeting.[542,658] The LI motif is similar to the double-Leu motif observed in the cytoplasmic tails of CIMPR and CDMPR. The Ile residue in the LI motif can be replaced by Leu without loss in the sorting efficiency, suggesting that the LI motif may contribute to sorting in a way similar to the double-Leu motif.

In addition to these resident membrane proteins of lysosomes, sorting signals for the lysosome have also been identified in the cytoplasmic tails of CD3 γ and δ chains.[405] One double-Leu motif and a Tyr-based motif have been implicated in lysosomal targeting. The double-Leu motif and the Tyr-based motif could individually sort proteins to the lysosome via the surface route. However, both sorting motifs can together mediate direct lysosomal sorting in the TGN.

Membrane proteins destined for the lysosome are sorted via the endosomal system. The direct route for lamp1 and LIMP II from the TGN to the lysosome may be mediated by the late endosome. The surface route for lysosomal delivery may be mediated by the early and subsequently the late endosome. In this respect, all these sorting signals must have an initial sorting information for the endosome and possible information for later delivery from the endosome to the lysosome. There are some proteins which are sorted to the endosome or endosome-like structures without further delivery to the lysosome, including the invarient (Ii) chain of MHC class II antigens and glucose transporter 4 (GLUT-4). These proteins may not contain signals for further transport to the lysosome. Alternatively, late-endosome to lysosome transport may be a signal-independent default process and these proteins may contain signals that prevent their transport to the lysosome. GLUT-4 is predicted to have 12 transmembrane domains with N-terminal and C-terminal tails both oriented on the cytoplasmic side of the membrane (Fig. 19). GLUT-4 is targeted to a unique intracellular compartment that may be equivalent to endosomes in fat and muscle cells. Insulin triggers a redistribution of GLUT-4 from this intracellular compartment to the plasma membrane. The property of intracellular sequestration of GLUT-4 also occurs in cells other than fat and muscle cells when expressed by transfection. Other forms of glucose transporters such as GLUT-1, GLUT-2 and GLUT-3 are transported to the plasma membrane irrespective of cell type. Detailed dissections of GLUT-4 have identified two signals involved in its intracellular sequestration and trafficking in response to insulin.[142,264,458,596,810,811] The first signal is located at the N-terminal cytoplasmic tail and is based on a critical Phe residue at position 5 and an Ile at position 8 (MPSGFQQIGS) in which Phe could be functionally replaced by a Tyr residue, while the other signal resides in the C-terminal cytoplasmic tail and is based on a double-Leu motif in the context of RRTPSLLEQEVK (Fig. 19). The p31 form of Ii chain is essential for targeting MHC class II antigens to the late endosome.[75,514,781] As a type II integral membrane protein, the N-terminal cytoplasmic tail of Ii chain contains this sorting information. There are several double-Leu-like motifs in its N-terminal cytoplasmic tail and various Ii forms contain pairs of a LI motif and either a ML, IL, or VL motif. Mutational studies of human Ii have established that the LI and ML motifs can each mediate endosomal targeting (Fig. 19), suggesting that both motifs are involved in efficient endosomal targeting of the wild-type molecule.[87]

VACUOLAR TARGETING IN YEAST *S. CEREVISIAE*

SIGNAL-MEDIATED SORTING OF SOLUBLE VACUOLAR ENZYMES

Similar to lysosomal enzymes, soluble vacuolar proteins in the yeast also travel the early secretory pathway from the ER to the Golgi, and in the Kex2p-containing late Golgi compartment, they are sorted from proteins destined for the plasma membrane and secretion.[306,366,620,813,814] Sorting of soluble enzymes to the yeast vacuole involves a process that is not mediated by the M6P signal. The sorting signals for vacuoles have been identified for carboxypeptidase Y (CPY) (encoded by the PRC1 gene)[335,803,804] as well as proteinase A (PrA) (encoded by the PEP4 gene).[364] The open reading frame of PRC1 encodes a polypeptide of 532 amino acids. Upon translocation into the ER, the N-terminal 20 residues, functioning as the signal peptide, are cleaved by the signal peptidase complex and the translocated polypeptide (residue 21-532) is modified by the attachment of four N-linked core-glycans, giving rise to a precursor termed p1CPY of about 67 kDa. As p1CPY transits through the Golgi, extension of the glycans results in the Golgi modified form designated p2CPY of about 69 kDa, which is sorted to the vacuole via the late endosome-like PVC. Upon arrival at the vacuole, the N-terminal prosequence (residues 21-111) of p2CPY is cleaved to yield the enzymatically active mature CPY (residues 112-432) of about 61 kDa and this cleavage pro-

cess is mediated by PrA and therefore is PEP4-dependent. When fused to the secretory enzyme invertase (encoded by the SUC2 gene), the N-terminal 50-residue sequence of CPY is sufficient to sort this chimeric protein to the vacuole. Since the first 20 residues of this 50-residue sequence is cleaved during ER translocation, the vacuolar targeting signal must be contained between amino acids 21 and 50. Consistently, a deletion of residues 21-50 from the wild-type CPY leads to a deficiency in its vacuolar targeting and the mutant CPY is secreted. A single residue mutation that converts Gln at position 24 into a Lys residue similarly affects vacuolar sorting of the mutant CPY. Further studies have established that four contiguous amino acids Gln[24]-Arg-Pro-Leu[27] represent the vacuolar sorting signal for CPY. The vacuolar sorting signal for PrA has also been mapped to the prosequence of this protein. PrA is synthesized as a 405-residue precursor, in which the signal peptide (residues 1-22) and the prosequence (residue 23-76) make up the N-terminal 76 amino acids. The signal peptide is removed during translocation across the ER and the translocated precursor is modified by the attachment of two N-linked glycans, which is further modified in the Golgi to give rise to a Golgi form of about 48-52 kDa. Upon arrival at the vacuole, the prosequence is proteolytically removed to give a mature active enzyme of 329 amino acids (42 kDa). When the N-terminal 90 amino acids or more of PrA are fused to invertase, more than 95% of the chimeric proteins are sorted to the vacuole, and about 90% of a chimeric protein consisting of the N-terminal 76 amino acids fused to the invertase is also correctly sorted. However, the N-terminal 23-, 39-, or 61- residue sequence of PrA is not sufficient to sort the resulting chimeric proteins to the vacuole, suggesting that residues 61-76 play a key role in vacuolar sorting. The vacuolar sorting signals for other soluble vacuolar enzymes remain to be experimentally established.

GENETIC SCREENS IDENTIFY MORE THAN 60 GENES INVOLVED IN VACUOLAR SORTING AND/OR BIOGENESIS

Several genetic screens have isolated a large number of mutants defective in vacuolar sorting and/or vacuolar biogenesis and these mutants fall into more than 60 complementation groups.[366,620,631,686,733] Most of them are identified as mutants that secrete vacuolar proteins (vps) and are defective in delivery of multiple vacuolar proteins. Morphological studies have divided vps mutants into six classes.[40,366,619] Class A mutants include vps8, vps10, vps13, vps29, vps30, vps35, vps38, vps44 and vps46 and display almost wild-type morphology and vacuoles often appear as 1-10 subcompartments that cluster to one region of the cytoplasm. During the cell cycle, portions of the mother vacuole in class A mutants appear to actively partition into developing buds. The characteristic mother to bud vacuolar extensions have been termed "vacuolar segregation structures" and the process whereby maternal vacuolar material is passed to the daughter cell has been designated "vacuole inheritance". In wild-type cells, vacuolar integral membrane protein alkaline phosphatase (ALP) encoded by the PHO8 gene and components of vacuolar H⁺-ATPase (V-ATPase) are localized in the vacuoles and this property is not affected in class A mutants. Class B mutants include vps5, vps17, vps39, vps41 and vps43 and have their vacuoles fragmented, exhibiting >20 vacuole-like compartments per cell. The fragmented vacuoles cluster within one region in some mutants such as vps5 and vps17 mutants. In contrast, the fragmented vacuoles are randomly dispersed throughout the cytoplasm in vps39, vps41 and vps43 mutants. vps11, vps16, vps18 and vps33 are examples of class C mutants that lack any identifiable vacuole structures. ALP labeling is distinctly punctate, suggesting that this vacuolar membrane protein is distributed among vesicle-like structures. V-ATPase staining is more diffuse, suggesting that class C mutants are at least partially defective for the assembly of V-ATPase. Class D mutants include vps3, vps6, vps9, vps15, vps19, vps21, vps34 and vps45 and are morphologically similar to class A mutants but have a large single vacuolar structure and exhibit defects in vacuole inheritance. Furthermore, V-ATPase fails to assemble properly onto the vacuolar membrane and there is a defect in vacuolar acidification in class D mutants. Class E mutants include vps2, vps4, vps20, vps22, vps23, vps24, vps25, vps27, vps28, vps31, vps32, vps36 and vps37 and contain a novel endosome-like PVC in addition to normal vacuoles.[814] ALP antibodies label vacuoles, whereas V-ATPase antibodies label the PVC. Significant portions of CPY and PrA and Golgi membrane proteins are also localized to the PVC, suggesting that class E mutants possess an exaggerated form of the PVC that is normally less obvious in wild-type cells. Class F mutants, such as vps1 and vps26, have a large central vacuole surrounded by a number of class B-like fragmented vacuolar structures. ALP and V-ATPase are colocalized in class E mutants. Examples of well-characterized proteins involved in vacuolar sorting are shown in Figure 20.

VPS10/PEP1 ENCODES THE MEMBRANE SORTING RECEPTOR FOR CPY

Cloned VPS10 is identical to PEP1 and encodes a 210 kDa polypeptide of 1577 amino acids.[457] The N-terminal 21 residues may represent the signal peptide for ER targeting and translocation, resulting in a type I membrane protein with a luminal domain of 1370 amino acids (residues 22-1391) that contains seven potential sites for N-linked glycosylation followed by a 22-residue transmembrane domain (residues 1392-1413) and a 164-residue cytoplasmic domain (residues 1414-1577). The role of Vps10p/Pep1p is investigated in a yeast strain in which more than 80%

of the VPS10/PEP1 coding region is deleted and replaced with a DNA fragment containing the HIS3 gene. Cells with disrupted VSP10 are viable and show normal vacuolar morphology, suggesting that Vps10p/Pep1p is not essential for growth. However, more than 90% of CPY is secreted from the mutant cell as the Golgi form (p2CPY), suggesting that Vps10p/ Pep1p is essential for vacuolar sorting of CPY. Furthermore, the mutant cell sorts other luminal vacuolar enzymes, including PrA and proteinase B (PrB) normally, suggesting that VPS10 disruption has a selective effect on vacuolar sorting of CPY. Biochemical characterizations suggest that Vps10p/Pep1p is indeed an integral membrane protein associated with the Golgi

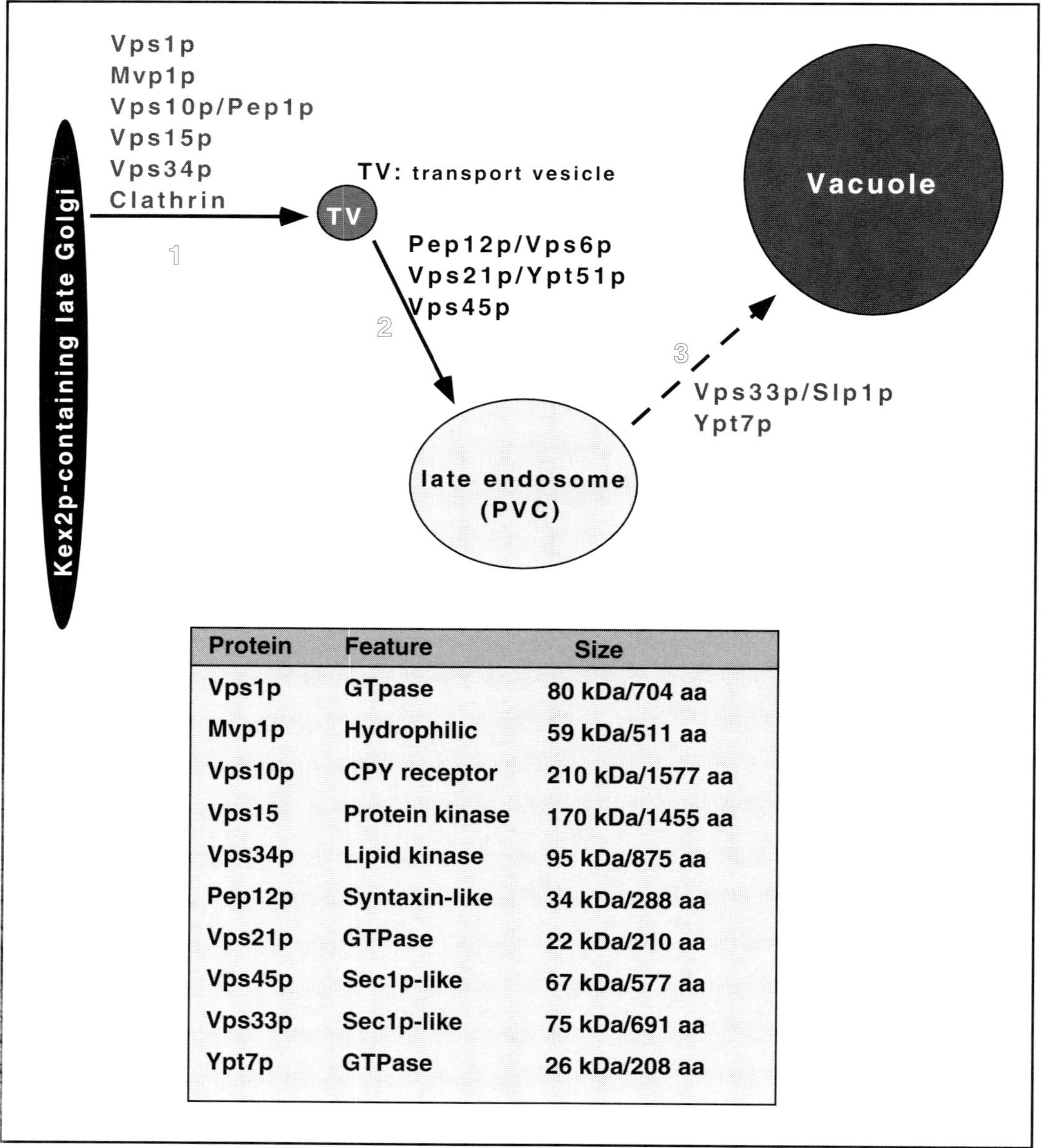

Protein	Feature	Size
Vps1p	GTpase	80 kDa/704 aa
Mvp1p	Hydrophilic	59 kDa/511 aa
Vps10p	CPY receptor	210 kDa/1577 aa
Vps15	Protein kinase	170 kDa/1455 aa
Vps34p	Lipid kinase	95 kDa/875 aa
Pep12p	Syntaxin-like	34 kDa/288 aa
Vps21p	GTPase	22 kDa/210 aa
Vps45p	Sec1p-like	67 kDa/577 aa
Vps33p	Sec1p-like	75 kDa/691 aa
Ypt7p	GTPase	26 kDa/208 aa

Fig. 20. Examples of proteins involved in vacuolar sorting in S. cerevisiae. The steps of action, structural features and sizes of the respective proteins are shown.

complex (most likely the Kex2p-containing late Golgi). Chemical crosslinking experiments demonstrate that Vps10p can interact specifically with the p2CPY (the Golgi form) and this interaction is abolished by a single amino acid change (Qln24 to Lys) in the vacuolar sorting signal of CPY. Furthermore, a chimeric protein containing the CPY vacuolar targeting signal fused to invertase also interacts with Vps10p. It is therefore suggested that Vps10/Pep1p is the "private" receptor for CPY and that interaction of p2CPY with Vps10p in the late Golgi compartment results in selective packaging of CPY into vesicles destined for the PVC.[457] Upon reaching the PVC, p2CPY dissociates from Vps10p and is transported to the vacuole, while Vps10p is recycled back to the late Golgi for subsequent sorting events.

INVOLVEMENT IN VACUOLAR SORTING OF A PROTEIN COMPLEX COMPOSED OF VPS15 PROTEIN KINASE AND VPS34 PHOSPHATIDYLINOSITOL 3-KINASE (PtdIns 3-KINASE)

Yeast cells with mutant vps15 alleles are characterized by the secretion of >90% of newly synthesized CPY as the p2CPY form. VPS15 encodes a 170 kDa protein of 1455 amino acids and the N-terminal 300 residues bear significant similarity to Ser/Thr protein kinases with best identity to the catalytic subunit of phosphorylase β kinase (25% identity) and *Schizo. pombe* Wee1p kinase (23% identity).[289] Ser/Thr protein kinases are characterized by the consensus element DLKPEN, in which the Lys residue is absolutely conserved, and Vps15p has the DIKTEN sequence in the corresponding region. Vps15p is also characterized by other conserved residues for protein kinases. The N-terminal sequence MGAQLSL fits a consensus sequence for the attachment of myristic acid to the N-terminal penultimate Gly (after removal of the initiation Met), suggesting that Vps15p could be myristylated. Biochemical characterizations of Vps15p suggest that it is a peripheral

protein associated on the cytoplasmic side of the late Golgi compartment. VPS15 disruption results in a ts growth defect in which mutant cells grow at 26°C but not at 37°C, suggesting that Vps15p is required for vegetative growth at an elevated growth temperature. Point mutations in Vps15 that alter highly conserved residues among protein kinases result in a defect in its kinase activity and defective sorting of multiple vacuolar proteins, suggesting that the kinase activity of Vps15 is required for vacuolar sorting. The role played by C-terminal 1100 amino acids is unknown, but a mutant Vps15 (vsp13/del30C) lacking the 30 amino acids at the C-terminus results in a temperature-conditional vacuolar sorting defect. A shift to the nonpermissive temperature in vsp13/del30C cells causes an immediate but reversible defect in vacuolar sorting, suggesting that Vps15 is directly involved in vacuolar sorting of luminal enzymes.

Mutations in VPS34 also lead to sorting defects for multiple soluble vacuolar enzymes and gene disruption causes a ts phenotype for vegetative growth. VPS34 encodes a 95 kDa hydrophilic protein of 875 amino acids that shares extensive sequence identity with the p110 catalytic subunit of mammalian phosphoinositide 3-kinase (PI 3-kinase). The mammalian PI 3-kinase (in conjunction with the regulatory subunit p85) phosphorylates membrane PtdIns and its more highly phosphorylated derivatives, PtdIns(4)P and PtdIns(4,5)P2. *S. cerevisiae* has PtdIns 3-kinase activity and this activity is extremely defective in cells with VPS34-disruption. Unlike the mammalian p110, Vps34p uses only PtdIns but not PtdIns-(4)P and PtdIns(4,5)P2 as the substrate, suggesting that it is a PtdIns-specific 3-kinase. Alterations of conserved residues in the lipid kinase domain of Vps34p result in severe defects in both PtdIns 3-kinase activity and vacuolar protein sorting, suggesting that its PtdIns 3-kinase activity and the production of the phosphoinositide PtdIns(3)p are required for correct sorting of soluble vacuolar enzymes.[674,734] Cell fractionation studies suggest that a significant fraction (50%) of Vps34p is associated with membrane structures. Characterization of a ts allele of VPS34 demonstrates that inactivation of Vps34p leads to an immediate defect in vacuolar sorting, indicating that, like Vps15p, Vps34p may be directly involved in vacuolar sorting.

Genetic and biochemical evidence suggest that Vps15p and Vps34p form a protein complex that may be directly involved in vacuolar sorting of soluble enzymes.[732,735] As mentioned, VPS15 kinase domain mutants are defective in vacuolar sorting of multiple soluble enzymes. Overexpression of Vps34p can suppress sorting defects exhibited by these mutants, indicating that Vps34p may function downstream of Vps15p. Biochemical analysis suggests that membrane association of Vps34p is mediated by Vps15p. In control cells, about 50% of Vps34p is associated with membrane fraction. However, less than 10% of Vps34p is membrane-associated in VPS15-disrupted cells, suggesting that membrane association of Vps34p requires Vps15p. Overexpression of Vps15p enhances membrane association of Vps34p to 90%. Sucrose density gradient analysis suggests that Vps15p and Vps34p are enriched in the same membrane fractions. Antibodies against either Vps15p or Vps34p can precipitate a protein complex that contains both proteins. Vps15p and Vps34p can also be chemically crosslinked. In addition, Vps15p kinase activity is required for activation of Vps34p PtdIns 3-kinase activity, suggesting that phosphorylation events catalyzed by Vps15p may regulate the PtdIns 3-kinase activity and therefore vacuolar sorting. Mutant Vps34p molecules with alterations in the conserved lipid kinase domain are defective in PtdIns 3-kinase activity and behave as dominant negative mutants, most likely through titration of Vps15p such that endogenous wild-type Vps34p cannot be recruited and activated. In contrast, protein kinase-negative mutant Vps15p molecules do not exhibit dominant-negative phenotype because they

are unable to associate with Vps34p, demonstrating that intact Vps15p kinase domain is required for recruitment of Vps34p and activation of its lipid kinase activity. Characterization of a ts allele of VPS15 reveals that a shift to the nonpermissive temperature leads to a decrease in cellular PtdIns 3-kinase activity, suggesting that functional Vps15p is indeed required for the activation of Vps34p. These results suggest that sorting of soluble vacuolar enzymes involves a process that results in the activation of the Vps15p protein kinase, which in turn leads to membrane recruitment of Vps34p and the activation of its PtdIns 3-kinase activity. These events result in the phosphorylation of PtdIns to PtdIns(3) and may trigger the budding process of vesicles destined for the PVC. How the kinase activity of Vps15p is regulated has not been established and how PtdIns(3) participates in the sorting event remains to be mechanistically investigated. Based on the finding that CPY has its own private receptor, it seems that there are multiple membrane receptors in the late Golgi that are involved in sorting of different vacuolar enzymes. The sorting receptors should somehow all activate the Vps15p-Vps34p complex to mediate the selective incorporation of vacuolar enzymes into vesicles destined for the PVC. Whether individual receptors can each directly activate the Vps15p-Vps34p complex or there exists an intermediate component that links the various receptors to the Vps15p-Vps34p complex remains to be resolved.

Based on the conserved regions between Vps34p and the bovine p110 catalytic subunit of mammalian phosphoinositide 3-kinase (PI 3-kinase), a PCR-based approach has identified a cDNA for a human protein (termed human PtdIns 3-kinase) composed of 887 amino acids with a high degree of homology with Vps34p over its entire length (37% identity and 58% similarity).[815] In contrast to the p110 catalytic subunit of PI 3-kinase, human PtdIns 3-kinase does not associate with the regulatory subunit p85.

Rather, human PtdIns 3-kinase is present in a complex with a protein of about 150 kDa (p150). Similar to Vps34p but distinct from p110 of PI 3-kinase, the human PtdIns only phosphorylates PtdIns but not PtdIns(4)P or PtdIns(4,5)P2. Protein sequence analysis and cDNA cloning both establish that p150 is highly homologous to Vps15p, suggesting that the human PtdIns 3-kinase-p150 protein complex may be a functional counterpart of yeast Vps34p-Vps15p complex and could be involved in lysosomal/endosomal sorting.[815] This hypothesis is supported by the observation that wortmannin and LY294002 at doses that inhibit human PtdIns 3-Kinase in vitro each cause a defect in lysosomal sorting of cathepsin D.[93,163,796] The CIMPR-containing late endosomes become highly swollen vacuoles that are easily detectable under phase contrast microscopy and these swollen vacuoles gradually become depleted of CIMPR. It was suggested that these drugs inhibit transport from the TGN to the late endosome and the depletion of CIMPR in the swollen vacuoles is due to the net effect of inhibiting transport from TGN to these swollen structures and apparent normal recycling of CIMPR from these swollen structures back to the TGN. The detailed functional aspects of the human PtdIns 3-kinase-p150 protein complex remain to be established.

OTHER PROTEINS INVOLVED IN VACUOLAR TARGETING

Vps1p

VPS1 is important for correct sorting of vacuolar enzymes and encodes an 80-kDa hydrophilic protein of 704 amino acids.[651,808,845] Vps1p is structurally related to dynamins and possesses consensus motifs for GTP binding and hydrolysis. Consistent with its role in vacuolar sorting, Vps1p is preferentially associated with the Golgi membrane. The 704-residue polypeptide of Vps1p can be divided into two functionally distinct domains. The N-terminal half habors the consensus mo-

tifs for GTP binding and hydrolysis and is a GTPase, even in the absence of the C-terminal portion. Several point mutations in the GTPase domain lead to dominant-negative mutant proteins that cause wild-type cells to mis-sort vacuolar proteins. The C-terminal domain of Vps1p may mediate an interaction with other proteins involved in vacuolar sorting because overexpression of this portion in wild-type cells can abolish vacuolar sorting. It was suggested that, by GTP binding and hydrolysis cycles, the GTPase domain could regulate the functional interaction of the C-terminal domain with other proteins. Genetic screening for multicopy suppressers of dominant-negative vps1 mutations has identified a novel gene, MVP1, that exhibits genetic interaction with VPS1 and can suppress several dominant alleles of VPS1. This suppression is dependent on the presence of wild-type Vps1p. MVP encodes a 59-kDa hydrophilic protein of 511 amino acids and its disruption results in the secretion of 65% of newly made CPY, demonstrating that Mvp1p is also involved in vacuolar sorting.[198]

Vps45p

VPS45 encodes a 67 kDa hydrophilic protein of 577 amino acids.[144a,596a] Vps45p is structurally related to Sec1p, Sly1p, Vps33p and their counterparts in other organisms. Since Sec1p and Sly1p are involved in docking/fusion of transport vesicles, Vps45p may function likewise. It was suggested that Vps45p may be involved in docking and fusion of vesicles containing vacuolar enzymes with the PVC. VPS45 disruption leads to cells that grow normally at permissive temperatures (such as 25°C) but not at elevated temperatures such as 38°C. Sorting to the vacuoles of CPY and PrA is affected, resulting in the secretion of these vacuolar proteins. Furthermore, vesicles or clusters of vesicles of 40-50 nm are seen to accumulate in VPS45-disrupted cells or cells with a ts allele of VPS45 at the restrictive temperature. Biochemical characterizations suggest that Vps45p is a peripheral protein associated with a high-speed membrane fraction that includes Golgi, transport vesicles and possibly the PVC. This Vps45p-containing membrane fraction lacks ER, vacuole and the plasma membrane, suggesting that Vps15 is associated with membrane structures between the late Golgi and the vacuole. The association of Vps45p with the membrane is saturable because Vps45p overexpression results in higher proportion of Vps45p that is not membrane-associated, suggesting the existence of a specific but limiting membrane receptor for Vps45p. Since Sec1p and Sly1p interact with members of the syntaxin family (Sso1p/Sso2p and Sed5p, respectively), the potential membrane receptor for Vps15p could be another member of the syntaxin family. The PEP12/VPS6 encodes a syntaxin-like protein involved in vacuolar sorting and it will be interesting to examine whether Vsp45p interacts with Pep12p.

Pep12p/Vps6p

Mutations in PEP12, which is identical to the VPS6, result in a sorting defect for soluble vacuolar enzymes, such as CPY, PrA and PrB.[619] The predicted Pep12p/Vps6p is a 34 kDa integral membrane protein with 288 amino acids anchored to the membrane by a C-terminal hydrophobic tail. Its amino acid sequence bears clear sequence homology with other members of the syntaxin family, suggesting that it is involved in vesicle docking/fusion. It has been suggested that Pep12p/Vsp6p may function as a t-SNARE of the PVC for vesicles derived from the late Golgi.

Vps21p/Ypt51p

VPS21 encodes a 22 kDa protein of 210 amino acids that is structurally most related to mammalian Rab5 (54% identical).[305,716] Mutations in VPS21 result in a defective vacuolar sorting of soluble enzymes and vps Class D phenotype. By analogy with Ypt1p and Sec4p, Vps21p

may be involved in docking/fusion with the PVC for vesicles derived from the late Golgi or from the endocytotic pathway. In this regard, Vps21p, Vps6p/Pep12p and Vps45p may act together in vesicle docking/fusion with the PVC.

Vps33p/Slp1p

Mutations in the VPS33/SLP1 result in a defect in vacuolar sorting and biogenesis. VPS33 encodes a 75 kDa hydrophilic protein of 691 amino acids,[40a,818a] which is related to members of Sec1p protein family,[1] suggesting that Vps33p/Slp1p may also be involved in vesicles docking/fusion during vacuolar sorting of soluble enzymes. It has been speculated that Vps33p/Slp1p may be involved in transport from the PVC to the vacuole. Furthermore, the amino acid sequence of Vps33p/Slp1p contains motifs that are characteristic of ATPases and ATP-binding proteins, suggesting that it may be an ATPase (although it remains to be biochemically established). Whether the ATPase motifs of Vps33p/Slp1p are involved in vacuolar sorting awaits further investigation.

Ypt7p

YPT7 encodes a 26 kDa protein of 208 amino acids with 63% identity with mammalian Rab7, which is associated with the late endosome.[669,840] Cells with disrupted YPT7 exhibit highly fragmented vacuoles and delayed transport of CPY, PrA, PrB to the vacuole as evidenced by their delayed maturation. In contrast to vps mutants, cells with disrupted YPT7 do not secrete soluble vacuolar enzymes, suggesting that sorting at the late Golgi and transport to the PVC are not affected. Furthermore, maturation of vacuolar membrane protein ALP is delayed to a greater extent. In YPT7 disrupted cells, endocytosis of the pheromone α-factor is not affected but their degradation is inhibited. The undegraded α-factor accumulates in the PVC, suggesting that Ypt7p is involved in transport from the PVC to the vacuole.

Evidence that Vacuolar Transport of Membrane Protein Occurs by Default

The best studied integral membrane proteins associated with vacuoles are ALP and dipeptidyl peptidase B (DADP B), both of which traverse the early secretory pathway and are then transported from the late Golgi to the vacuole via the PVC. This direct intracellular pathway has been clearly established by the fact that mutations of genes (such as SEC1) involved in transport from the late Golgi to the plasma membrane have no effect on their transport to the vacuole. DADP B, encoded by DAP2, is a 120 kDa type II integral glycoprotein with 841 amino acids. The N-terminal 29 residues form the cytoplasmic domain, followed by a 16-residue transmembrane domain and 796-residue luminal domain. Newly made DADP B is about 110 kDa and is converted into the 120 kDa form by Golgi modification of the glycans. The 120 kDa Golgi form is transported to the vacuole without detectable PEP4-dependent proteolytic cleavage. ALP, encoded by PHO8, is composed of 566 amino acids and is synthesized as an inactive precursor; its enzymatic activation being dependent on the removal of a C-terminal prosequence in a PEP4-dependent manner. Newly made ALP is about 76 kDa with two attached N-glycans. This 76 kDa form is modified in the Golgi without a detectable change in its apparent size. Upon delivery to the vacuole, its C-terminal prosequence is removed by PrA, resulting in the enzymatically active mature ALP of about 72 kDa.[365,840] Experiments designed to identify potential vacuolar sorting signals for these two integral membrane proteins have led to the hypothesis that their vacuolar transport involve no signal and occurs by a default pathway. Furthermore, late Golgi proteins (such as Kex1p, DADP A and Kex2p), when overexpressed, are transported to the vacuole. Mutant versions of these proteins that are not Golgi-retained are also delivered to the vacuole, further supporting the hypothesis that vacuolar transport of integral membrane proteins in *S. cerevisiae* may occur by a signal-independent process.[366,534,620] This is in marked contrast to the signal-mediated lysosomal/endosomal targeting of integral membrane proteins in higher eukaryotic cells.

CHAPTER 7

POLARIZED SURFACE TARGETING

DISTINCT SURFACE DOMAINS OF POLARIZED CELLS

Polarized cells, such as epithelial cells and neurons, organize their plasma membrane into morphologically, biochemically and functionally distinct domains.[497,517,639,662,711,713] In simple epithelial cells, such as the epithelial cells of the intestine and kidney tubules, the plasma membrane is differentiated into the apical domain, which faces the external environment (or its equivalent) and the basolateral domain, which faces the internal milieu of the organism. In hepatocytes, the plasma membrane is divided into several bile canalicular domains, from which bile materials are secreted into the gut and are equivalent to the apical domain of simple epithelial cells; and a few sinusoidal and lateral domains that are equivalent to the basolateral domain. The apical/bile canalicular domain is separated by tight junctions from the basolateral/sinusoidal-lateral domain, and the tight junctions function as a diffusion barrier for proteins and lipids. In neurons, the plasma membrane is similarly differentiated into the axon, cell body and dendrites. Even in yeast *S. cerevisiae*, polarity development is essential for cell cycle progression. During the growth of buds, the secretory pathway is directed to the growing buds only. The differentiation of the plasma membrane polarity is associated with the development of cytoskeletal polarity. Many plasma membrane proteins have been identified and characterized that are localized to specific surface domains.[302,432,497,639,662,711] Examples of integral membrane proteins enriched in the apical/bile canalicular domain include aminopeptidase N (ApN), dipeptidyl peptidase IV (DPPIV), sucrose-isomaltase (SIM), ecto-ATPase (CAM105), neutral endopeptidase and γ-glutamyl transpeptidase. The transferrin receptor, low density lipoprotein (LDL) receptor, NaK+ ATPase, EGF receptor, hepatic asialoglycoprotein receptor and the cell adhesion molecule uvomorulin are examples of integral membrane proteins associated preferentially with the basolateral domains. TGN38/41, M6P receptors and lysosomal integral membrane proteins also use preferentially the basolateral surface for recycling.[505,605,607,613] The intact polymeric IgA receptor is initially targeted to the sinusoidal/

basolateral domain. Upon binding with polymeric IgA, this protein complex is selectively transcytosed to the bile canalicular/apical domain. In addition to studies using intact organs or organisms, the majority of polarity studies are performed in cell lines that can differentiate their plasma membrane into apical and basolateral surfaces in culture. MDCK cells have been the major cell line used for polarity studies.[497,639,711] In addition, human intestinal Caco2 cells,[497,629,793] pig kidney LLC-PK1 cells,[437,440,497] and Fischer rat thyroid (FRT) cells[872,873] have also been used. Early studies on plasma membrane polarity were greatly facilitated by the observation that influenza virus (HV) buds solely from the apical surface, whereas vesicular stomatitis virus (VSV) buds only from the basolateral domain of infected MDCK cells. Furthermore, the envelope proteins, hemagglutinin (HA) of HV and G-protein of VSV have been shown to be transported respectively to the apical and basolateral domains in the absence of other viral proteins.

Furthermore, proteins that are preferentially secreted from either the apical or the basolateral surfaces have been characterized.[72,244,439,440,497,517,629,639,662,711,793] The majority of serum proteins such as albumin and transferrin are secreted from the sinusoidal domain of hepatocytes. An 80 kDa (gp80) protein and several others have been shown to be secreted preferentially from the apical surface of MDCK cells,[244,439,799] while extracellular matrix protein laminin, heparan sulfate proteoglycans (HSPG) and several other proteins have been shown to be released preferentially from the basolateral surface of MDCK cells. Some proteins, such as the extracellular matrix protein fibronectin are secreted equally from both surfaces of MDCK cells.[440] The majority of secretory proteins, including apolipoproteins apoA-I, apoA-IV, apoE, apoB and fibronectin are preferentially secreted from the basolateral surface of Caco2 cells. Similarly, the majority of secretory proteins such as fibronectin and HSPG are released predomi-

nantly from the basolateral surface of LLC-PK1 cells.[440] gp80 in transfected Caco2 cells is secreted predominantly from the apical surface, although no endogenous proteins have been shown to be preferentially secreted from the apical surface of Caco2 cells.[15,440,629]

PROTEIN TARGETING TO SURFACE DOMAINS

How proteins are transported to these distinct surface domains has been a central question in the study of polarity. Proteins destined for the basolateral or equivalent domain are sorted intracellularly in the TGN and transported directly from the TGN to the basolateral surface. Two pathways have been uncovered for proteins destined for the apical surface.[395,420,466,497,517,639,713] One involves direct transport from the TGN to the apical surface. In the other pathway, proteins are initially delivered to the basolateral surface followed by their selective transcytosis from the basolateral to the apical surface. These two pathways are not mutually exclusive and can be utilized by the same protein in the same cell. In MDCK cells, the majority of exogenous as well as endogenous domain-specific proteins are seen to be transported directly from the TGN to the apical or the basolateral surfaces, respectively.[396,420] However, in hepatocytes, all the apical proteins thus far studied (including DPPIV, ApN and 5'-nucleotidase) have been shown to be first delivered to the sinusoidal surface and then selectively retrieved and transcytosed to the bile canalicular domain.[51,665] In Caco2 cells, SIM, DPPIV and ApN are all enriched in the apical surface by different transport routes. The majority of SIM is efficiently transported directly from the TGN to the apical surface, while DPPIV and ApN use both the direct and transcytotic pathways.[395,466] When expressed in transfected MDCK cells, the majority of DPPIV and ApN is directly targeted to the apical surface.[438] However, a significant portion of DPPIV is delivered to the apical surface by the transcytotic pathway when expressed in LLC-PK1 cells.[437] These results demonstrate that the same protein could use different routes for apical surface targeting in different cells and that different proteins in the same cell type could also use different pathways. The trafficking pathway for IgA receptor is faithfully reconstituted when expressed in MDCK cells and the newly made receptor is initially transported to the basolateral surface and is efficiently transcytosed to the apical surface. In addition to the direct sorting and indirect surface retrieval pathways, there is another mechanism for selective expression of domain-specific proteins. In MDCK cells, Na^+, K^+-ATPase is confined to the basolateral domain, although newly made Na^+, K^+-ATPase is equally transported to both domains. The basolateral restriction of this protein is mediated by the selective retention/stabilization at the basolateral domain and inactivation/degradation at the apical domain.[258]

The apical and basolateral targeting events are believed to involve different machineries. In MDCK cells, it is seen that brefeldin A can selectively inhibit apical delivery of both secretory as well as membrane proteins, leading to their enchanced transport to the basolateral surface.[434,439] Furthermore, activators of protein kinase A stimulate apical but not basolateral transport in MDCK cells and a Gs class of heterotrimeric G proteins has been implicated in the apical transport.[593,594] To biochemically dissect these transport events, transport vesicles that mediate transport to the apical and basolateral domains of MDCK cells have been isolated and characterized.[833] For isolation of these vesicles, MDCK cells are simultaneously infected with HV and VSV and vesicles derived from the TGN in perforated cells are isolated and it is found that the apical vesicles marked by HA and the basolateral vesicles marked by G-protein contain a set of proteins that are common to both types of vesicles. More importantly, proteins enriched in either apical or basolateral vesicles have also been

identified. When the purified vesicles are solubilized with the detergent CHAPS, the apical marker protein HA is found to exist in a large CHAPS-insoluble complex with several other integral membrane proteins. Furthermore, a similar set of CHAPS-insoluble proteins is found after solubilization of a total cellular membrane fraction, allowing a rapid purification of these proteins directly from the total cellular membrane fraction. This CHAPS-insoluble complex is enriched in glycosphingolipids (which are enriched in the apical membrane), apical cargo and a set of proteins of TGN-derived transport vesicles. A protein of about 21 kDa (referred to as VIP21 for vesicular integral-membrane protein of 21kD) that is common to both apical and basolateral vesicles has been molecularly-cloned. When an epitope-tagged VIP21 is expressed, it is detected on the Golgi apparatus, plasma membrane as well as vesicular structures.[384] VIP21 is identical to caveolin, the major component of caveolae.[646] Caveolae were originally identified as tiny, flask-shaped membrane invaginations on the plasma membrane of endothelial cells and now of most other cells.[10,422] Originally proposed to be involved in transcytosis in epithelial cells, it is now established that they are also the surface sites where small molecules (such as folate) are concentrated and internalized through a process termed potocytosis. The small molecules flow into the cytoplasm through carriers or channels embedded in the caveolae membrane. Current studies also suggest that caveolae are enriched in GPI-anchored proteins as well as molecules (such as Src kinase) involved in signal transduction, suggesting that they may play a critical role in the cellular signaling pathway.[10,422,638,663] Biochemically, the formation of caveolae is regulated through a protein coat distinct from clathrin, and the 21 kDa caveolin is found to be the major structural component of the caveolae coat. Formation of caveolae is dependent on cholesterol because its depletion leads to the disruption of caveolae structure.

Immunolocalization of endogenous VIP21/caveolin has confirmed that it is preferentially associated with the TGN, TGN-derived vesicles and the plasma membrane caveolae.[192] VIP21/caveolin may play an important role in the formation of glycosphingolipid clusters in the TGN for sorting GPI-anchored and other proteins destined for the apical surface in polarized cells.[192,873]

Further biochemical and molecular analysis has led to the identification and characterization of VIP36, another protein associated with the CHAPS-insoluble complex.[211] The predicted amino acid sequence of VIP36 suggests that it is a type I integral membrane protein with an N-terminal 31 kDa luminal/extracellular domain, followed by a 23-residue transmembrane domain and a short C-terminal cytoplasmic tail with 11 amino acids (QKRQERNKRFY). The luminal domain of VIP36 is structurally related to leguminous plant lectins and ERGIC53, which is a mannose-specific lectin.[212] Detailed subcellular localization suggests that VIP36 is enriched in the Golgi apparatus, endosomal and vesicular structures as well as the plasma membrane, consistent with its nature as a transport vesicle protein between the TGN and the plasma membrane. The luminal domain of VIP36 may bind sugar residues of glycosphingolipids and/or the GPI anchor of GPI-anchored proteins and thus may provide a link between the luminal/extracellular face of the glycolipid rafts/clusters and the cytoplasmic protein sorting machinery for apical delivery. Consistent with this proposal, it was revealed that selective apical sorting of a GPI-anchored protein is abolished in a ConA resistant MDCK cell line, which has altered sugar structures in glycolipids and glycoproteins, although this mutant cell can still cluster GPI-anchored proteins and glycolipids. These studies indicate that some sugar structure is required for apical sorting of GPI-anchored proteins clustered in the glycolipid rafts and that VIP36 might be involved in this process. Stud-

ies on VIP21 and VIP36 provide significant support for the proposal that sorting of apically-destined proteins in epithelial cells is mediated by coclustering of apical cargo protein and glycosphingolipids into microdomains, rafts or clusters followed by their selective apical transport.[873]

As discussed previously, the C-terminal cytoplasmic tail of ERGIC53 contains a KKXX-based ER retrieval signal in its last four residues (KKFF). When ERGIC53 is overexpressed to levels that saturate the KKXX-mediated retrieval, it is transported to the plasma membrane. The surface ERGIC53 was found to be internalized rapidly from the cell surface and this internalization is dependent on the KKFF sequence. Detailed mutagenesis has defined a novel internalization signal composed of a C-terminal K-K/R-F/Y-F/Y motif.[322] The C-terminal tail (KRFY) of VIP36 fits perfectly with this consensus and has been shown to promote internalization, suggesting that this motif may play a role in recycling surface VIP36 back to the TGN.[211,322]

A protein of about 46 kDa has been identified that is 38-fold more enriched in apical transport vesicles as compared to the basolateral vesicles. Cloning this protein established that it is a new isoform of annexin XIII, which includes a previously described intestine-specific annexin XIIIa. This new isoform, termed annexin XIIIb, consists of 357 amino acids with a unique insert of 41 amino acids in the N-terminus as compared to annexin XIIIa. The annexin protein family contains at least 13 distinct members and each member is characterized by its Ca^{2+}-dependent lipid-binding activity. Immunofluorescence microscopy showed that annexin XIIIb is localized to the apical membrane and underlying punctate structures. Antibodies specific for annexin XIIIb inhibit in vitro transport of influenza HA to the apical surface but have no effect on in vitro basolateral delivery of VSV G-protein, suggesting that annexin XIIIb may play a role in vesicular transport to the apical domain of MDCK cells or other polarized cells.[210]

When transport from the TGN to the apical or the basolateral surface is examined in permeabilized MDCK cells,[592] it was found that transport from the TGN to the basolateral surface is inhibited by antibodies against NSF and stimulated by exogenous α-SNAP. In marked contrast, transport from the TGN to the apical domain is not affected by antibodies against NSF or exogenous α-SNAP, indicating that transport from the TGN to the basolateral but not the apical domain is mediated by NSF and SNAPs.[321] Furthermore, basolateral but not apical transport is inhibited by RabGDI and botulinum neurotoxins.[321] Whether apical transport involves a different set of NSF- and SNAPs-like proteins or a totally different mechanism that is NSF- and SNAPs-independent remains to be resolved.

GPI-ANCHOR MEDIATES APICAL SEGREGATION

Endogenous GPI-anchored proteins are all targeted preferentially to the apical domain of MDCK cells, suggesting that GPI-anchored proteins may have a common structure that mediates their apical transport.[417,421,497,517,713,853] When expressed in MDCK cells, exogenous GPI-anchored proteins are also selectively transported to the apical surface. These exogenous GPI-anchored proteins include Thy-1, decay-accelerating factor (DAF) and placental alkaline phosphatase (PLAP). When chimeric proteins are constructed between each of these GPI-anchored proteins and others that are not apically-transported, it was revealed that the sequence responsible for the attachment of a GPI-anchor is also sufficient for apical delivery of these chimeric proteins, suggesting that the GPI-anchor is responsible for apical transport.[97,418] When the C-terminal 37 amino acids of DAF were fused to the ectodomain of herpes simplex glycoprotein D (which is basolaterally sorted) or to human growth hormone and expressed in transfected MDCK cells, both of these chimeric proteins were transported to the apical domain. Since nine residues of this 37 amino acid sequence remain associated with the GPI-anchored chimeric proteins after signal cleavage and GPI-attachment, there is a possibility that this 9-residue sequence could be involved in apical localization. The proposal that the GPI-anchor is solely responsible for apical expression was validated by the fact that when eight of these nine residues were removed from the chimeric growth hormone, the GPI-anchored protein is still targeted to the apical surface.[419] Furthermore, when the Ser residue for GPI attachment is replaced by a Gly in this 29-residue minimal GPI attachment signal, the chimeric protein is similarly GPI-anchored and apically targeted. Similarly, the GPI attachment signal of Thy-1, when fused to the ectodomain of VSV G-protein (which is expressed basolaterally) is able to confer apical localization to the GPI-anchored chimeric protein. When the biogenesis of GPI-anchored proteins was examined, it was revealed that newly made proteins became associated with detergent-insoluble glycosphingolipid-enriched complexes upon reaching the Golgi apparatus.[99] This suggests that segregation of GPI-anchored proteins into glycosphingolipid-enriched membrane microdomains, rafts or clusters may be responsible for apical delivery of these and other proteins (such as HA) that can incorporate into these glycosphingolipid-enriched structures.[156,209,417,713]

CYTOPLASMIC SIGNALS MEDIATE BASOLATERAL TARGETING

Signals responsible for the basolateral sorting have been identified in the cytoplasmic domains of several proteins (Fig. 21). The existence of a basolateral signal was first demonstrated in the cytoplasmic domain of the polymeric IgA receptor (pIgR).[22,113,497] When expressed in MDCK cells, the majority of the newly

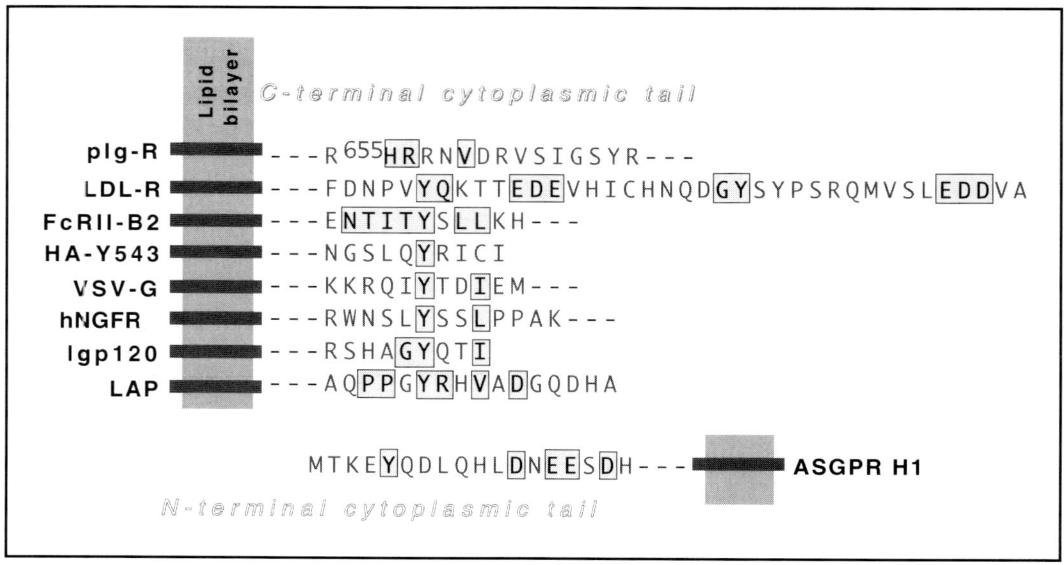

Fig. 21. A compilation of cytoplasmic sorting signals for the basolateral surface of epithelial cells.

made receptor is initially transported to the basolateral domain, from which the receptor is internalized and transcytosed to the apical domain. This transcytosis occurs both in the presence and absence of the ligand. Upon arrival at the apical domain, the receptor is cleaved and the major portion of the extracellular domain (known as the secretory component) is released into the apical medium.[498] In contrast, the majority of a mutant receptor (plgR654t) retaining only two amino acids ($R^{653}A^{654}$) proximal to the membrane of its 103-residue cytoplasmic domain is directly transported to the apical domain and cleaved, suggesting that the cytoplasmic tail is necessary for basolateral sorting and subsequent transcytosis.[499] Another mutant receptor (plgR669t) retaining C-terminal additional 15 amino acids (R^{655}HRRNVDRVSIGSYR^{669}) as compared to plgR654t is sorted to the basolateral surface. This additional 15-residue sequence may therefore contain a basolateral sorting signal. When 14 (R^{655}-Y^{668}) of the 15 residues are deleted from the wild-type receptor, the mutant receptor (plgRdel655-668) is transported to the apical domain, suggesting that this 14-residue sequence

is indeed necessary for basolateral sorting. When the polypeptide portion of PLAP, which is GPI-anchored and apically targeted, is fused to the transmembrane domain of plgR with different cytoplasmic sequences, it was found that this 14-residue sequence is sufficient for basolateral targeting. Site-directed mutagenesis of this putative basolateral sorting signal has identified that residues H656, R657 and V660 are most essential for basolateral sorting.[22] When endocytosis is measured, it is found that plgRdel655-668 has a similar endocytosis rate as the intact receptor, while plgR669t has no detectable endocytosis. These observations suggest that this 14-residue basolateral sorting signal is not involved in rapid endocytosis but cytoplasmic sequences other than these 14 amino acids are involved. Indeed, the cytoplasmic tail of IgA receptor contains two internalization signals that are distinct from the basolateral sorting signal.[546] Furthermore, the basolateral signal of plgR has been demonstrated to function also in the endocytotic pathway for basolateral targeting.[23]

Signals responsible for basolateral sorting have been identified in the cytoplasmic domains of several other proteins. The LDL receptor is sorted to the basolateral domain of MDCK cells and two signals for basolateral sorting have been identified in its cytoplasmic domain: a proximal signal (FD<u>NPVYQ</u>KTTEDEVH) and a distal one (QDGYSYPSRQMVSLEDDVA).[470] Both of these two signals can independently mediate basolateral sorting with different capacities; the distal signal being more potent. Both signals contain essential Tyr residues and the proximal signal is colinear with, but distinct from, the endocytosis signal (NPVYQ). Both signals contain an essential cluster of three acidic residues at the C-terminal end and the residue Q and G adjacent to the essential Y in the second and the first signal, respectively, are also involved in basolateral sorting.[468] The basolateral signals of LDL receptor also sort proteins basolaterally in the endocytotic pathway.[469]

Two isoforms of Fc receptors, designated FcRII-B1 and FcRII-B2, have been identified in mouse macrophage and lymphocytes. These two isoforms are identical except for an in-frame insertion of an additional 47 amino acids in the cytoplasmic domain of FcRII-B1. These additional amino acids in the cytoplasmic tail of FcRII-B1 inhibit its ability to be internalized by clathrin-mediated endocytosis. When these two isoforms are expressed in transfected MDCK cells, it is found that the majority (80-90%) of internalization-competent FcRII-B2 is selectively transported to the basolateral surface, while internalization-incompetent FcRII-B1 is targeted equally to both surfaces.[316] When mutant FcRII-B2 molecules with different deletions in the cytoplasmic tail are expressed in MDCK cells, one mutant FcRII-B2 (CT31) with the C-terminal 16 residues deleted is targeted almost exclusively (97% as compared to 80-90% of original FcRII-B2) to the basolateral surface and this mutant internalizes slightly better than wild-type FcRII-B2. Another mutant FcRII-B2 (CT18) with a truncation of the C-terminal 29 residues, which removes a potential Tyr-based endocyto-

sis signal and/or a double-Leu motif (ITYSLLK), is incapable of internalization and is predominantly (88%) transported to the apical surface, suggesting that the signal for clathrin-mediated endocytosis is required for the basolateral targeting of FcRII-B2.[315] Further dissections of the cytoplasmic tail of FcRII-B2 reveal that this double-Leu motif plays a critical role in both basolateral sorting and endocytosis.[314,468]

HA is preferentially delivered to the apical surface and is incapable of internalization via clathrin-mediated endocytosis. HA has a short cytoplasmic tail (NGSLQCRICI). When C^6 but not S^3 or C^9 is replaced by a Y residue, the resulting mutant (HA-Tyr543) is rapidly internalized via a clathrin-mediated process,[393] suggesting that replacement of C^6 by a Y has generated a Tyr-based endocytosis signal. More significantly, the HA-Tyr543 is directly sorted to the basolateral surface, suggesting that a Tyr-based endocytosis signal is sufficient for basolateral targeting.[98] The human nerve growth factor receptor (NGFR) is mainly sorted to the apical surface of MDCK cells. A mutant NGFR (NGFR$_m$) in which a cytoplasmic Tyr-based signal (SWNSLYSSLPPAK) is moved closer to the membrane by an internal 58-residue deletion is sorted to the basolateral domain and is capable of rapid internalization, further supporting the proposal that a Tyr-based endocytosis signal is sufficient for basolateral sorting.[397] A Tyr[19]-based basolateral targeting signal (KKRQIYTDIEM) has also been identified in the 29-residue cytoplasmic tail of VSV G-protein.[783,784] Ile[22], 3 residues carboxyl terminal to this Tyr is also important but can be replaced by other aliphatic residues. The major subunit of human asialoglycoprotein receptor is sorted to the basolateral domain and this process is similarly dependent on Y residue at position 5 in its N-terminal cytoplasmic tail and this Y residue is also critical for endocytosis.[227,834]

When expressed in MDCK cells, the majority of lgp120/lamp1 is directly delivered to the lysosome.[304] When the G^7 residue is replaced by an A residue, the mutant protein is still capable of internalization and is targeted to the lysosome via a surface internalization route from the basolateral domain. About 75% of newly made G^7-A mutant of lgp120 is targeted to the basolateral domain. However, mutants in which Y^8 is replaced by a C residue or with the cytoplasmic domain deleted are predominantly transported to the apical surface, suggesting that basolateral sorting of lgp120 requires a functional internalization signal. Lysosomal acid phosphatase (LAP) is transported to the lysosome via the plasma membrane. When expressed in MDCK cells, LAP is sorted to the lysosome via the basolateral domain, suggesting that it harbors a signal for basolateral targeting. A mutant with its cytoplasmic tail deleted is expressed on the apical domain, while a chimeric protein in which the extracellular and transmembrane domains of HA is fused to the cytoplasmic domain is targeted to the basolateral domain, suggesting that the cytoplasmic domain contains a basolateral sorting signal. The basolateral signal of LAP is within the first 12 residues of its cytoplasmic domain and overlaps with the endocytosis signal. Although the structures responsible for basolateral sorting and endocytosis share some common properties, they do exhibit other distinct features.[605] A basolateral sorting signal has also been implied in the N-terminal cytoplasmic domain of the transferrin receptor but this signal is independent of the endocytosis signal.[157]

It is currently impossible to derive a common structure for basolateral sorting signals. Known signals can be divided into two classes. The first type of signal resembles those involved in rapid endocytosis via clathrin-coated vesicles and is either Tyr-based and/or double-Leu-based. This type of signal may be colinear or overlapping with the endocytosis signal but may involve distinct structural features for basolateral sorting.[605] The other type of signal is clearly different from the endocy-

tosis signals. Examples include those of the LDL receptor which contain an essential Tyr followed by a cluster of acidic residues. The molecular mechanism responsible for basolateral sorting mediated by these signals is unknown. Since they are all located in the cytoplasmic domain and resemble, to different extents, signals that are recognized by TGN AP1 adaptin complex or surface AP2 adaptin complex, it is possible that a different but related adaptin complex may be involved in recognizing these basolateral signals.

SURFACE POLARITY OF NEURONS

In addition to the cell body, a neuronal cell has two different types of neurites: a single axon that is typically involved in transmitting signals and several dendrites that are generally involved in receiving signals transmitted from axons of other neurons. Rat hippocampal neurons in culture undergo a well-defined sequence of events from being nonpolarized to a polarized stage with distinct axons and dendrites.[185] When polarized hippocampal neurons are infected with the influenza Fowl Plague virus (FPV), the envelope glycoprotein HA, which is apically sorted in epithelial cells, is preferentially delivered to the axon. In contrast, the envelope glycoproteins of Semiliki Forest virus (SFV) and VSV are selectively transported to the dendrites and cell body.[185] Furthermore, like in epithelial cells in which GPI-anchored proteins are apically sorted, the GPI-anchored protein Thy-1 is exclusively localized to the axonal surface of hippocampal neurons in culture.[186] These results suggest that the axon may be equivalent to the apical domain, whereas the dendrite-cell body may be equivalent to the basolateral domain of epithelial cells. The apical and basolateral domains in epithelial cells are separated by tight junctions and the presence of similar structures that segregate the axonal membrane from that of the cell body and dendrites has been investigated. These studies show that there is a functional barrier in the axonal hillock/initial segment and that this barrier prevents the intermixing of membrane constituents, and may therefore play a key role in maintaining the unique protein and lipid compositions of the axon and dendrite-cell body.[371] Whether this barrier is biochemically similar to tight junctions or is formed by a different mechanism remains to be investigated. Expression and analysis of wild-type plgR and its mutants in hippocampal neurons in culture have demonstrated the existence of a transcytotic pathway from the dendrite-cell body to the axon and this transcytotic pathway shares similarities with the basolateral to apical transcytotic pathway of epithelial cells.[166]

REGULATED EXOCYTOSIS

REGULATED EXOCYTOSIS MEDIATED BY SECRETORY GRANULES

BIOGENESIS OF SECRETORY GRANULES

Exocrine cells, endocrine cells and peptidergic neurons possess a regulated pathway for timely and controlled release of zymogens, peptide hormones and neuropeptides, in addition to the constitutive secretory pathway.[24,101,102,354,555,729,788,858] Proteins destined for the regulated pathway traverse the early secretory pathway from the ER to the TGN, together with other proteins of the constitutive secretory pathway. Upon arrival in the TGN, proteins destined for the regulated pathway are sorted from other proteins and selectively incorporated into dense-core aggregates (or induced to form aggregates) and incorporated into vesicles known as dense-core immature secretory granules, which then undergo a maturation process to become dense-core mature secretory granules (Fig. 22). During the process of aggregation, budding and maturation, proteins in the mature secretory granules become concentrated many fold (up to several hundred fold) over their concentration in the lumen of the Golgi apparatus. Although variable, the secretory granules are generally about 100 nm in diameter and have a characteristic dense core under electron microscopy due to their high luminal protein concentrations. The mature secretory granules are stored adjacent to the plasma membrane. Upon proper stimuli, they fuse with the plasma membrane (or its restricted regions), resulting in the release of the luminal material to the cell exterior. Other cell types that have a secretory granule pathway include mast cells, platelets, neutrophiles and large granular lymphocytes.

Like other proteins destined for the secretory pathway, all proteins destined for the regulated exocytotic pathway are targeted and translocated into the ER lumen and this process is mediated by a N-terminal signal peptide, which is cleaved by the ER signal peptidase complex and is referred to as the presequence. In addition, the majority of proteins destined for the regulated secretory pathway in endocrine cells and peptidergic neurons travel the early secretory pathway as inactive precursors (proproteins) that contain a prosequence. The prosequence is then proteolytically removed during the process of sorting, budding and/or maturation, giving rise to the biologically active mature protein in the mature secretory granule.[729,858] The proteolytic cleavage occurs usually at sites characterized by pairs of basic amino acids. The Kex2p-like endopeptidases PC2 and PC3 have been proposed to be responsible for this proteolytic processing. The processing events for somatostatin are shown in Figure 23.[402] In contrast, exocrine cells secrete their enzymes via the regulated pathway as inactive precursors

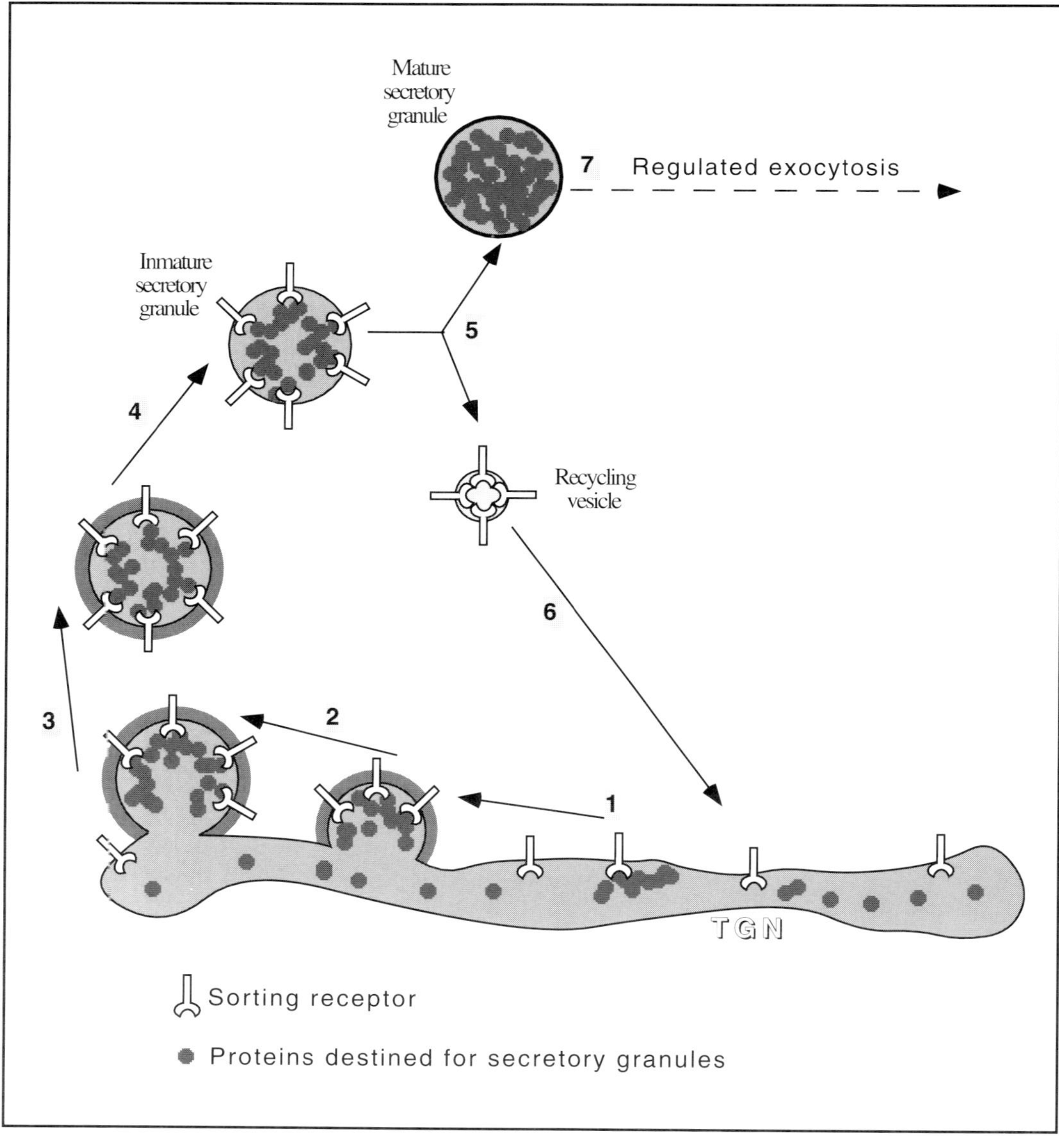

Fig. 22. A model for the biogenesis of dense-core secretory granules. Upon arrival at the TGN, proteins destined for secretory granules are induced to form aggregates, or incorporated into existing aggregates, which are then recognized by a sorting receptor (step 1), leading to the formation of coated buds and budding vesicles (step 3). The coated vesicles pinch off to form the coated (step 3) and then uncoated (step 4) immature secretory granules, which undergo a maturation step (5) to form the mature secretory granules and some recycling vesicles that transport the sorting receptor back to the TGN (6). The mature secretory granules undergo controlled exocytosis upon appropriate stimulation (step 7).

which are subsequently activated by proteolytic processing outside the cell. The sorting process for proteins destined for the secretory granule occurs in the TGN and may involve two events. The first sorting event involves the TGN milieu (such as its low pH and high Ca^{2+})-induced selective aggregation of precursors of peptide hormones and neuropeptides together with granins, the general components of secretory granules.[112,116,117,730] This process, which gives rise to the dense cores observed in the TGN by electron microscopy, may reflect the intrinsic property of proteins destined for the secretory granule. The second sorting event is the selective packaging of the aggregated proteins into coated buds and vesicles that eventually pinch off from the TGN membrane to give rise to immature secretory granules. The second sorting event may involve a specific membrane protein in the TGN that functions as a sorting receptor for recognizing the aggregated proteins and for initiating the budding process. After budding and uncoating, immature secretory granules undergo a maturation process in which further sorting events can occur, including further proteolytic processing of precursors by PC2, PC3 and/or

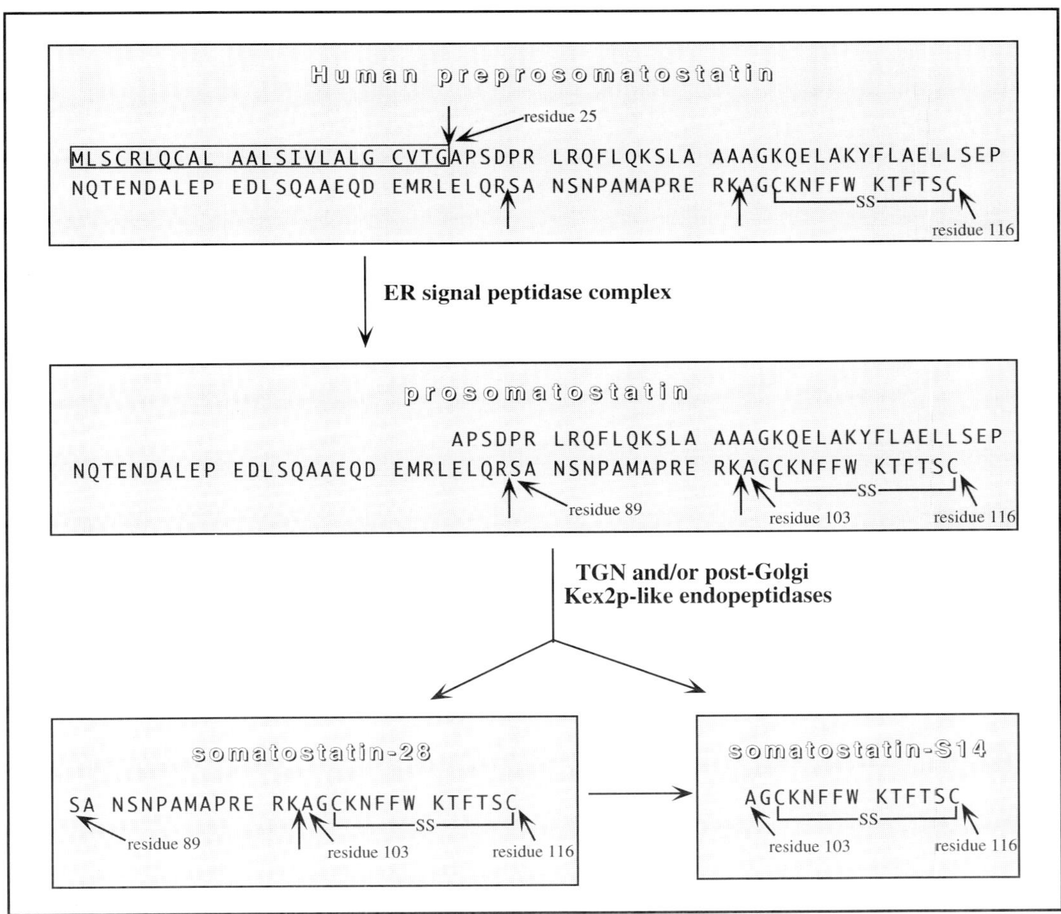

Fig. 23. The processing pathway of somatostatin. Newly made human preprosomatostatin is composed of 116 residues; the N-terminal presequence/signal peptide (residues 1-24) is removed during the translocation process, leading to the generation of prosomatostatin, which further undergo a series of proteolytic cleavages in the TGN and in the immature secretory granules to give rise to somatostatins that will be stored in the mature secretory granules.

similar endoproteases, segregation of mature peptide hormones and neuropeptides from prosequences and the sorting receptor and condensation of the contents. The maturation process leads to the formation of mature secretory granules that will fuse with the plasma membrane upon proper stimulation and recycling vesicles that may be involved in cycling sorting receptor and other potential components (endoproteases?) back to the TGN (Fig. 22). Although immature secretory granules can also undergo regulated exocytosis, the mature secretory granule is believed to be the one that mediates physiological regulated exocytosis.

GRANINS AS MAJOR COMPONENTS OF SECRETORY GRANULES

Granins, also known as chromogranins/secretogranins, are a family of acidic, heat-stable proteins of secretory granules in a wide variety of endocrine cells and peptidergic neurons.[318,319,788] Chromogranin A, chromogranin B (secretogranin I) and secretogranin II (chromogranin C) are three classic granins that have been well characterized. In addition, there are three other proteins that belong to the granin protein family and include secretogranin III (1B1075), secretogranin IV (HISL-19) and secretogranin V(7B2). These proteins have apparent sizes of 75 kDa for human chromogranin A (439 aa after removal of the signal peptide), 120 kDa for human chromogranin B (657 aa after the removal of the signal peptide), 85 kDa for human secretogranin II (587 aa after removal of the signal peptide), 57 kDa for rat secretogranin III (533 aa inclusive of the signal peptide), 35 kDa for secretogranin IV and 24 kDa for human secretogranin V(185 aa after the signal peptide removal). In addition to the acidic nature (about 21%, 18% and 13% negatively charged Glu and Asp residues in the primary sequences of chromogranin A, chromogranin B and secretogranin II, respectively), all granins are also characterized by the existence of multiple sites with two or more adjacent basic residues that are potential sites for

proteolytic processing by PC2 and/or PC3. It is known that chromogranin A can give rise to pancreastatin, chromostatin and betagranin, chromogranin B to GAWK and secretogranin II to secretoneurin, respectively, after proteolytic processing. Furthermore, chromogranin A, chromogranin B and secretogranin II have been shown to bind Ca^{2+} and to aggregate in the presence of this divalent cation. Although these granins can be precursors for biological peptides such as pancreastatin and secretoneurin, the major physiological function of these proteins has been proposed to act as helper proteins in the segregation, aggregation, packaging of peptide hormones and neuropeptides into the secretory granules as well as to modulate the proteolytic processing of these peptide hormones and neuropeptides.[318,319,642]

SORTING SIGNALS FOR SECRETORY GRANULES

Studies have been performed to understand the molecular features of the putative sorting signals harbored by proteins destined for the secretory granules and it seems that the signal for secretory granules may not be mediated by linear amino acid sequences but rather by three-dimensional signal patches. Several cell lines that possess the secretory granule pathway have been useful for the studies of secretory granule biogenesis (and the potential sorting signal for this pathway). These include ACTH-secreting corticotrophic AtT-20 cells derived from mouse anterior pituitary tumor, growth hormone- and prolactin-producing GH3 cells derived from rat pituitary, mouse neuroblastoma Neuro2a cells and PC12 cells derived from rat adrenal medulla tumor. In addition to the secretory granule pathway, PC12 cells also possess the synaptic vesicle-mediated exocytosis pathway.[149,175] When the mature luminal domain of VSV G-protein is fused to the C-terminus of human growth hormone, the chimeric protein (hGH-TG) is correctly sorted to the secretory granules in AtT-20 cells.[489] When a chimeric protein composed of the signal peptide (25 residues) and prosequence (82 residues) of anglerfish somatostatin I fused to the α-globin is expressed in GH3 cells, a significant amount of the chimeric protein is delivered to the secretory granules, suggesting that the 82-residue prosequence of anglerfish somatostatin I contains sufficient information for sorting to the secretory granule.[748] The prosequence of rat preprosomatostain has been similarly shown to possess the sorting information for the secretory granules.[692] The best studied sorting signal for secretory granules may be the one found in pro-opiomelanocortin (POMC) which are processed at paired basic residues by PC2 and/or PC3 endopeptidase in the TGN and the immature secretory granule to give rise to many biologically important peptides, including adrenocorticotropin (ACTH), α-melanocyte stimulating hormone (α-MSH), β-lipotropin (β-LPH), β-endorphin, γ-MSH and others.[141,767] The prohormone is mainly synthesized in the anterior pituitary corticotroph cells. During translocation across the ER, the N-terminal 26-residue presequence (the signal peptide) is removed to give rise to the pro-hormone.[141,767] Upon reaching the TGN, the prohormone is sorted to the secretory granule along with proteolytic processing during its sorting in the TGN and its maturation in the immature secretory granule. By fusing various N-terminal regions of the mouse pro-opiomelanocortin to chloramphenicol acetyltransferase (CAT) and expressing the chimeric proteins in AtT-20 or Neuro-2a cells, the N-terminal 26-residue sequence of mouse pro-opiomelanocortin was shown to contain sufficient sorting information for the secretory granule. This 26-residue sequence (WC^{2}LESSQC^{8}QD LTTESNLLAC^{20} IRAC^{24}KP) contains four Cys residues that form two disulfide bonds (C^2-C^4 and C^8-C^{20}). Importantly, these two disulfide bonds (particularly the C^8-C^{20} one) cause the formation of an amphipathic hairpin loop structure between C^8 and C^{20}. Disruption of both disulfide bridges or the C^8-C^{20} one or the removal of the amphipathic hairpin loop results

in the failure of these mutant POMC to be sorted to the secretory granule, and they are instead secreted by the constitutive pathway. It is proposed that four highly conserved hydrophobic and acidic residues (D^{10}-L^{11}, E^{14} and L^{18}) in the amphipathic loop may play a key role in sorting to the secretory granule. The POMC sorting signal for the secretory granule thus seems to be encoded by conformational information in the 13-residue amphipathic loop stabilized by the C^8-C^{20} disulfide bond.[141] Sorting motifs similar to the amphipathic loop of POMC have been proposed for other proproteins destined for the secretory granule.[141]

Fusion of Secretory Granules in Endocrine Cells May Be Mediated by Proteins that Are Used for Docking/Fusion of Synaptic Vesicles

The fusion process of mature secretory granules with the plasma membrane has been studied both in intact cells as well as in permeabilized cells and it seems that proteins involved in the docking/fusion of synaptic vesicles with the presynaptic plasma membrane (see later section) may similarly participate in the fusion of secretory granules with the plasma membrane. Originally identified to be neuron-specific, SNAP-25 has recently been shown to be present in the plasma membrane of endocrine but not the exocrine cells of the pancreas.[654] When the β cell line (HIT) or primary rat islet cells are permeabilized and treated with botulinum neurotoxins (BoNT) A and E which cleave SNAP-25 at different sites, Ca^{2+}-stimulated insulin release is inhibited, suggesting that SNAP-25 is involved in the fusion of secretory granules with the plasma membrane in endocrine cells such as the pancreatic β-cells. Another synaptic docking/fusion protein, synaptobrevin-2 (VAMP-2) is also expressed in the pancreatic β-cells and similarly implicated in the Ca^{2+}-induced insulin secretion.[623] In addition, syntaxin 1A and 1B as well as the syntaxin-binding protein, n-Sec1/rb-sec1/Munc-18, have been shown to be expressed on the plasma membrane of endocrine chromaffin cells of adrenal medulla[301] and synaptobrevin is a component of chromaffin granules, highlighting the involvement of these proteins in the fusion of chromaffin granules with the plasma membrane. The general soluble fusion proteins NSF and SNAPs have also been revealed to participate in the fusion of secretory granules with the cell surface in adrenal chromaffin cells.[115,491,493] Interestingly, the trimeric G-protein Gi has been shown to be associated with the insulin secretory granules of pancreatic β-cells and to participate in insulin release.[374] Other proteins, such as Rab3, a rabphilin-3A isoform, annexin II, 14-3-3 proteins and calmodulin have been also implicated in the fusion of secretory granules with the plasma membrane in endocrine cells.[115,130]

REGULATED EXOCYTOSIS MEDIATED BY SYNAPTIC VESICLES

SYNAPTIC VESICLES AND PRESYNAPTIC PLASMA MEMBRANE

In addition to regulated release of neuropeptides from the dense-core secretory granules in peptidergic neurons, all neurons possess another regulated exocytosis pathway mediated by synaptic vesicles.[26,45,57,60,101,175,354,355,666,682,753,755] Communication among different neurons and between the peripheral neurons and the target cells (such as the muscle cells) is mediated by regulated release of neurotransmitters from the nerve terminal. The presynaptic nerve terminal is filled with small translucent synaptic vesicles that are about 40-50 nm in diameter and homogeneous in morphology. It is estimated that synaptic vesicle-associated proteins account for about 7% of total brain proteins. In contrast to secretory granules, which arise from the biosynthetic pathway via the TGN and contain peptide-based materials, synaptic vesicles are primarily derived from the endocytotic pathway

and contain small chemical neurotransmitters such as acetylcholine, glutamate, γ-aminobutyric acid (GABA), glycine and the biogenic amines. No soluble proteins are present in the lumen of synaptic vesicles. At the resting state, some vesicles are docked at the active zone, a specialized region of the presynaptic membrane facing the synaptic cleft. Upon proper stimulation, the docked synaptic vesicles fuse with the presynaptic plasma membrane, resulting in the release of their luminal neurotransmitters. The synaptic vesicle proteins are then retrieved from the presynaptic plasma membrane via endocytosis and the endocytosed vesicles are then refilled with neurotransmitters by its associated membrane channels and carriers. Because of their abundance and homogeneity, synaptic vesicles can be relatively easily purified for biochemical and molecular studies. The major proteins associated with the synaptic vesicle have been molecularly and functionally characterized. Similarly, major proteins associated with the presynaptic plasma membrane that are involved in synaptic vesicle docking/fuion have also been characterized in significant detail.

CLOSTRIDIAL NEUROTOXINS

An important aspect of synaptic studies has been to unravel the mechanism of action of the tetanus and botulinum neurotoxins that cause neuroparalytic syndromes of tetanus and botulism respectively.[280,527,667] Bacteria of the genus *Clostridium* produce one tetanus neurotoxin (TeNT) and seven different serotypes of botulinum neurotoxins (BoNT/A, B, C, D, E, F and G). These clostridial neurotoxins are made as an inactive 150 kDa precursor that is activated by cleavage via bacterial or host proteases into two disulfide-linked polypeptide chains of about 100 and 50 kDa, respectively. Once inside the body, both TeNT and BoNTs act on motorneurons at the neuromuscular junctions by binding through the 100 kDa heavy chain to the presynaptic plasma membrane. The neurotoxins are then endocytosed into vesicles. BoNT-containing vesicles remain in the nerve termini of the motor neurons, while TeNT-containing vesicles are transported retrogradely along the axon of the motor neuron and discharged via transcytosis into the intersynaptic space between the motor neuron and the inhibitory interneurons of the spinal cord, from where TeNT is internalized into the nerve terminal of the interneuron. The 50 kDa light chain of these neurotoxins is then translocated into the cytosol of the respective nerve termini, where it displays a zinc-dependent endopeptidase activity for proteins associated with synaptic vesicles (TeNT, BoNT/B, BoNT/D, BoNT/F and BoNT/ G for synaptobrevin) and/or with presynaptic plasma membrane (BoNT/C for syntaxins and BoNT/A and BoNT/E for SNAP-25)[64,65,66,668] (Fig. 24). In this way, BoNTs act at the motor neuron to inhibit the exocytotic release of acetylcholine contained in the synaptic vesicles and to cause the clinical manifestation of botulism: flaccid paralysis of all peripheral muscles. TeNT acts at the inhibitory interneuron to block the release of inhibitory neurotransmitters and leads to a failure to counterbalance excitatory stimuli, resulting in the spastic paralysis of all striated muscles that is characteristic of tetanus. The molecular and neurological effects of these clostridial neurotoxins have thus established the functional importance of synaptobrevins, syntaxins and SNAP-25 in synaptic vesicles docking/fusion and neurotransmitter release.

PROTEINS ASSOCIATED WITH SYNAPTIC VESICLES OR THE PRESYNAPTIC PLASMA MEMBRANE

Synaptobrevins

Synaptobrevins, also termed VAMPs (vesicle-associated membrane proteins), have been molecularly characterized in many species and have apparent sizes of about 18 kDa.[52,202,754,861] There are two highly related forms, referred to as VAMP-1 and VAMP-2. VAMP-1 was originally cloned from the marine ray Torpedo california and is composed of 120 amino acids.[794] Rat VAMP-1 and VAMP-2 were cloned by screening cDNA libraries with the Torpedo VAMP-1 cDNA.[202] While VAMP-2 is expressed at a relatively higher level than VAMP-1 in whole brain, VAMP-1 is more abundant in the spinal cord. Rat VAMP-1 and VAMP-2 are of 118 and 116 amino acids (Fig. 24), respectively, and are 77% identical to each other, and 84% and 75% identical to Torpedo VAMP-1, respectively. The N-terminal 28 residues of rat VAMP-1 and 26 residues of VAMP-2 have been referred to as the Pro-rich head domain. Following this Pro-rich head, the central 68 amino acids define the hydrophilic core that can potentially form a coiled-coil domain, while following C-terminal 22 hydrophobic residues form the C-terminal membrane anchor domain. The sequences of the central coiled coil domains of several members of VAMP family are aligned in Figure 25. VAMPs are initially targeted to the ER and transported to the synaptic vesicle via the biosynthetic and endocytotic pathway.[385] Residues 41-50 of rat VAMP-2 have been shown to play a critical role in its targeting to the synaptic vesicles.[255] VAMPs are substrates for TeNT, BoNT B, D, F and G, which cleave at different positions along the VAMP polypeptide (Fig. 24). VAMPs can interact specifically with syntaxin 1 in solution.[107] However, the affinity of VAMPs for syntaxin 1 is increased if it is complexed with SNAP-25, suggesting that VAMP on synaptic vesicles interacts with the syntaxin 1-SNAP-25 complex at the presynaptic plasma membrane.[586] The synaptic vesicle protein synaptophysin can also associate with VAMPs and this association inhibits the interaction of VAMP with the syntaxin 1-SNAP-25 complex, suggesting that synaptophysin has a regulatory role in synaptic vesicle docking/ fusion.[108,197] Using the yeast two-hybrid

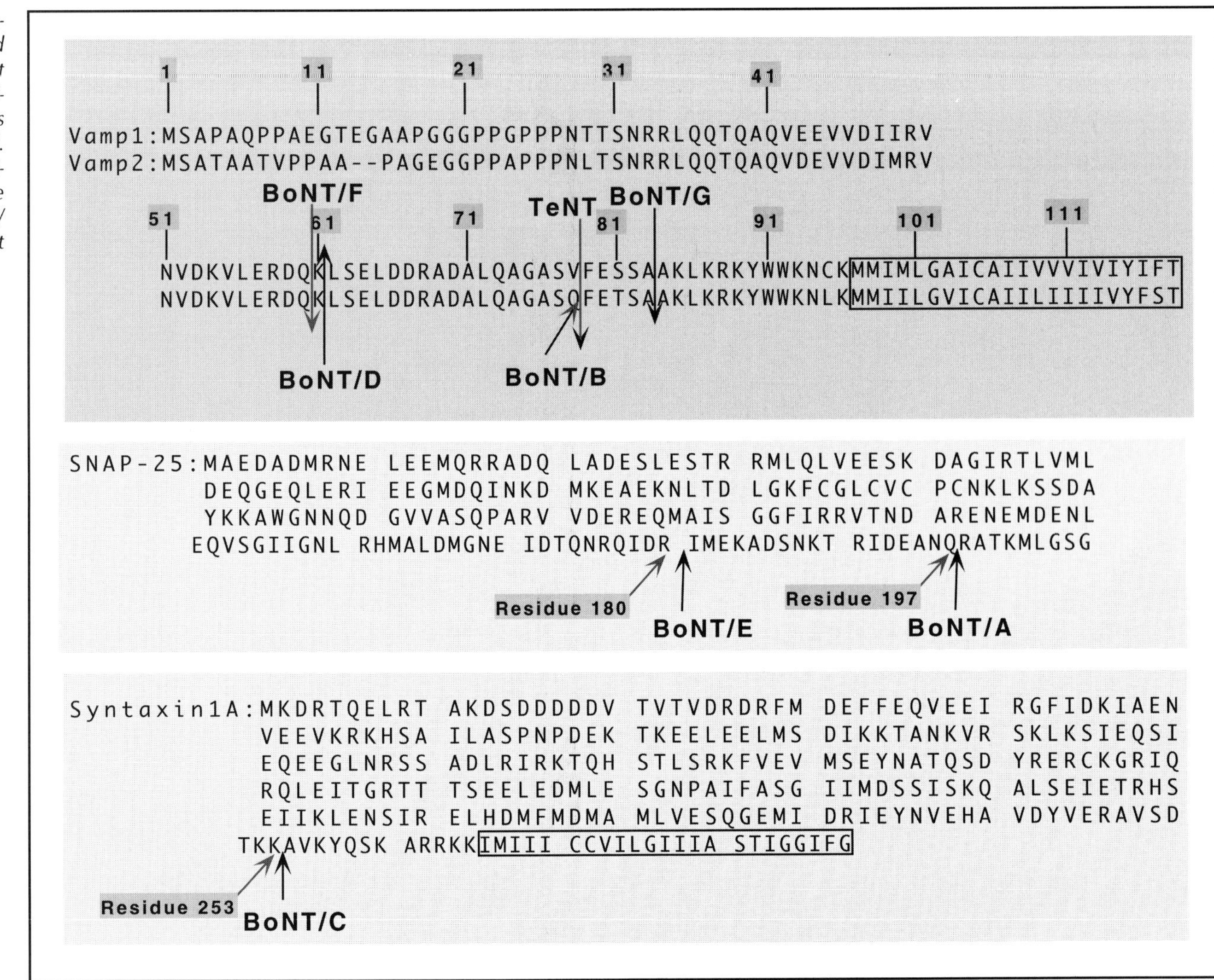

Fig. 24. The primary sequences of rat VAMP1 and 2, mouse SNAP-25 and rat syntaxin 1A. The C-terminal tail anchors of VAMPs and syntaxin 1A are boxed. The cleavage site for various clostridial toxins are indicated. Note that BoNT/B cleave only VAMP2 but not VAMP1.

system for protein-interaction screening, another C-terminal-tail-anchored protein of 33 kDa (termed VAP-33 for VAMP-associated protein of 33 kDa) which interacts with VAMPs has recently been identified in Aplysia.[718] VAP-33 is composed of 260 amino acids with the C-terminal 23 amino acids forming a hydrophobic membrane anchor so that the major portion of the polypeptide (residue 1-237) is oriented on the cytoplasmic side of the membrane. The deduced 99 amino acid sequence of a partial nucleotide sequence of a human cDNA clone (Gene-Bank accession number of F06472) is 65% identical to VAP-33, suggesting that VAP-33 is a conserved protein. As presynaptic injection of antibodies specific for VAP-33 inhibits synaptic transmission, VAP-33 may be involved in the exocytosis of neurotransmitters. VAP-33 has been proposed to function as a t-SNARE due to its association with plasma membrane of neurites. VAMPs have been proposed to be important for synaptic vesicle docking and fusion. Recent studies using a synaptobrevin-deficient mutant Drosophila strain have shown that although synaptobrevin is essential for synaptic transmission, it is dispensable for synaptic vesicle docking. It has thus been suggested that synaptobrevin may be required somewhere between receipt of the evoked fusion signal and the actual fusion event of the docked vesicles.[95]

Synaptophysin

Synaptophysin is an integral membrane protein of about 38 kDa. The human protein is composed of 313 amino acids and has four transmembrane domains with its N-terminal and C-terminal tails oriented on the cytoplasmic side of the synaptic vesicle. Synaptoporin is a synaptophysin isoform and the rat protein is about 30 kDa composed of 265 amino acids. Synaptophysin and synaptoporin can interact with VAMPs[108,197] and prevent interaction between VAMPs and the syntaxin 1-SNAP-25 complex, suggesting that they

play a regulatory role in synaptic vesicle docking/fusion.

Rab3A and Rabphilin-3A

Rab3A is composed of 220 amino acids and anchored to the synaptic vesicle membrane via geranylgeranyl lipid anchors attached to C-terminal Cys residues. Rab3A undergoes a membrane association-dissociation cycle that is associated with neurotransmitter release.[424,541] Synaptic vesicle bound Rab3A is in the GTP form whereas cytosolic Rab3A contains only GDP, suggesting that nucleotide exchange of GDP for GTP may be associated with synaptic vesicle recruitment of Rab3A and that hydrolysis of bound GTP by Rab3A may play an important role in its dissociation from the membrane.[467,736] Presynaptic injection of the nonhydrolyzable GTP analog GTPγS, or GDP analog GDPβS, inhibits neurotransmitter release, suggesting that guanine nucleotide exchange and/or GTP hydrolysis play a critical role in synaptic vesicle docking/fusion.[290] Similar to other Rabs, cytosolic Rab3A is in a complex with RabGDI and this property has actually led to the identification of RabGDI. Rab3A:GTP interacts with Rabphilin-3A and this interaction is involved in the recruitment of Rabphilin-3A onto the synaptic vesicle membrane and may be important for Rab3A to effect its role in neurotransmitter release. The detailed properties of Rabphilin-3A have been described in chapter 3. Its N-terminal domain binds Rab3A, while its C-terminal domain is structurally related to synaptotagmin and contains two copies of Ca^{2+}/phospholipid-binding C2 domains originally described in protein kinase C.[702] Since a transient increase of Ca^{2+} plays an essential role in neurotransmitter release, Rab3A:GTP-Rabphilin-3A complex may be involved in coupling the Ca^{2+} increase to synaptic vesicle fusion. Although the exact role played by Rab3A and rabphilin-3A is unknown, it is believed that they may regulate the formation of the 7S docking complex and/or the 20S fusion complex between proteins of the synaptic vesicle with those of the presynaptic plasma membrane.[307,736]

Synaptotagmins

Synaptotagmin I was originally identified as 65 kDa abundant integral membrane protein (p65) of synaptic vesicles.[577 578,579] The primary sequence was first established for rat synaptotagmin I and subsequently for human and other species. The rat and human proteins of synaptotagmin I are 97% identical and are composed of 421 and 422 amino acids respectively. Synaptotagmin I is a type I integral membrane protein with a intravesicular/extracellular domain composed of the N-terminal 53 residues (residues 1-53 for human synaptotagmin I) or 52 residues (residues 1-52 for rat synaptotagmin I), followed by a transmembrane domain of 27 residues (residues 54-80 for human and residues 53-79 for rat synaptotagmin I, respectively) that functions as a type I signal/anchor sequence. The C-terminal 342 amino acids (residues 81-422 for human and residues 80-421 for rat synaptotagmin I) form the large C-terminal cytoplasmic domain, within which there are two copies (residues 136-249 and residues 267-382 for human synaptotagmin I; residues 135-248 and residues 266-381 for rat synaptotagmin I) of about 115 amino acid repeats that share 41% identity with each other. Furthermore, these two repeats are related to the regulatory C2 domain of protein kinase C (32% to 40% identical depending on the isoforms of protein kinase C) and capable of Ca^{2+} and phospholipid binding.[578] Synaptotagmin I interacts with syntaxin 1 and this formed the basis for the original identification of syntaxin 1A and 1B. In addition, synaptotagmin I interacts with the complex of VAMP, syntaxin 1 and SNAP-25 and effectively prevents the binding of α-SNAP and NSF to this protein complex. Synaptotagmin I also interacts with the cytoplasmic domain of neurex-

ins[274,585,801] which are a family of highly polymorphic neuronal cell surface proteins, including the α-latrotoxin receptor. α-latrotoxin (from black widow spider venom) is a potent neurotoxin that causes massive transmitter release. It was recently demonstrated that the second C2 domain in the cytoplasmic region of synaptotagmin I is a high affinity binding site for AP2, suggesting that synaptotagmin I may also be involved in biogenesis and regeneration of synaptic vesicles.[870] When synaptotagmin I gene is disrupted in mice,[516] heterozygotes are phenotypically normal, but homozygotes die within 48 hours after birth. When hippocampal neurons from homozygous mutant mice are studied, it is seen that Ca^{2+}-dependent neurotransmitter release is severely impaired. The fast synchronous component of Ca^{2+}-triggered release is inhibited, while the slow asynchronous component is unimpaired. In these synaptotagmin I-negative neurons, neurotransmitter release can still be triggered normally by agents that act by Ca^{2+}-independent mechanisms, such as hypertonic sucrose or α-latrotoxin. Other genetic evidence also suggest that synaptotagmin I plays a fundamental role in Ca^{2+}-triggered synchronous neurotransmitter release and that synaptotagmin I may function as a Ca^{2+} sensor and/or Ca^{2+}-sensitive fusion clamp for docked synaptic vesicles. In addition to synaptotagmin I,[228,423] there are at least seven other synaptotagmins referred to as synaptotagmin II, III, IV, V, VI, VII and VIII, respectively.[410] Synaptotagmin I, II, III and V are brain-specific whereas synaptotagmin IV, VI, VII and VIII are ubiquitously expressed in most tissues. Some synaptotagmins appear to be independent of Ca^{2+} (IV, VI and VIII), others exhibit a similar high Ca^{2+} affinity for syntaxin and for phospholipid binding (III and VII) and a third class binds syntaxin only at high Ca^{2+} concentrations but phospholipids at low Ca^{2+} concentrations (I and II). The exact role played by these other synaptotagmins remains to be established.

Dynamin

Dynamin has been described in detail in chapter 3 and is a high molecular weight GTPase involved in the fission process of clathrin-coated synaptic vesicles in presynaptic nerve terminals or endocytotic vesicles in other cells.[636,765,802] It is believed that dynamin plays a fundamental role in biogenesis and regeneration of synaptic vesicles.

Synapsins

Synapsins are peripheral proteins associated with the cytoplasmic face of synaptic vesicles. Synapsin Ia (74 kDa), Ib (70 kDa), IIa (64 kDa) and IIb (53 kDa) are alternatively spliced products of two genes (synapsin I and II) and together account for about 9% of the synaptic vesicle proteins. Synapsins are the major phosphoproteins in nerve terminals. The N-terminal regions of synapsins are phosphorylated by Ca^{2+}, calmodulin-dependent protein kinase I (CaMKI) and cAMP-dependent protein kinase. The C-terminal portions of synapsin Ia and Ib but not synapsin II are Pro-rich and phosphorylated by Ca^{2+}, calmodulin-dependent protein kinase II (CaMKII). Synapsins can interact with actin filaments and neurofilaments and have been proposed to guide the migration of synaptic vesicles to the active zones in nerve terminals. Dephosphorylated synapsin I can anchor synaptic vesicles to actin cytoskeleton.[114] In squid neurons, injection of synapsin I that is unphosphorylated at the C-terminal portion results in inhibition of neurotransmitter release. Using mutant mice whose synapsin I and/or synapsin II genes are disrupted, it was revealed that synapsins play an essential function in recruitment of synaptic vesicles to the active zone and/or maturation of synaptic vesicles in the active zone.[643]

Syntaxins

Syntaxin 1A and 1B are about 84% identical to each other and were originally identified by analyzing proteins that were coimmunoprecipitated with synaptotagmin I.[58] Associated with the presynaptic plasma membrane (particularly the active zone), syntaxins are integral membrane proteins of about 35 kDa (288 amino acids for rat syntaxin 1A, Fig. 24) anchored to the cytoplasmic side of the membrane by a C-terminal 23- (syntaxin 1A) or 24-residue (syntaxin 1B) hydrophobic tail. Preceding the C-terminal tail is a 71-residue region that can potentially form coiled-coil structures for protein-protein interaction and these regions for several members of syntaxin family have been aligned in Figure 26. Syntaxin 1A and 1B are substrates of BoNT/C (Fig. 24)[66,668] and can interact with SNAP-25 that is associated with the presynaptic plasma membrane. As mentioned, the syntaxin 1-SNAP-25 complex appears to be recognized by VAMPs on the synaptic vesicle. Syntaxin 1 also interacts with n-Sec1/rb-Sec1/Munc-18 and the brain specific Sec1p-like protein.[223,586] Association of n-Sec1 with syntaxin 1 can inhibit interaction of syntaxin 1 with both SNAP-25 and VAMP, suggesting that n-Sec1 may regulate the functional state of syntaxin 1. A protein complex composed of syntaxin 1, SNAP-25 and VAMP can interact either with synaptotagmin the synaptic vesicle to form a tetra-protein complex referred to as the 7S docking complex, or with α-SNAP and subsequently NSF to form a 20S SNARE complex or fusion complex composed of at least five distinct proteins. Mutational studies of syntaxin 1A have defined the coiled-coil domain (residue 191-266) preceding the C-terminal membrane anchor to be responsible for binding with SNAP-25 (residues 191-221), α-SNAP (residues 191-240) and VAMP (residues 191-266).[350,473] n-Sec1 binding requires an extended structure from residues 4 to 240.[350] These observations suggest that unique, yet overlapping domains of syntaxin are required for distinct protein-protein interactions and conformational changes of syntaxin, resulting from distinct protein-protein interactions, may mediate synaptic vesicle docking/fusion.[265] The interaction of

```
VAMP1:TSNRRLQQTQAQVEEVVDIIRVNVDKVLERDQKLSELDDRADALQAGASVFESSAAKLKRKYWWKNCKMM
VAMP2:TSNRRLQQTQAQVDEVVDIMRVNVDKVLERDQKLSELDDRADALQAGASQFETSAAKLKRKYWWKNLKMM
Celbr:GSNRRLQQTQNQVDEVVDIMRVNVDKVLERDQKLSELDDRADALQAGASQFETSAAKLKRKYWWKNCKMW
Snc1p:QSKSRTAELQAEIDDTVGIMRDNINKVAERGERLTSIEDKADNLAVSAQGFKRGANRVRKAMWYKDLKMK
Snc2p:QSQNKTAALRQEIDDTVGIMRDNINKVAERGERLTSIEDKADNLAISAQGFKRGANRVRKQMWWKDLKMR
Conse: S              I R N   KV ER   L      D AD L    A   F    A         W  K  KM
```

```
 1A:IETRHSEIIKLENSIRELHDMFMDMAMLVESQGEMIDRIEYNVEHAVDYVERAVSDTKKAVKYQSKARRKKIMIII
 1B:IETRHNEIIKLETSIRELHDMFVDMAMLVESQGEMIDRIEYNVEHSVDYVERAVSDTKKAVKYQSKARRKKIMIII
  2:IESRHKDIMKLETSIRELHEMFMDMAMFVETQGEMVNNIERNVVNSVDYVEHAKEETKKAIKYQSKARRKKWIIAA
  3:IEGRHKDIVRLESSIKELHDMFMDIAMLVENQGEMLDNIELNVMHTVDHVEKARDETKRAMKYQGQARKKLIIIIV
  4:ISARHSEIQQLERTIRELHEIFTFLATEVEMQGEMINRIEKNILSSADYVERGQEHVKIALENQKKARKKKVMIAI
  5:IQSRADTMQNIESTIVELGSIFQQLAHMVKEQEETIQRIDENVLGAQLDVEAAHSEILKYFQSVTSNRWLMVKIFL
SSD1:VQARHQELLKLEKSMAELTQLFNDMEELVIEQQENVDVIDKNVEDAQLDVEQGVGHTDKAVKSARKARKNKIRCWL
SSD2:VQARHQELLKLEKTMAELTQLFNDMKELVIEQQENVDVIDKNVEDAQQDVEQGVGHTNKAVKSARKARKNKIRCLI
SED5:LQERNRAVETIESTIQEVGNLFQQLASMVQEQGEVIQRIDANVDDIDLNISGAQRELLKYFDRIKSNRWLAAKVFF
PEP12:IEQRDQEISNIERGITELNEVFKDLGSVVQQQGVLVDNIEANIYTTSDNTQLASDELRKAMRYQKRTSRWRVYLLI
Cons:    R       E     EL   F    V    Q      I  N
```

synaptotagmin or α-SNAP with the syntaxin 1-containing complex is mutually exclusive. The in vivo interaction of α-SNAP with syntaxin 1 and the formation of the 20S fusion particle can be triggered by the dissociation of synaptotagmin from the 7S complex. The formation of the 7S docking complex is dependent on several prior events (prior-7S events), including dissociation of VAMP from synaptophysin in the synaptic vesicle, dissociation of syntaxin 1 from n-Sec1 and association of the dissociated syntaxin 1 with SNAP-25 to form the syntaxin 1-SNAP-25 complex on the presynaptic plasma membrane. Other processes that can potentially affect these prior-7S events will have a regulatory role in the 7S complex formation and therefore vesicle docking/fusion. Rab3A and rabphilin-3A can have a regulatory role in events prior to and/or associated with the 7S or 20S complex formation. Studies using a syntaxin-negative Drosophila strain have established the essential role of syntaxin in synaptic transmission. However, docking of synaptic vesicles onto the active zone occurs normally, suggesting that syntaxin may be dispensable for docking. It is believed that syntaxin may participate in the actual fusion event between the synaptic vesicle and the presynaptic plasma membrane.[95]

SNAP-25

SNAP-25 (synaptosomal associated protein of 25 kDa) was originally identified in mouse as a brain-specific cDNA that encodes a 25 kDa protein associated with the presynaptic plasma membrane in the nerve terminal and it is palmitoylated.[561] The deduced mouse SNAP-25 is composed of 206 amino acids and lacks any membrane spanning domain. Its association with the presynaptic plasma membrane is most likely mediated by its attached palmitate and by its interaction with syntaxin 1. Interaction of syntaxin 1 with SNAP-25 is essential for the formation of the high affinity t-SNARE (syntaxin 1-SNAP-25) that interacts with the v-SNARE (VAMP) of the synaptic vesicle.[586] SNAP-25 is the substrate for BoNT/A and BoNT/E (Fig. 24).[64,65] SNAP-25 is involved in docking/fusion of synaptic vesicles by functioning as a part of the t-SNARE. Yeast Sec9p, involved in docking/fusion of Golgi-derived vesicles with the plasma membrane, is structurally related to SNAP-25 in its essential C-terminal portion.

n-Sec1/Munc-18/rbSec1

This brain-specific Sec1p-like protein was identified independently by biochemical isolation of syntaxin 1-binding proteins as well as by the genetic isolation of mammalian homologues of Sec1p, Drosophila Rop and C. elegans UNC8.[223,587] n-Sec1/Munc-18/rbSec1 is a hydrophilic protein of about 67 kDa and composed of 594 amino acids. n-Sec1/Munc-18/rbSec1 can bind syntaxin 1 and this interaction prevents interaction of syntaxin 1 with SNAP-25, suggesting that synaptic vesicle docking requires prior dissociation of syntaxin 1 from n-Sec1/Munc-18/rbSec1 and association with SNAP-25. n-Sec1/Munc-18/rbSec1 may thus have a regulatory role in synaptic vesicle docking/fusion. Whether n-Sec1/Munc-18/rbSec1 is also directly involved in vesicle docking/fusion remains to be established.

NSF and SNAPs

Synaptic vesicle fusion requires soluble general fusion proteins NSF and SNAPs, which are described in detail in chapter 3. Interaction of VAMP on the synaptic vesicle with syntaxin 1-SNAP-25 on the presynaptic plasma membrane results in the recruitment of SNAPs and subsequently NSF to form a 20S fusion complex composed of at least five proteins; and this 20S fusion complex is directly involved in neurotransmitter release.[170]

In addition to these proteins, there are many other proteins that are involved directly or indirectly in neurotransmitter release and these include neurotransmitter transporters involved in charging newly-generated synaptic vesicles with neurotransmitters, cytochromeB561, channel proteins, proton pump proteins associated with the synaptic vesicles as well as neurexins and voltage-gated Ca^{2+} channels associated with the presynaptic plasma membrane.[753,755,801]

A Model for Docking/Fusion of Synaptic Vesicles

A working model[26,45,60,281,307,666,682,753] for neurotransmitter release is presented in Figure 27. First, VAMP in the synaptic vesicle must dissociate from synaptophysin (step 1) and this process can be regulated by Rab3A. Next, syntaxin 1 must dissociate from n-Sec1 and become associated with SNAP-25 to form the syntaxin 1-SNAP-25 complex. Docking of synaptic vesicles with presynaptic plasma membrane is achieved by interaction of VAMP with syntaxin 1-SNAP-25 complex to form a 7S docking complex. Synaptotagmin may bind to this 7S complex to function as a Ca^{2+}-sensitive fusion clamp by preventing association of SNAPs with this complex (step 2). Upon the arrival of an action potential which triggers an influx of Ca^{2+} and transient increase of Ca^{2+} concentration in the nerve terminal, synaptotagmin becomes dissociated from the 7S complex and this dissociation results in the association of SNAPs,[473] and subsequently NSF, with the 7S complex to form the 20S fusion complex (step 3). ATP hydrolysis by NSF in the 20S complex triggers a series of poorly-defined events that lead to the fusion of synaptic vesicles with the presynaptic plasma membrane, the release

of neurotransmitter and dissociation of the 20S complex (step 4). Synaptic vesicles are then endocytosed by a process that can be initiated by an interaction between the clathrin AP2 complex and the second C2 domain within the cytoplasmic region of synaptotagmin I. The fission of the endocytotic vesicle is mediated by dynamins (step 5). The endocytosed vesicles fuse with an early endosome-like compartment[816,822] which is involved primarily in the generation of new synaptic vesicles (step 6). The newly-generated synaptic vesicles are then charged with their respective neurotransmitters, a process mediated by neurotransmitter transporters. The charged synaptic vesicles are guided to the active zone of the nerve terminal through interaction of synapsin I and II with the cytoskeletal elements (step 7). The synaptic vesicles are now ready for a new cycle of neurotransmitter release. Recent studies using syntaxin- or synaptobrevin-negative Drosophila mutants have suggested that synaptobrevin and syntaxin may be dispensable for the docking of synaptic vesicles.[95] Synaptobrevin may play a role in events required somewhere between the receipt of the evoked fusion signal and the actual fusion, while syntaxin may be directly involved in the actual fusion event. Future biochemical, molecular, genetic and cell biological studies will certainly provide insights that will modify and/or extend the working model.

Fig. 27 (opposite). A model for synaptic vesicle-mediated neurotransmitter release. VAMPs associated with synaptophysin on the synaptic vesicle first become dissociated from synaptophysin (Step 1). Syntaxin 1 in the active zone of presynaptic membrane becomes dissociated from n-Sec1 and associates instead with SNAP-25 to form the syntaxin-SNAP-25 complex which then interact with VAMPs and synaptotagmin on the synaptic vesicle to form the 7S docking complex (step 2). Upon arrival of an action potential and a transient Ca^{2+} increase at the nerve terminal, synaptotagmin dissociates from this 7S complex, resulting in the association of SNAPs, and subsequently NSF, onto the VAMP-syntaxin-SNAP-25 complex to form the 20 S fusion (SNARE) complex (step 3). ATP hydrolysis by NSF triggers the fusion of the synaptic vesicle membrane with the presynaptic membrane and release of the neurotransmitters (step 4). The synaptic vesicle components are then internalized by a process involving an interaction between the AP2 adaptin complex with the C2 domain in the cytoplasmic region of synaptotagmin (step 5). Through an early-endosome-like compartment, new synaptic vesicles are generated (step 6) and the vesicles are then charged with neurotransmitters (step 7). The charged synaptic vesicles are translocated to the active zone and become docked (step 1-2) and are ready for the next cycle of neurotransmitter release.

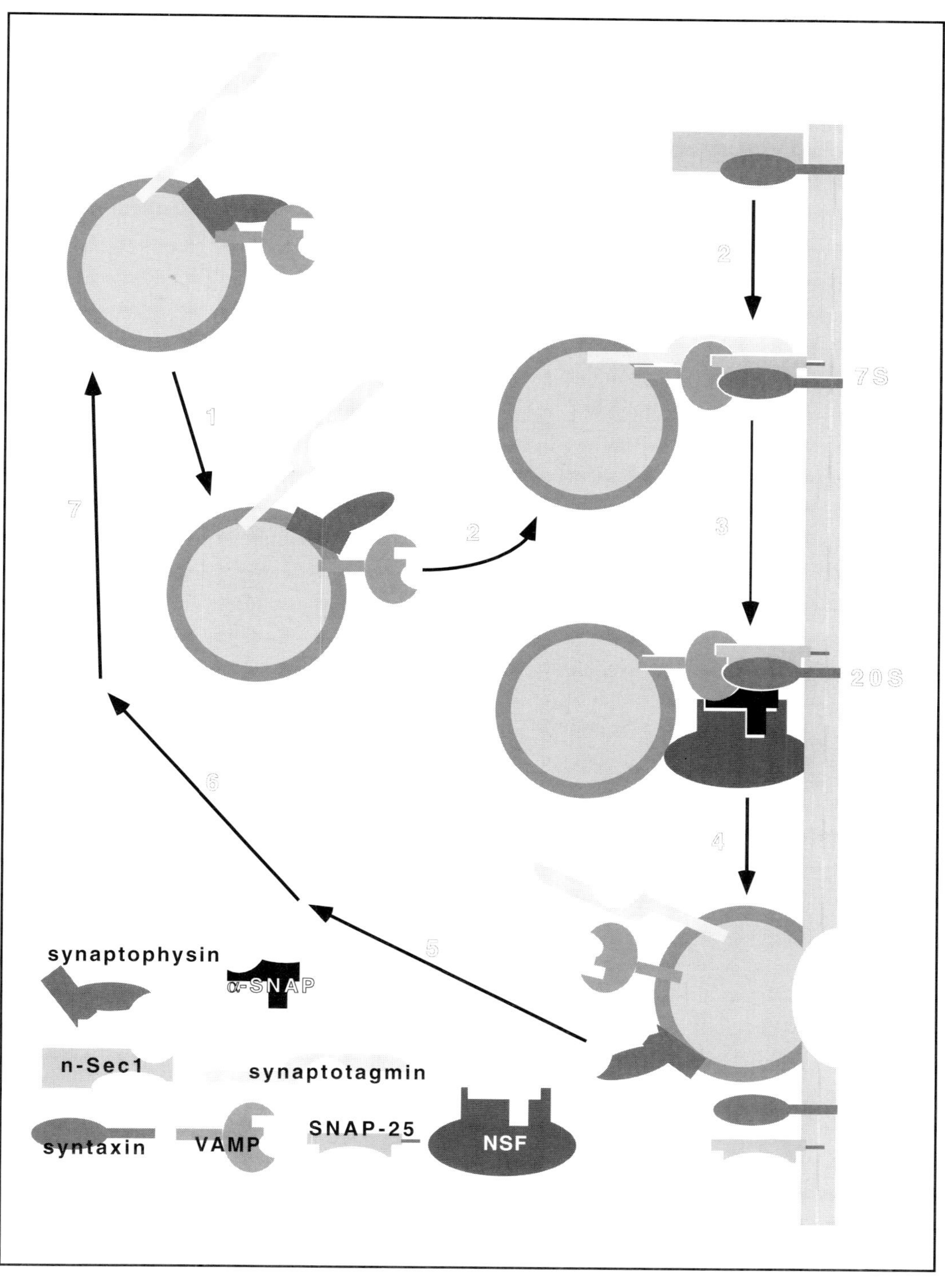
synaptophysin
α-SNAP
n-Sec1
synaptotagmin
syntaxin
VAMP
SNAP-25
NSF
7S
20S

REFERENCES

1. Aalto MK, Keranen S, Ronne H. A family of proteins involved in intracellular transport. Cell 1992; 68:181-182.
2. Aalto MK, Ronne H, Keranen S. Yeast syntaxins Sso1p and Sso2p belong to a family of related mammalian proteins that function in vesicular transport. EMBO J 1993; 12:4095-4104.
3. Aalto MK, Ruohonen L, Hosono K, Keranen S. Cloning and sequencing of the yeast Saccharomyces cerevisiae SEC1 gene localized on chromosome IV. Yeast 1992; 8:587-588.
4. Acharya U, Jacobs R, Peters JM et al. The formation of Golgi stacks from vesiculated Golgi membranes requires two distinct fusion events. Cell 1995; 82:895-904.
5. Achstetter T, Franzusof A, Field C, Schekman R. SEC7 encodes an unusual, high molecular weight protein required for membrane traffic from the yeast Golgi apparatus. J Biol Chem 1988; 263:11711-11717.
6. Ahluwalia N, Bergeron JJM, Wade I et al. The p88 molecular chaperone is identical to the endoplasmic reticulum membrane protein, calnexin. J Biol Chem 1992; 267:10914-10918.
7. Alexandrov K, HoriuchiH, Steele-Mortimer O et al. Rab escort protein-1 is a multifunctional protein that accompanies newly-prenylated rab proteins to their target membranes. EMBO J 1994; 13:5262-5273.
8. Allan VJ, Kreis T. A microtubule-binding protein associated with membranes of the Golgi apparatus. J Cell Biol 1986; 103:2229-2239.
9. Althoff SM, Stevens SW, Wise JA. The Srp54 GTPase is essential for protein export in the fisson yeast Schizosaccharomyces pombe. Mol Cell Biol 1994; 14:7839-7854.
10. Anderson RGW. Caveolae: where incoming and outgoing messengers meet. Proc Natl Acad Sci 1993; 90:10909-10913.
11. Andres DA, Milatovich A, Ozcelik T et al. cDNA cloning of the two subunits of human CAAX fanesyltransferase and chromosomal mapping of FNTα and FNTβ loci and related sequences. Genomics 1993; 18:105-112.
12. Andres DA, Seabra MC, Brown MS et al. cDNA cloning of component A of Rab geranylgeranyltransferase and demonstration of its role as a Rab escort protein. Cell 1993; 73:1091-1099.
13. Antony C, Cibert C, Geraud G et al. The small GTP-binding protein rab6p is distributed from medial Golgi to the trans Golgi network as determined by a confocal microscopic approach. J Cell Sci 1992; 103:785-796.
14. Aoki D, Lee N, Yamaguchi N et al. Golgi retention of a trans Golgi membrane protein, galactosyltransferase, requires cysteine and histidine residues within the transmembrane domains. Proc Natl Acad Sci 1992; 89:4319-4323.
15. Appel D, Kock-Brandt C. Sorting of a secretory protein (gp80) to the apical surface of Caco-2 cells. J Cell Sci 1994; 107:553-559.
16. Arar C, Carpentier V, Le Caer J-P et al. ERGIC-53, a membrane protein of the endoplasmic reticulum-Golgi intermediate compartment, is identical to MR60, an intracellular mannose-specific lectin of myelomonocytic cells. J Biol Chem 1995;

270:3551-3553.

17. Armstrong J, Craighead MW, Watson R et al. Schizosaccharomyces pombi ypt5: a homologue of the rab5 endosome fusion regulator. Mol Biol Cell 1993; 4:583-592.

18. Armstrong J, Patel S. The Golgi sorting domain of coronavirus E1 protein. J Cell Sci 1991; 98:567-575.

19. Armstrong J, Patel S, Riddle P. Lysosomal sorting mutants of coronavirus E1 protein, a Golgi membrane protein. J Cell Sci 1990; 95:191-197.

20. Armstrong SA, Hannah VC, Goldstein JL, Brown MS. CAAX geranylgeranyltransferase transfers farnesyl as efficiently as geranylgeranyl to RhoB. J Biol Chem 1995; 270:7864-7868.

21. Armstrong SA, Seabra MC, Sudhof TC et al. cDNA cloning and expression of the α and β subunits of rat Rab geranylgeranyl transferase. J Biol Chem 1993; 258:12221-12229.

22. Aroeti B, Kosen PA, Kuntz ID, et al.. Mutational and secondary structural analysis of the basolateral sorting signal of the polymeric immunoglobulin receptor. J Cell Biol 1993; 123:1149-1160.

23. Aroeti B, Mostov K. Polarized sorting of the polymeric immunoglobulin receptor in the exocytotic and endocytotic pathways is controlled by the same amino acids. EMBO J 1994; 13:2297-2304.

24. Arvan P, Castle D. Protein sorting and secretion granule formation in regulated secretory cells. Trends Cell Biol 1992; 2:327-331.

25. Bacon RA, Salminen A, Ruohola H et al. The GTP-binding protein Ypt1 is required for transport in vitro: the Golgi apparatus is defective in ypt1 mutants. J Cell Biol 1989; 109:1015-1022.

26. Bajjalieh SM, Scheller RH. The biochemistry of neurotransmitter secretion. J Biol Chem 1995; 270:1971-1974.

27. Baker D, Hicke L, Rexach M, et al. Reconstitution of SEC gene product-dependent intercompartment protein transport. Cell 1988; 54:335-344.

28. Baker D, Wuestehube L, Schekman R et al. GTP-binding Ypt1 protein and Ca^{2+} function independently in a cell-free protein transport system. Proc Natl Acad Sci 1990; 87:355-359.

29. Balch WE. Small GTP-binding proteins in vesicular transport. Trends Biochem Sci 1990; 15:473-477.

30. Balch WE, Dunphy, WG, Braell WA, Rothman JE. Reconstitution of transport of protein between successive compartments of the Golgi measured by the coupled incorporation of N-acetylglucosamine. Cell 1984; 39:405-416.

31. Balch WE, Glick BS, Rothman JE. Sequential intermediates in the pathway of intercompartmental transport in a cell-free system. Cell 1984; 39:525-536.

32. Balch WE, Kahn RA, Schwaninger R. ADP-ribosylation factor is required for vesicular trafficking between the endoplasmic reticulum and the cis-Golgi compartment. J Biol Chem 1992; 267: 13053-13061.

32a. Balch WE, McCaffery JM, Plunter H, Farquhar MG. Vesicular stomatitis virus glycoprotein is sorted and concentrated during export from the endoplasmic reticulum Cell 1994; 77:841-852.

33. Balch WE, Wagner KR, Keller DS. Reconstitution of transport of vesicular stomatitis virus G protein from the endoplasmic reticulum to the Golgi complex using a cell-free system. J Cell Biol 1987; 104:749-760.

34. Baldini G, Hohl T, Lin HY, Lodish HF. Cloning of a Rab3 isotype predominantly expressed in adipocytes. Proc Natl Acad Sci 1992; 89:5049-5052.

35. Banfield DK, Lewis MJ, Pelham HRB. A SNARE-like protein required for traffic through the Golgi complex. Nature 1995; 375:806-809.

36. Banfield DK, Lewis MJ, Rabouille C et al. Localization of Sed5, a putative vesicle targeting molecule, to the cis-Golgi network involves both its transmembrane and cytoplasmic domains. J Cell Biol 1994; 127:357-371.

37. Bankaitis VA, Aitken JR, Cleves AE, Dowhan W. An essential role for a phopholipid transfer protein in yeast Golgi function. Nature 1990; 347: 561-562.

38. Bankaitis VA, Malehorn DE, Emr SD, Greene R. The Saccharomyces cerevisiae SEC14 gene encodes a cytosolic factor that is required for transport of secretory protein from the yeast Golgi complex. J Cell Biol 1989; 108:1271-1281.

39. Bansal A, Gierasch LM. The NPXY internalization signal of the LDL receptor adopts a reverse-turn conformation. Cell 1991; 67:1195-1201.

40. Banta LM, Robinson JS, Klionsky DJ, Emr SD. Organelle assembly in yeast: characterization of yeast mutants defective in vacuolar biogenesis and protein sorting. J Cell Biol 1988; 107: 1369-1383.

40a. Banta LM, Vida TA, Herman PK, Emr SD. Characterization of yeast Vps33p, a protein required for vacuolar protein sorting and vacuole biogenesis. Mol Cell Biol 1990; 10:4638-4649.

41. Baranski TJ, Cantor AB, Kornfeld S. Lysosomal enzyme phosphorylation. I. protein recognition determinants in both lobes of procathepsin D mediates its interaction with UDP-GlcNac:lysosomal enzyme N-acetylglucosamine-1-phospho-transferase. J Biol Chem 1992; 267: 23342-23348.

42. Baranski TJ, Faust PL, Kornfeld S. Generation of a lysosomal enzyme targeting signal in the secretory protein pepsinogen. Cell 1990; 63:281-291.

43. Baranski TJ, Koelsch G, Hartsuck JA, Kornfeld S. Mapping and molecular modeling of a recognition domain for lysosomal enzyme targeting. J Biol Chem 1991; 266:23365-23372.

44. Barbieri MA, Li G, Colombo MI, Stahl PD. Rab5, an early acting endosomal GTPase, supports in vitro endosome fusion without GTP hydrolysis. J Biol Chem 1994; 269:18720-18722.

45. Bark IC, Wilson MC. Regulated vesicular fusion in neurons: snapping together the details. Proc Natl Acad Sci 1994; 91:4621-4624.

46. Barlowe C, d'Enfert C, Schekman R. Purification and characterization of SAR1p, a small GTP-binding protein required for transport vesicle formation from the endoplasmic reticulum. J Biol Chem 1993; 268:873-879.

47. Barlowe C, Orci L, Yeung T et al. COPII: a membrane coat formed by Sec proteins that drive vesicle budding from the endoplasmic reticulum. Cell 1994; 77:895-907.

48. Barlowe C, Schekman R. SEC12 encodes a guanine-nucleotide-exchange factor essential for transport vesicle budding from the ER. Nature 1993; 365:347-349.

49. Barr FA, Leyte A, Huttner WB. Trimeric G proteins and vesicle formation. Trends Cell Biol 1992; 2:91-94.

50. Barroso M, Nelson, DS, Sztul E. Transcytosis-associated protein (TAP)/p115 is a general fusion factor required for binding of vesicles to acceptor membranes. Proc Natl Acad Sci 1995; 92:527-531.

51. Bartles JR, Feracci HM, Stieger B, Hubbard AL. Biogenesis of the rat hepatocyte plasma membrane in vivo: comparison of the pathways taken by apical and basolateral proteins using subcellular fractionation. J Cell Biol 1987; 105:1241-1251.

52. Baumert M, Maycox PR, Navone F et al. Synaptobrevin: an integral membrane protein of 18000 daltons present in small synaptic vesicles of rat brain. EMBO J 1989; 8:379-384.

53. Becker J, Tan TJ, Trepte H-H, Gallwitz D. Mutational analysis of the putative effector domain of the GTP-binding Ypt1 protein in yeast suggests specific regulation by a novel GAP activity. EMBO J 1991; 10:785-792.

54. Beckers CJM, Block MR, Glick BS et al. Vesicular transport between the endoplasmic reticulum and the Golgi stack requires the NEM-sensitive fusion protein. Nature 1989; 339:397-400.

55. Beckers CJM, Keller DS, Balch WE. Semi-intact cells permeable to macromolecules: use in reconstitution of protein transport from the endoplasmic reticulum to the Golgi complex. Cell 1987; 50: 523-534.

56. Beltzer JP, Spiess M. In vitro binding of the asialoglycoprotein receptor to the β-adaptin of plasma membrane coated vesicles. EMBO J 1991; 10:3735-3742.

57. Bennett MK, Calakos N, Kreiner T, Scheller RH. Synaptic vesicle membrane proteins interact to form a multimeric complex. J Cell Biol 1992; 116:761-775.

58. Bennett MK, Calakos N, Scheller RH. Syntaxin: a synaptic protein implicated in docking of synaptic vesicles at presynaptic active zones. Science 1992; 257: 255-259.

59. Bennett MK, Garcia-Arraras JE, Elferink LA et al. The syntaxin family of vesicular transport receptors. Cell 1993; 74: 863-873.

60. Bennett MK, Scheller RH. A molecular description of synaptic vesicle membrane trafficking. Annu Rev Biochem 1994; 63:63-100.

61. Bennett MK, Wandinger-Ness A, Simons K. Release of putative exocytotic vesicles from perforated MDCK cells. EMBO J 1988; 7:4075-4085.

62. Beranger F, Cadwallader K, Porfiri E et al. Determination of structural requirements for the interation of Rab6 with RabGDI and Rab geranylgeranyltransferase. J Biol Chem 1994; 269: 13637-13643.

63. Bergeron JJM, Brenner MB, Thomas DY, Williams DB. Calnexin: a membrane-bound chaperone of the endoplasmic reticulum. Trends Biochem Sci 1994; 19:124-128.

64. Binz T, Blasi J, Yamasaki S et al. Proteolysis of SNAP-25 by types E and A botulinal neurotoxins. J Biol Chem 1994; 269:1617-1620.

65. Blasi J, Chapman ER, Link E et al. Botulinum neurotoxin A selectively cleaves the synaptic protein SNAP-25. Nature 1993;

365:160-163.

66. Blasi J, Chapman ER, Yamasaki S et al. Botulinum neurotoxin C1 blocks neurotransmitter release by means of cleaving HPC-1/syntaxin. EMBO J 1993; 12:4821-4828.

67. Block MR, Glick BS, Wilcox CA et al. Purification of an N-ethylmaleimide-sensitive protein catalyzing vesicular transport. Proc Natl Acad Sci 1988; 85: 7852-7858.

68. Blond-Elguindi S, Fourie AM, Sambrook JF, Gething M-J. Peptide-dependent stimulation of the ATPase activity of the molecular chaperone BiP is the result of conversion of oligomers to active monomers. J Biol Chem 1993; 268: 12730-12735.

69. Bobak DA, Nightingale MS, Murtagh J et al. Molecular cloning, characterization, and expression of human ADP-ribosylation factors: two guanine nucleotide-dependent activators of cholera toxin. Proc Natl Acad Sci 1989; 86:6101-6105.

70. Boehm J, Ulrich HD, Ossig R, Schmitt HD. Kex2-dependent invertase secretion as a tool to study the targeting of transmembrane proteins which are involved in ER-Golgi transport in yeast. EMBO J 1994; 13:3696-3710.

71. Bohni PC, Deshaies RJ, Schekman RW. SEC11 is required for signal peptide processing and yeast cell growth. J Cell Biol 1988; 106:1035-1042.

72. Boll W, Partin JS, KAtz AI, et al.. Distinct pathways for basolateral targeting of membrane and secretory proteins in polarized epithelial cells. Proc Natl Acad Sci 1991; 88: 8592-8596.

73. Boman AL, Kahn RA. Arf proteins: the membrane traffic police? Trends Biochem Sci 1995; 20:147-150.

74. Bonatti S, Migliaccio G, Simons K. Palmitylation of viral membrane glycoproteins takes place after exit from the endoplasmic reticulum. J Biol Chem 1989; 264:12590-12595.

75. Bonnerot C, Marks MS, Cosson P et al. Association of BiP and aggregation of

class II MHC molecules synthesized in the absence of invariant chain. EMBO J 1994; 13:934-944.

76. Bos K, Wraight C, Stanley KK. TGN38 is maintained in the trans Golgi network by a tyrosine-containing motif in the cytoplasmic domain. EMBO J 1993; 12:2219-1128.

77. Bosshart H, Humphrey J, Deignan E et al. The cytoplasmic domain mediates localization of furin to the trans Golgi network en route to the endosomal/lysosomal system. J Cell Biol 1994; 126: 1157-1172.

78. Boulianne GL, Trimble WS. Identification of a second homolog of N-ethyl-maleimide-sensitive fusion protein that is expressed in the nervous system and secretory tissues of Drosophila. Proc Natl Acad Sci 1995; 92:7095-7099.

79. Bourne HR, Sanders DA, McCormick F. The GTPase superfamily: conserved structure and molecular mechanism. Nature 1991; 349:117-127.

80. Bowser R, Muller H, Govindan B, Novick P. Sec8p and Sec15p are component of a plasma membrane-associated 19.5S particle that may function downstream of Sec4p to control exocytosis. J Cell Biol 1992; 118:1041-1056.

81. Bowser R, Novick P. Sec15 protein, an essential component of the exocytotic apparatus, is associated with the plasma membrane and with a soluble 19.5S particle. J Cell Biol 1991; 112:1117-1131.

82. Braakman, I, Helenius J, Helenius A. Manipulating disulfide bond formation and protein folding in the endoplasmic reticulum. EMBO J 1992; 11: 1717-1722.

83. Braakman, I, Helenius J, Helenius A. Role of ATP and disulfide bonds during protein folding in the endoplasmic reticulum. Nature 1992; 356:260-262.

84. Braell WA, Balch WE, Dobbertin DC, Rothman JE. The glycoprotein that is transported between successive compartments of the Golgi in a cell-free system resides in stacks of cisternae. Cell 1984; 39:511-524.

85. Braun M, Waheed A, von Figura K. Lysosomal acid phosphatase is transported to lysosomes via the cell surface. EMBO J 1989; 8:3633-3640.

86. Brennwald P, Novick P. Interactions of three domains distinguishing the ras-related GTP-binding proteins Ypt1 and Sec4. Nature 1993; 362:560-563.

87. Bremnes B, Madsen T, Gedde-Dahl M, Bakke O. An LI and ML motif in the cytoplasmic tail of the MHC-associated invariant chain mediate rapid internalization. J Cell Sci 1994; 107:2021-2032.

88. Rodsky FM, Hill BL, Acton SL et al. Clathrin light chains: arrays of protein motifs that regulate coated vesicle dynamics. Trends Biochem Sci 1991; 16:208-213.

89. Brodsky JL, Schekman R. A Sec63p-Bip complex from yeast is required for protein translocation in a reconstituted proteoliposome. J Cell Biol 1993; 123: 1355-1363.

90. Brown HA, Gutowski S, Moomaw CR et al. ADP-ribosylation factor, a small GTP-dependent regulatory protein, stimulates phospholipase D activity. Cell 1993; 75:1137-1144.

91. Brown JD, Hann BC, Medzihradszky KF, et al. Subunits of Saccharomyces cerevisiae signal recognition particle required for its functional expression. EMBO J 1994; 13:4390-4400.

92. Brown MS, Goldstein JL. Mad bet for Rab. Nature 1993; 366:14-15.

93. Brown WJ, DeWald DB, Emr SD, et al.. Role of phosphatidylinositol 3-kinase in sorting and transport of newly-synthesized lysosomal enzymes in mammalian cells. J Cell Biol 1995; 130:781-796.

94. Brennwald P, Kearns B, Champion K, et al. Sec9 is a SNAP-25-like component of a yeast SNARE complex that may be the effector of Sec4 function in exocytosis. Cell 1994; 79:245-258.

95. Broadie K, Prokop A, Bellen HJ et al. Syntaxin and synaptobrevin function sownstream of vesicle docking in Droso-

phila. Cell 1995; 15:663-673.

96. Brondyk WH, McKiernan CJ, Fortner KA et al. Interaction cloning of Rabin3, a novel protein that associates with the ras-like GTPase Rab3A. Mol Cell Biol 1995; 15:1137-1143.

97. Brown DA, Crise B, Rose JK. Mechanism of membrane anchoring affects polarized expression of two proteins in MDCK cells. Science 1989; 245:1499-1501.

98. Brewer CB, Roth MG. A single amino acid change in the cytoplasmic domain alters the polarized delivery of influenza virus hemagglutinin. J Cell Biol 1991; 114:413-421.

99. Brown DA, Rose JK. Sorting of GPI-anchored proteins to glycolipid-enriched membrane subdomains during transport to the apical cell surface. Cell 1992; 68:533-544.

100. Bucci C, Parton RG, Mather IH et al. The small GTPase rab5 functions as a regulatory factor in the early endocytotic pathway. Cell 1992; 70:715-728.

101. Burgess TL, Kelly RB. Constitutive and regulated secretion of proteins. Annu Rev Cell Biol 1987; 3:243-293.

102. Burgoyne RD, Morgan A. Ca^{2+} and secretory-vesicle dynamics. Trends Neurosci 1995; 18:191-196.

103. Burke J, Pettitt JM, Humphris D et al. The transmembrane and flanking sequences of β1,2-N-acetylglucosaminyltransferase I specify medial-Golgi localization. J Biol Chem 1992; 267:24433-24440.

104. Burke J, Pettitt JM, Humphris D, Glesson PA. Medial-Golgi retention of N-acetylglucosaminyltransferae I: contribution from all domains of the enzyme. J Biol Chem 1994; 269:12049-12059.

105. Burton JL, Burns ME, Gatti E et al. Specific interactions of Mss4 with members of the Rab GTPase subfamily. EMBO J 1994; 13:5547-5558.

106. Burton JL, Roberts D, Montaldi M et al. A mammalian guanine-nucleotide releasing protein enhances function of yeast secretory protein Sec4p. Nature 1993; 261:464-467.

107. Calakos N, Bennett MK, Peterson KE, Scheller R. Protein-protein interactions contributing to the specificity of intracellular vesicular trafficking. Science 1994; 263:1146-1149.

108. Calakos N, Scheller R. Vesicle-associated membrane protein and synaptophysin are associated on the synaptic vesicle. J Biol Chem 1994; 269:24534-24537.

109. Canfield WM, Johnson KF, Ye RD et al. Localization of the signal for rapid internalization of bovine cation-independent mannose 6-phosphate/insulin-like growth factor-II receptor to amino acids 24-29 of the cytoplasmic tail. J Biol Chem 1991; 266:5682-5688.

110. Cantor AB, Baranski TJ, Kornfeld S. Lysosomal enzyme phosphorylation. II. protein recognition determinants in either lobe of procathepsin D are sufficient for phosphorylation of both the amino and carboxyl lobe oligosaccharides. J Biol Chem 1992; 267:23349-23356.

111. Cao X, Zhou Y, Lee AS. Requirement of tyrosine- and serine/threonine kinases in the transcriptional activation of the mammalian grp78/BiP promoter by thapsigargin. J Biol Chem 1995; 270; 494-502.

112. Carnell L, Moore HPH. Transport via the regulated secretory pathway in semi-intact PC12 Cells: role of intra-cisternal calcium and pH in the transport and sorting of secretogranin II. J Cell Biol 1994; 127:693-705.

113. Casanova JE, Apodaca G, Mostov KE. An autonomous signal for basolateral sorting in the cytoplasmic domain of the polymeric immunoglobulin receptor. Cell 1991; 66:65-75.

114. Ceccaldi P-E, Grohovaz F, Benfenati F et al. Dephosphorylated synapsin I anchors synaptic vesicles to actin cytoskeleton: an analysis by videomicroscopy. J Cell Biol 1995; 128:905-912.

115. Chamberlain LH, Roth D, Morgan A, Burgoyne RD. Distinct effects of α-

SNAP, 14-3-3 proteins, and calmodulin on priming and triggering of regulated exocytosis. J Cell Biol 1995; 130: 1063-1070.

116. Chanat E, Huttner WB. Milieu-induced, selective aggregation of regulated secretory proteins in the trans Golgi network. J Cell Biol 1991; 115:1505-1519.

117. Chanat E, Weiss U, Huttner WB, Tooze SA. Reduction of the disulfide bond of chromogranin B (secretogranin I) in the trans Golgi network causes its misorting to the constitutive secretory pathway. EMBO J 1993; 12:2159-2168.

118. Chang MP, Mallet WG, Mostov KE, Brodsky FM. Adaptor self-aggregation, adaptor-receptor recognition and binding of α-adaptin subunits to the plasma membrane contribute to recruitment of adaptor (AP2) components of clathrin-coated pits. EMBO J. 1993; 12: 2169-2180.

119. Chapman RE, Munro S. Retrieval of TGN proteins from the cell surface requires endosomal acidification. EMBO J 1994; 13:2305-2312.

120. Chapman RE, Munro S. The functioning of the yeast Golgi apparatus requires an ER protein encoded by ANP1, a member of a new family of genes affecting the secretory pathway. EMBO J 1994; 13:4896-4907.

121. Chavrier P, Gorvel J-P, Stelzer E et al. Hypervariable C-terminal domain of rab proteins acts as a targeting signal. Nature 1991; 353:769-772.

122. Chavrier P, Parton RG, Hauri HP et al. Localization of low molecular weight GTP binding proteins to exocytic and endocytic compartments. Cell 1990; 62:317-319.

123. Chavrier P, Vingron M, Sander C et al. Molecular cloning of YPT1/SEC4-related cDNAs from an epithelial cell line. Mol Cell Biol 1990; 10:6578-6585.

124. Chen HJ, Remmler J, Delaney JC et al. Mutational analysis of the cation-independent mannose 6-phosphate/insulin-like growth factor II receptor: a consensus casein kinase II site followed by 2 leucines near the carboxyl terminus is important for intracellular targeting of lysosomal enzymes. J Biol Chem 1993; 268:22338-22346.

125. Chen MS, Obar RA, Schroeder CC et al. Multiple forms of dynamin are encoded by shibire, a Drosophila gene involved in endocytosis. Nature 1991; 351:583-586.

126. Chen W, Helenius J, Braakman I, Helenius A. Cotranslational folding and calnexin binding during glycoprotein synthesis. Proc Natl Acad Sci 1995; 92: 6229-6233.

127. Chen WJ, Andres DA, Goldstein JL et al. cDNA cloning and expression of the peptide-binding β-subunit of rat p21^{ras} farneyltransferase, the counterpart of yeast DPR1/RAM1. Cell 1991; 66: 327-334.

128. Chen WJ, Andres DA, Goldstein JL, Brown MS. Cloning and expression of a cDNA encoding the α subunit of rat p21^{ras} farneyltransferase. Proc Natl Acad Sci 1991; 88:11368-11372.

129. Chen YT, Holcomb C, Moore HPH. Expression and localization of two low molecular weight GTP-binding proteins, Rab8 and Rab10, by epitope tag. Proc Natl Acad Sci 1993; 90:6508-6512.

130. Chung S-H, Takai Y, Holz R. Evidence that the Rab3a-binding protein, Rabphilin3a, enhances regualted secretion. Studies in adrenal chromaffin cells. J Biol Chem 1995; 270:16714-16718.

131. Clark J, Moore L, Krasinskas A et al. Selective amplification of additional members of the ADP- ribosylation factor (ARF) family: Cloning of additional human and Drosophila ARF-like genes. Proc Natl Acad Sci 1993; 90:8952-8956.

132. Clary DO, Griff IC, Rothman JE. SNAPs, a family of NSF attachment proteins involved in intracellular membrane fusion in animals and yeast. Cell 1990; 61:709-721.

133. Clary DO, Rothman JE. Purification of three related peripheral membrane pro-

teins needed for vesicular transport. J Biol Chem 1990; 265:10109-10117.

134. Cockcroft S, Thomas GMH, Fensome A et al. Phospholipase D: a downstream effector of ARF in granulocytes. Science 1994; 263:523-526.

135. Colley KJ, Lee EU, Adler B et al. Conversion of a Golgi apparatus sialyltransferase to a secretory protein by replacement of the NH2-terminal signal anchor with a signal peptide. J Biol Chem 1989; 264:17619-17622,

136. Colley KJ, Lee EU, Paulson JC. The signal anchor and stem regions of β-galactoside α-2,6-sialyltransferase may each act to localize the enzyme to the Golgi apparatus. J Biol Chem 1992; 267: 7784-7793.

137. Colombo MI, Mayorga LS, Casey PJ, Stahl PD. Evidence of a role for heterotrimeric GTP-binding proteins in endosome fusion. Science 1992; 255: 1695-1697.

138. Connolly T, Rapiejko PJ, Gilmore R. Requirement of GTP hydrolysis for dissociation of signal recognition particle from its receptor. Science 1991; 252: 1171-1173.

139. Connolly T, Gilmore R. The signal recognition particle receptor mediates the GTP-dependent displacement of SRP from the signal sequence of the nascent polypeptide. Cell 1989; 57:599-610.

140. Connolly T, Gilmore R. GTP hydrolysis by complex of the signal recognition particle and the signal recognition particle receptor. J Cell Biol 1993; 123: 799-807.

141. Cool DR, Fenger M, Snell CR, Loh YP. Identification of the sorting signal motif within pro-opiomelanocortin for the regulated secretory pathway. J Biol Chem 1995; 270:8723-8729.

142. Corvera S, Chawla A, Chakrabarti R et al. A double-leucine within the GLU4 glucose transporter COOH-terminal domain functions as an endocytosis signal. J Cell Biol 1994; 126:979-989

143. Cosson P, Letourneur F. Coatomer interaction with di-lysine endoplasmic reticulum retention motifs. Science 1994; 263: 1629-1631.

144. Couve A, Gerst JE. Yeast Snc proteins complex with Sec9. J Biol Chem 1994; 269:23391-23394.

144a. Cowles CR, Emr S, Horazdovsky BF. Mutations in the VPS45 gene, a SEC1 homologue, result in vacuolar protein sorting defects and accumulation of membrane vesicle. J Cell Sci 1994; 107:3449-3459.

145. Cox JS. Shamu CE, Walter P. Transcriptional induction of genes encoding endoplasmic reticulum resident proteins requires a transmembrane protein kinase. Cell 1993; 73:1197-1206.

146. Craighead MW, Bowden S, Watson R, Armstrong J. Function of the ypt2 gene in the exocytotic pathway of Schizosaccharomyces pombe. Mol Biol Cell 1993; 4:1069-1076.

147. Cremers FPM, Armstrong SA, Seabra MC et al. REP-2, a rab escort protein encoded by the choroideremia-like gene. J Biol Chem 1994; 269:2111-2117.

148. Crowley KS, Reinhart GD, Johnson AE. The signal sequence moves through a ribosomal tunnel into a noncytoplasmic aqueous envionment at the ER membrane early in translocation. Cell 1993; 73:1101-1115.

149. Cutler DF, Cramer LP. Sorting during transport to the surface of PC12 cells: divergence of synaptic vesicle and secretory granule proteins. J Cell Biol 1990; 110:721-730.

150. Dahdal RY, Colley KJ. Specific sequences in the signal anchor of β-galactoside α-2,6-sialyltransferase are not essential for Golgi localization. J Biol Chem 1993; 268:26310-26319.

151. Dahms NM, Kornfeld S. The cation-dependent mannose 6-phosphate receptor. structural requirements for mannose 6-phosphate binding and oligomerization. J Biol Chem 1989; 264:11458-11467.

152. Dahms NM, Lobel P, Breitmeyer J et al. 46 kd mannose 6-phosphate receptor:

cloning, expression, and homology to the 215 kd mannose 6-phosphate receptor. Cell 1987; 50:181-192.

153. Dahms NM, Lobel P, Kornfeld S. Mannose-6-phosphate receptors and lysosomal enzyme targeting. J Biol Chem 1989; 264:12115-12118.

154. Dalbey RE. Positively charged residues are important determinants of membrane protein topology. Trends Biochem Sci 1990; 15:253-257.

155. Damke H, Baba T, Warnock DE, Schmid SL. Induction of mutant dynamin specifically blocks endocytotic vesicle formation. J Cell Biol 1994; 127:915-934.

156. Danielsen EM. Involvement of detergent-insoluble complex in the intracellular transport of intestine brush border enzymes. Biochemistry 1995; 34:1596-1605.

157. Dargemont C, Le Bivic A, Rothenberger S, et al.. The internalization signal and phosphorylation site of transferrin receptor are distinct from the main basolateral sorting information. EMBO J 1993; 12:1713-1721.

158. Dascher C, Balch WE. Dominant inhibitory mutants of ARF1 block Endoplasmic Reticulum to Golgi transport and trigger disassembly of the Golgi apparatus. J Biol Chem 1994; 269:1437-1448.

159. Dascher C, Matteson J, Balch WE. Syntaxin 5 regulates endoplasmic reticulum to Golgi transport. J Biol Chem 1994; 269:29363-29366.

160. Dascher C, Ossig R, Gallwitz D, Schmitt HD. Identification and structure of four yeast genes (SLY) that are able to suppress the functional loss of YPT1, a member of the RAS superfamily. Mol Cell Biol 1991; 11:872-885.

161. Dashaies RJ, Sanders SL, Feldheim DA, Schekman R. Assembly of yeast Sec proteins involved in translocation into the endoplasmic reticulum into a a membrane-bound multisubunit complex. Nature 1991; 349:806-808.

162. Dashaies RJ, Schekman R. SEC62 en-

codes a putative membrane protein required for protein translocation into the yeast endoplasmic reticulum. J Cell Biol 1989; 109:2653-2664.

163. Davidson HW. Wortmannin causes mistargeting of procathepsin D. evidence for the involvement of a phosphatidylinositol 3-kinase in vesicular transport to lysosomes. J Cell Biol 1995; 130:797-805.

164. Davidson HW, Balch WE. Differential inhibition of multiple vesicular transport steps between the endoplasmic reticulum and trans Golgi network. J Biol Chem 1993; 268:4216-4226.

165. De Curtis I, Simons K. Isolation of exocytotic carrier vesicles from BHK cells. Cell 1989; 58:719-727.

166. de Hoop M, von Poser C, Lange C et al. Intracellular routing of wild-type and mutated polymeric immunoglobulin receptor in hippocampal neurons in culture. J Cell Biol 1995; 130:1447-1459.

167. de Silva A, Braakman I, Helenius A. Posttranslational folding of vesicular stomatitis virus G protein in the ER: involvement of noncovalent and covalent complexes. J Cell Biol 1993; 120:647-655.

168. Dean N, Pelham HRB. Recycling of proteins from the Golgi compartment to the ER in yeast. J Cell Biol 1990; 111:369-377.

169. Deleted during revision.

170. DeBello WM, O'Conner V, Dresbath T et al. SNAP-mediated protein-protein interactions essential for neurotransmitter release. Nature 1995; 373:626-630.

171. Delbruck R, Desel C, von Figura K, Hill-Rehfeld A. Proteolytic processing of cathepsin D in prelysosomal organells. Eur J Cell Biol 1994; 64:7-14.

172. d'Enfert C, Barlowe C, Nishikawa S-I, et al. Structural and functional dissection of a membrane glycoprotein required for vesicle budding from the endoplasmic reticulum. Mol Cell Biol 1991; 11:5729-5734.

173. d'Enfert C, Gensse M, Gaillardin C. Fis-

sion yeast and a plant have functional homologues of the Sar1 and Sec12 proteins involved in ER to Golgi traffic in budding yeast. EMBO J 1992; 11: 4205-4211.

174. d'Enfert C, Wuestehube LJ, Lila T, Schekman R. Sec12p-dependent membrane binding of the small GTP-binding protein Sar1p promotes formation of transport vesicles from the ER. J Cell Biol 1991; 114:663-670.

175. Desnos C, Clift-O'Grady L, Kelly RB. Biogenesis of synaptic vesicles in vitro. J Cell Biol 1995; 130:1041-1049.

175a. Diatloff-Zito C, Gordon AJE, Duchaud E, Merlin G. Isolation of an ubiquitously expressed cDNA encoding human dynamin II, a member of the large GTP-binding protein family. Gene 1995; 163:301-306.

176. Dintzis SM, Pfeffer SR. The mannose 6-phosphate receptor cytoplasmic domain is not sufficient to alter the cellular distribution of a chimeric EGF receptor. EMBO J 1990: 9:77-84.

177. Dintzis SM, Velculescu VE, Pfeffer SR. Receptor extracellular domains may contain trafficking information: studies of the 300-kDa mannose 6-phosphate receptor. J Biol Chem 1994; 269: 12159-12166.

178. Dirac-Svejstrup AB, Soldati T, Shapiro AD, Pfeffer SR. Rab-GDI presents functional Rab9 to the intracellular transport machinery and contributes selectivity to Rab9 membrane attachment. J Biol Chem 1994; 269:15427-15430.

179. Dobberstein B. On the beaten pathway. Nature 1994; 367:599-560.

180. Donaldson JG, Cassel D, Kahn RA, Klausner RD. ADP- ribosylation factor, a small GTP-binding protein, is required for binding of the coatamer protein βCOP to Golgi membranes. Proc. Natl. Acad. Sci 1992; 89:6408-6412.

181. Donaldson JG, Finazzi D, Klausner RD. Brefeldin A inhibits Golgi membrane-catalysed exchange of guanine nucleotide onto ARF protein. Nature 1992; 360: 350-352.

182. Donaldson JG, Lippincott-Schwartz J, Bloom GS et al. Dissociation of a 110-kD peripheral membrane protein from the Golgi apparatus is an early event in brefeldin A action. J Cell Biol 1990; 111:2295-2306.

183. Donaldson JG, Lippincott-Schwartz J, Klausner R. Guanine nucleotides modulate the effects of brefeldin A in semi-permeabilized cells: regulation of the association of a 110-kD peripheral membrane protein with the Golgi apparatus. J Cell Biol 1991; 112:579-588.

184. Donaldson JG, Kahn RA, Lippincott-Schwartz J, Klausner RD. Binding of ARF and β-COP to Golgi membranes: possible regulation by a trimeric G protein. Science 1991; 254:1197-1199.

185. Dotti CG, Simons K. Polarized sorting of viral glycoproteins to the axon and dendrites of hippocampal neurons in culture. Cell 1990; 62:63-72.

186. Dotti CG, Parton RG, Simons K. Polarized sorting of glypiated proteins in hippocampal neurons. Nature 1991; 349:158-161.

187. D'Souza-Schorey C, Li G, Colombo MI, Stahl PD. A regulatory role for ARF6 in receptor-mediated endocytosis. Science 1995; 267:1175-1178.

188. Duden R, Allan V, Kreis T. Involvement of β-COP in membrane traffic through the Golgi complex. Trends Cell Biol 1991; 1:14-19.

189. Duden R, Griffiths G, Frank R et al. β-COP, a 110 kd protein associated with non-clathrin-coated vesicles and the Golgi complex, shows homology to β-adaptin. Cell 1991; 64:649-665.

190. Duden R, Hosobuchi M, Hamamoto S et al. Yeast β- and β'-coat proteins (COP). two coatomer subunits essential for endoplasmic reticulum-to-Golgi protein traffic. J Biol Chem 1994; 269: 24486-24495.

191. Dunn B, Stearns T, Botstein D. Specifying domains distinguish the ras-related GTPases Ypt1 and Sec4. Nature 1993; 362:563-565.

192. Dupree P, Parton RG, Raposo G, et al.. Caveolae and sorting in the trans Golgi network of epithelial cells. EMBO J 1993: 12:1597-1605.

193. Duronio RJ, Reed SI, Gordon JI. Mutations of human myristoyl-CoA:protein N-myristoyltransferase cause temperature-sensitive myristic acid auxotrophy in Saccharomyces cerevisiae. Proc Natl Acad Sci 1992; 89:4129-4133.

194. Dustin ML, Baranski TJ, Sampath S, Kornfeld S. A novel mutagenesis strategy identifies distinctly spaced amino acid sequences that are required for the phosphorylation of both the oligosaccharides of procathepsin D by N-acetylglucosamine 1-phosphotransferase. J Biol Chem 1995; 270:170-179.

195. Eakle KA, Bernstein M, Emr SD. Characterization of a component of the yeast secretion machinery: identification of the SEC18 gene product. Mol Cell Biol 1988; 8:4098-4109.

196. Eberie W, Sander C, Klaus W. The essential tyrosine of the internalization signal in lysosomal acid phosphatase is part of a β turn. Cell 1991; 67:1203-1209.

197. Edelmann L, Hanson PI, Chapman ER, Jahn R. Synaptobrevin binding to synaptophysin: a potential mechanism for controlling the exocytotic fusion machine. EMBO J 1995; 14:224-231.

198. Ekena K, Stevens TH. The Saccharomyces cerevisiae MVP1 gene interacts with VPS1 and is required for vacuolar protein sorting. Mol Cell Biol 1995; 15: 1671-1678.

199. Elazar Z, Mayer T, Rothman JE. Removal of Rab GTP-binding proteins from Golgi membranes by GDP dissociation inhibitor inhibits inter-cisternal transport in the Golgi stacks. J Biol Chem 1994; 269:794-797.

200. Elazar Z, Orci L, Sotermann J et al. ADP-ribosylation factor and coatomer couple fusion to vesicle budding. J Cell Biol 1994; 124:415-424.

201. Elferink LA, Anzai K, Scheller RH. Rab15, a novel low molecular weight GTP-binding protein specifically expressed in rat brain. J Biol Chem 1992; 267:5768-5775.

202. Elferink LA, Trimble WS, Scheller RH. Two vesicle-associated membrane protein genes are differentially expressed in the rat central nervous system. J Biol Chem 1989; 264:11061-11064.

203. Ercolani L, Stow JL, Boyle JF et al. Membrane localization of the pertussis toxin-sensitive G-protein subunits αi-2 and αi-3 and expression of a metallo-thionein-αi-2 fusion gene in LLC-PK1 cells. Proc Natl Acad Sci 1990; 87: 4635-4639.

204. Esmon B, Novick P, Schekman R. Compartmentalized assembly of oligosaccharides on exported glycoproteins in yeast. Cell 1981; 25:451-460.

205. Esnault Y, Blondel M-O, Deshaies RJ, et al. The yeast SSS1 gene is essential for secretory protein translocation and encodes a conserved protein of the endoplasmic reticulum. EMBO J 1993; 12:4083-4093.

205a. Espenshade P, Gilmeno RE, Holzmacher E et al. Yeast SEC16 gene encodes a multidomain vesicle coat protein that interacts with Sec23p. J Cell Biol 1995; 131:311-324.

205b. Farquhar MG, Palade GE. The Golgi apparatus (complex)-(1954-1981)-from artifact to center stage. J Cell Biol 1981; 91:77s-103s.

206. Feldheim D, Schekman R. Sec72p contributes to the selective recognition of signal peptides by the secretory polypeptide translocation complex. J Cell Biol 1994; 126:935-943.

207. Ferro-Novick S, Jahn R. Vesicle fusion from yeast to man. Nature 1994; 370: 191-193.

208. Ferro-Novick S, Novick P. The role of GTP-binding proteins in transport along the exocytic pathway. Annu Rev Cell Biol 1993; 9:575-599.

209. Fiedler K, Kobayashi T, Kuzchalia TV, Simons K. Glycosphingolipid-enriched, detergent-insoluble complexes in protein sorting in epithelial cells. Biochemistry 1993; 32:6365-6373.

210. Fiedler K, Lafont F, Parton RG, Simons K. Annexin XIIIb: A novel epithelial specific annexin is implicated in vesicular traffic to the apical plasma membrane. J Cell Biol 1995; 128:1043-1053.

211. Fiedler K, Parton RG, Kellner R, et al. VIP36, a novel component of glycolipid rafts and exocytotic carrier vesicles in epithelial cells. EMBO J 1994; 13:1729-1740.

212. Fiedler K, Simons K. A putative novel class of animal lectins in the secretory pathway homologous to Leguminous lectins. Cell 1994; 77:625-626.

213. Finegold AA, Johnson DI, Farnsworth CC et al. Protein geranylgeranyltransferase of Saccharomyces cerevisiae is specific for Cys-Xaa-Xaa-Leu motif proteins and requires the CDC43 gene product but not the DPR1 gene product. Proc Natl Acad Sci 1991; 88:4448-4452.

214. Fleischer B, McIntyre O, Kempner ES. Target sizes of galactosyltransferae, sialyltransferase, and uridine diphosphatase in Golgi apparatus of rat liver. Biochemistry 1993; 32:2076-2081.

215. Franzusof A, Schekman R. Functional compartments of the yeast Golgi apparatus are defined by the sec7 mutation. EMBO J 1989; 8:2695-2702.

216. Fujimura K, Tanaka K, Nakano A, Tohe A. The Saccharomyces cerevisiae MSI4 gene encodes the yeast counterpart of component A of Rab geranylgeranyltransferase. J Biol Chem 1994; 269: 9205-9212.

217. Fukuda M. Lysosomal membrane glycoproteins. J Biol Chem 1991; 266: 21327-21330.

218. Freedman RB. Protein disulfide isomerases: multiple roles in the modification of nascent secretory proteins. Cell 1989; 57:1069-1072.

219. Galli T, Chilcote T, Mundigl O et al. Tetanus Toxin-mediated cleavage of cellubrevin impairs exocytosis of transferrin receptor-containing vesicles in CHO cells. J Cell Biol 1994; 125:1015-1024.

220. Gallusser A, Kirchhausen T. The beta 1 and beta 2 subunits of the AP complexes are the clathrin coat assembly components. EMBO J 1993; 13:5237-5244.

221. Gallwitz D, Donath C, Sander C. A yeast gene encoding a protein homologous to the human c-has/bas proto-oncogene product. Nature 1983; 306:704-707.

222. Gammie AE, Kurihara LJ, Vallee RB, Rose MD. DNM1, a dynamin-related gene, participates in endosomal trafficking in yeast. 1995; J Cell Biol 130: 553-566.

223. Garcia EP, Gatti E, Butler M, et al. A rat brain Sec1 homologue related to Rop and UNC18 interacts with syntaxin. Proc Natl Acad Sci 1994; 91:2003-2007.

224. Garrett MD, Kabcenell AK, Zahner JE et al. Interaction of Sec4 with GDI proteins from bovine brain, Drosophila melanogaster and Saccharomyces cerevisiae. FEBS Letters 1993; 331:233-238.

225. Garrett MD, Zahner JE, Cheney CM, Novick PJ. GDI1 encodes a GDP dissociation inhibitor that plays an essential role in the yeast secretory pathway. EMBO J 1994; 13:1718-1728.

226. Gaynor EC, te Heesen S, Graham TR, et al. Signal-mediated retrieval of a membrane protein from the Golgi to the ER in yeast. J Cell Biol 1994; 127:653-665.

227. Geffen I, Fuhrer C, Leitinger B, et al. Related signals for endocytosis and basolateral sorting of the asialoglycoprotein receptor. J Biol Chem 1993; 268:20772-20777.

228. Geppert M, Goda Y, Hammer RE et al. Synaptotagmin I: a major Ca²⁺ sensor for transmitter release at a central synapse. Cell 1994; 79:717-727.

229. Gerich, B, Orci L, Tschochner H et al. Non-clathrin-coat protein α is a conserved subunit of coatomer and in Saccharomyces cerevisiae is essential for growth. Proc Natl Acad Sci 1995; 92: 3229-3233.

230. Gerst JE, Rodgers L, Riggs M, Wigler M. SNC1, a yeast homolog of the synaptic vesicle-associated membrane protein/synaptobrevin gene family: genetic interactions with the RAS and CAP genes. Proc Natl Acad Sci 1992; 89:4338-4342.

231. Gething MJ, Sambrook J. Protein folding in the cell. Nature 1992; 355:33-45.

231a. Gilmeno RE, Espenshade P, Kaiser CA. SED4 encodes a yeast endoplasmic reticulum protein that binds Sec16p and participates in vesicle formation. J Cell Biol 1995; 131:325-338.

232. Gilmore R. Protein translocation across the endoplasmic reticulum: a tunnel with toll booths at entry and exit. Cell 1993; 75:589-592.

233. Glick BS, Rothman JE Possible role if fatty acyl-coenzyme A in intracellular protein transport. Nature 1987; 326:309-312.

234. Glickman JN, Conibear E, Pearse BMF. Specificity of binding of clathrin adaptors to signals on the mannose-6-phosphate/insulin-like growth factor II receptor. EMBO J 1989; 8:1041-1047.

235. Goda Y, Pfeffer S. Cell-free systems to study vesicular transport along the secretory and endocytotic pathways. FASEB J 1989; 3:2488-2495.

236. Goepfert PA, Wang G, Mulligan MJ. Identification of an ER retrieval signal in a retrovirus glycoprotein. Cell 1995; 82:543-544.

237. Goldenring JR, Shen KR, Vaughan HD, Modlin IM. Identification of a small GTP-binding protein, Rab25, expressed in the gastrointestinal mucosa, kidney, and lung. J Biol Chem 1993; 268:18419-18422.

238. Goodman LE, Judd SR, Farnsworth CC et al. Mutants of Saccharomyces cerevisiae defective in the farnesylation of ras proteins. Proc Natl Acad Sci 1990; 87:9665-9669.

239. Gordon JI, Duronio RJ, Rudnick DA et al. Protein N-myristoylation. J Biol Chem 1991; 266:8647-8650.

240. Gorlich D, Hartmenn E, Prehn S, Rapoport TA. A protein of the endoplasmic reticulum involved early in polypeptide translocation. Nature 1992; 357:47-52.

241. Gorlich D, Prehn S, Hartmenn E et al. A mammalian homolog of SEC61p and SECYp is associated with ribosomes and nascent polypeptides during translocation. Cell 1992; 71:489-503.

242. Gorlich D, Rapoport TA. Protein translocation into proteoliposomes reconstituted from purified components of the endoplasmic reticulum. Cell 1993; 75:615-630.

243. Gorvel J-P, Chavrier P, Zerial M, Gruenberg J. Rab5 controls early endosome fusion in vitro. Cell 1991; 64:915-925.

244. Gottlieb TA, Beaudry G, Rizzolo L, et al. Secretion of endogenous and exogenous proteins from polarized MDCK cell monolayers. Proc Natl Acad Sci 1986; 83:2100-2104.

245. Goud B, Salminen A, Walworth NC, Novick PJ. A GTP-binding protein required for secretion rapidly associates with secretory vesicles and the plasma membrane in yeast. Cell 1988; 53:753-768.

246. Graham TR, Emr SD. Compartmental organization of Golgi-specific protein modification and vacuolar protein sorting events defined in a yeast sec18 (NSF) mutant. J Cell Biol 1991; 114:207-218.

247. Graham TR, Seeger M, Payne GS et al. Clathrin-dependent localization of α1,3-mannosyltransferase to the Golgi complex of Saccharomyces cerevisiae. J Cell Biol 1994; 127:667-678.

248. Graham TR, Krasnov VA. Sorting of yeast α1,3-mannosyltransferase is mediated by a luminal domain interaction, and a transmembrane domain signal that can confer clathrin-dependent Golgi localization to a secreted protein. Mol Cell Biol 1995; 6:809-824.

249. Gravotta D, Adesnik M, Sabatini DD. Transport of influenza HA from the trans Golgi network to the apical surface of MDCK cells permeabilized in their basolateral plasma membranes: energy dependence and involvement of GTP-binding proteins. J Cell Biol 1990; 111:2893-2908.

250. Greenburg G, Blobel G. cDNA-derived primary structure of the 25-kDa subunit of canine microsomal signal peptidase

complex. J Biol Chem 1994; 269: 25354-25358.

251. Griff IC, Schekman R, Rothman JE, Kaiser CA. The yeast SEC17 gene product is functionally equivalent to mammalian α-SNAP protein. J Biol Chem 1992; 267:12106-12115.

252. Griffiths G, Ericsson M, Krijnse-Locker J et al. Localization of the lys, asp, glu, leu tetrapeptide receptor to the Golgi complex and the intermediate compartment in mammalian cells. J Cell Biol 1994; 127:1557-1574.

253. Griffiths G, Pepperkok R, Krijnse Locker J, Kreis TE. Immunocytochemical localization of β-COP to the ER-Golgi boundary and the TGN. J Cell Sci 1995; 108:2839-2856.

253a. Griffiths G, Simons K. The trans Golgi network: sorting at the exit site of the Golgi complex. Science 1986; 234: 438-443.

254. Groesch ME, Ruohola H, Bacon R, et al. Isolation of a functional vesicular intermediate that mediates ER to Golgi transport in yeast. J Cell Biol 1990; 111:45-53.

255. Grote E, Hao JC, Bennett MK, Kelly RB. A targeting signal in VAMP regulating transport to synaptic vesicle. Cell 1995; 81:581-589.

256. Guo Q, Vasile E, Krieger M. Disruption in Golgi structure and membrane traffic in a conditional lethal mammalian cell mutant are corrected by ε-COP. J Cell Biol 1994; 125:1213-1224.

257. Hamburger D, Egerton M, Riezman H. Yeast Gaa1p is required for attachment of a completed GPI anchor onto proteins. J Cell Biol 1995; 129:629-639.

258. Hammerton RW, Krzeminski, KA, Mays RW, et al. Mechanism for regulated cell surface distribution of Na⁺, K⁺-ATPase in polarized epithelial cells. Science 1991; 254: 847-850.

259. Hall A. Ras-related GTPases and the cytoskeleton. Science 1990; 249:635-640.

260. Hall A. The cellular functions of small GTP-binding proteins. Mol Biol Cell 1992; 3:475-479.

261. Hammond C, Helenius A. A chaperone with a sweet tooth. Cur Biol 1993; 3:884-886.

262. Hammond C, Helenius A. Folding of VSV G protein: sequential intraction with BiP and calnexin. Science 1994; 266:456-458.

263. Hammond C, Helenius A. Quality control in the secretory pathway. Curr Opini Cell Biol 1995:7:523-529.

264. Haney PM, Levy MA, Strube MS, Mueckler M. Insulin-sensitive targeting of the GLU4 glucose transporter in L6 myoblasts is conferred by its COOH-terminal cytoplasmic tail. J Cell Biol 1995; 129:641-658.

265. Hanson PI, Otto H, Barton N, Jahn R. The N-ethylmaleimide-sensitive fusion protein and α-SNAP induce a conformational change in syntaxin. J Biol Chem 1995; 270:16955-16961.

266. Hara-Kuge S, Kuge O, Orci L et al. En Bloc incorporation of coatomer subunits during the assembly of COP-caoted vesicles. J Cell Biol 1994; 124:883-892.

267. Hardwick KG, Boothroyd JC, Rudner AD, Pelham HRB. Genes that allows yeast cells to grow in the absence of the HDEL receptor. EMBO J 1992; 11: 4187-4195.

268. Hardwick KG, Lewis MJ, Semenza J, et al. ERD1, a yeast gene required for the retention of luminal endoplasmic reticulum proteins, affects glycoprotein processing in the Golgi. EMBO J 1990; 9:623-630.

269. Hardwick KG, Pelham HRB. SEC5 encodes a 39-kD integral membrane protein required for vesicular transport between the ER and the Golgi complex. J Cell Biol 1992; 119:513-521.

270. Harrison-Lavoie KJ, Lewis VA, Hynes GM et al. A 102 kDa subunit of a Golgi-associated particle has homology to β-subunit of trimeric G proteins. EMBO J 1993; 12:2847-2853.

271. Harter C, Mellman I. Transport of the lysosomal membrane glycoprotein lgp120

(lgp-A) to lysosomes does not require appearance on the plasma membrane. J Cell Biol 1992; 117:311-325.

272. Hartmann E, Gorlich D, Kostka S et al. A tetrameric complex of membrane proteins in the endoplasmic reticulum. Eur J Biochem 1993; 214:375-381.

273. Hartmann E, Sommer T, Prehn S et al. Evolutionary conservation of components of the protein translocation complex. Nature 1994; 367:654-657.

274. Hata Y, Davletov B, Petrenko AG et al. Interaction of synaptotagmin with the cytoplasmic domains of neurexins. Neuron 1993; 10:307-315.

275. Hata Y, Slaughter CA, Sudhof TC. Synaptic vesicle fusion complex contains unc-18 homologue bound to syntaxin. Nature 1993; 366:347-351.

276. Hata Y, Sudhof TC. A novel ubiquitous form of Munc-18 interacts with multiple syntaxins. use of the yeast two-hybrid system to study interactions between proteins involved in membrane traffic. J Biol Chem 1995; 270:13022-13028.

277. Haubruck H, Engelke U, Mertins P, Gallwitz D. Structural and functional analysis of ypt2, an essential ras-related gene in the fission yeast Schizosaccharomyces cerevisiae encoding a Sec4 protein homologue. EMBO J 1990; 9:1957-1962.

278. Haubruck H, Prange R, Vorgias C, Gallwitz D. The ras-related mouse ypt1 protein can functionally replace the YPT1 gene product in yeast. EMBO J 1989; 8:1427-1432.

279. Haun RS, Tsai SC, Adamik R et al. Effect of myristoylation on GTP-dependent binding of ADP-ribosylation factor to Golgi. J Biol Chem 1993; 268:7064-7068.

280. Hayashi T, McMahon H, Yamasaki S et al. Synaptic vesicle membrane fusion complex: action of clostridial neurotoxins on assembly. EMBO J 1994; 13:5051-5061.

281. Hayashi T, Yamasaki S, Nauenburg S et al. Disassmbly of the reconstituted synaptic vesicle membrane fusion complex in vitro. EMBO J 1995; 14:2317-2325.

282. He B, Chen P, Chen SY et al. RAM2, an essential gene of yeast, and RAM1 encode the two polypeptide components of the fanesyltransferase that prenylates α-factor and ras proteins. Proc Natl Acad Sci 1991; 88:11373-11377.

283. Hebert DN, Foellmer B, Helenius A. Glucose trimming and reglucosylation determine glycoprotein association with calnexin in the endoplasmic reticulum. Cell 1995; 81:425-433.

284. Helenius A, Marquardt T, Braakman I. The endoplasmic reticulum as a protein-folding compartment. Trends in Cell Biol 1992; 2:227-231.

285. Helms JB, Palmer DJ, Rothman JE. Two distinct populations of ARF bound to Golgi membranes. J Cell Biol 1993; 121:751-760.

286. Helms JB, Rothman JE. Inhibition by Brefeldin A of a Golgi membrane enzyme that catalyses exchange of guanine nucleotide bound to ARF. Nature 1992; 360:352-354.

287. Hengst L, Lehmeier T, Gallwitz D. The ryh1 gene in the fission yeast Schiosaccharomyces pombe encoding a GTP-binding protein related to ras, rho, and ypt: structure, expression and identification of its human homologue. EMBO J 1990; 9:1949-1955.

288. Herman PK, Emr SD. Characterization of VPS34, a gene required for vacuolar protein sorting and vacuole segregation in Saccharomyces cerevisiae. Mol Cell Biol 1990; 10:6742-6754.

289. Herman PK, Stack JH, DeModena JA, Emr SD. A novel protein kinase homolog essential for protein sorting to the yeast lysosome-like vacuole. Cell 1991; 64:425-437.

290. Hess SD, Doroshenko PA, Augustine GJ. A functional role for GTP-binding proteins in synaptic vesicle cycling. Science 1993; 259:1169-1172.

291. Hicke L, Schekman R. Yeast Sec23p acts in the cytoplasm to promote protein

transport from the endoplasmic reticulum to the Golgi complex in vivo and in vitro. EMBO J 1989; 8:1677-1684.

292. Hicke L, Yoshihisa T, Schekman R. Sec23p and a novel 105-kDa protein function as a multimeric complex to promote vesicle budding and protein transport from the endoplasmic reticulum. Mol Biol Cell 1992; 3:667-676.

293. High S. Membrane protein insertion into the endoplasmic reticulum-another channel tunnel? BioEssay 1992; 14:535-540.

294. High S, Andersen SSL, Gorlich D et al. Sec61p is adjacent to nascent type I and type II signal-anchor proteins during their membrane insertion. J Cell Biol 1993; 121:743-750.

295. High S, Dobberstein B. The signal sequence interacts with the methionine-rich domain of the 54-kD protein of signal recognition particle. J Cell Biol 1991; 113:229-233.

296. High S, Martoglio B, Gorlich D et al. Site-specific photocross-linking reveals that Sec61p and TRAM contact different regions of a membrane-inserted signal sequence. J Biol Chem 1993; 268: 26745-26751.

297. Hill KJ, Stevens TH. Vma21p is a yeast membrane protein retained in the endoplasmic reticulum by a di-lysine motif and is required for the assembly of the vacuolar H$^+$-ATPase complex. Mol Biol Cell 1994; 5:1039-1050.

298. Hinshaw JE, Schmid SL. Dynamin self-assemble into rings suggesting a mechanism for coated vesicle budding. Nature 1995; 374:190-192.

299. Hobman TC, Woodward L, Farquhar MG. The rubella virus E2 and E1 spike glycoproteins are targeted to the Golgi complex. J Cell Biol 1993; 121:269-281.

300. Hochstenbach, F, David V, Watkins S, Brenner MB. Endoplasmic reticulum resident protein of 90 kilodaltons associates with T- and B-cell antigen receptors and major histocompatibility complex antigens during their assembly. Proc Natl Acad Sci 1992; 89:4734-3738.

301. Hodel A, Schafer T, Gerosa D, Burger MM. In chromaffin cells, the mammalian Sec1p homologue is a syntaxin 1A-binding protein associated with chromaffin granules. J Biol Chem 1994; 269: 8623-8626.

302. Hong W. The plasma membrane and intercellular junction. In: LeBouton AV, ed. Molecular and Cell Biology of the Liver. CRC press, 1993:85-115.

303. Hong W, Tang BL. Protein trafficking along the exocytotic pathway. Bioassay 1993; 15:231-238.

304. Honing S, Hunziker W. Cytoplasmic determinants involved in direct lysosomal sorting, endocytosis, and basolateral targeting of rat lgp120 (lamp-I) in MDCK cells. J Cell Biol 1995; 128:321-332.

305. Horazdovsky BF, Busch GR, Emr SD. VPS21 encodes a rab5-like GTP binding protein that is required for the sorting of yeast vacuolar proteins. EMBO J 1994; 13:1297-1309.

306. Horazdovsky BF, DeWald DB, Emr SD. Protein transport to the yeast vacuole. Cur Opi Cell Biol 1995; 7:544-551.

307. Horikawa HPM, Saisu H, Ishizuka T et al. A complex of rab3A, SNAP-25, VAMP/synaptobrevin-2 and syntaxins in brain presynaptic terminals. FEBS Letters 1993; 330:236-240.

308. Hosobuchi M, Kreis T, Schekman R. SEC21 is a gene required for ER to Golgi protein transport that encodes a subunit of a yeast coatomer. Nature 1992; 360:603-605.

309. Hsu VW, Shah N, Klausner RD. A brefeldin A-like phenotype is induced by overexpression of a human ERD-2-like protein, ELP-1. Cell 1992; 69:625-635.

310. Huber LA, Dupree P, Dotti CG. A deficiency of the small GTPase rab8 inhibits membrane traffic in developing neurons. Mol Cell Biol 1995; 15:918-924.

311. Huber LA, Pimplikar S, Parton RG et al. Rab8, a small GTPase involved in vesicular traffic between the TGN and the basolateral plasma membrane. J Cell Biol 1993; 123:35-45.

312. Huber LA, Ullrich O, Takai Y et al. Mapping of Ras-related GTP-binding proteins by GTP overlay following two-dimensional gel electrophoresis. Proc Natl Acad Sci 1994; 91:7874-7878.

313. Humphrey JS, Peters PJ, Yuan LC, Bonifacino JS. Localization of TGN38 to the trans Golgi network: involvement of a cytoplasmic tyrosine-containing sequence. J Cell Biol 1993; 120: 1123-1135.

314. Hunziker W, Fumey C. A di-leucine motif mediates endocytosis and basolateral sorting of macrophage IgG Fc receptors in MDCK cells. EMBO J 1994; 13:2963-2969.

315. Hunziker W, Harter C, Matter K, Mellman I. Basolateral sorting in MDCK cells requires a distinct cytoplasmic domain determinant. Cell 1991; 66: 907-920.

316. Hunziker W, Mellman I. Expression of mcarophage-lymphocyte Fc receptors in Madin-Darby canine kidney cells: polarity and transcytosis differ for isoforms with or without coated pit localization domains. J Cell Biol 1989; 109: 3291-3302.

317. Hurtley SM, Helenius A. Protein oligomerization in the endoplasmic reticulum. Annu Rev Cell Biol 1989; 5:277-307.

318. Huttner WB, Gerdes H-H, Rosa P. The granin (chromogranin/secretogranin) family. Trends Biochem Sci 1991; 16:27-30.

319. Huttner WB, Natori S. Helper proteins for neuroendocrine secretion. Cur Biol 1995; 5:242-245.

320. Hwang C, Sinskey AJ, Lodish HF. Oxidized redox state of glutathione in the endoplasmic reticulum. Science 1992; 257:1496-1502.

321. Ikonen E, Tagaya M, Ullrich O, et al. Different requirements for NSF, SNAP, and Rab proteins in apical and basolateral transport in MDCK cells. Cell 1995; 81: 571-580.

322. Itin C, Kappeler, F, Linstedt, AD, Hauri H-P. A novel endocytosis signal related to the KKXX ER-retrieval signal. EMBO J 1995; 14:2250-2256.

323. Itin C, Schindler R, Hauri H-P. Targeting of protein ERGIC-53 to the ER/ERGIC/cis-Golgi recycling pathway. J Cell Biol 1995; 131:57-67.

324. Jackson AP, Parham P. Structure of human clathrin light chains. J Biol Chem 1988; 263; 16688-16695.

325. Jackson, AP, Seow HF, Holmes N et al. Clathrin light chains contain brain-specific insertion sequences and a region of homology with intermediate filaments. Nature 1987; 326:154-159.

326. Jackson MR, Nilsson T, Peterson PA. Identification of a consensus motif for retention of transmembrane proteins in the endoplasmic reticulum. EMBO J 1990; 9:3153-3162.

327. Jackson MR, Nilsson T, Peterson PA. Retrieval of transmembrane proteins to the endoplasmic reticulum. J Cell Biol 1993; 121:317-333.

328. Jannatipour M, Rokeach LA. The Schizosaccharomyces pombe homologue of the chaperone calnexin is essential for viability. J Biol Chem 1995; 270:4845-4853.

329. Jiang Y, Rossi G, Ferro-Novick S. Bet2p and Mad2p are components of a prenyltransferase that adds geranylgeranyl onto Ypt1p and Sec4p. Nature 1993; 84-86.

330. Jiang Y, Ferro-Novick S. Identification of yeast component A: reconstitution of the geranylgeranyltransferase that modifies Ypt1p and Sec4p. Proc Natl Acad Sci 1994; 91:4377-4381.

331. Joberty G, Tavitian A, Zahraoui A. Isoprenaylation of Rab proteins possessing a C-terminal CaaX motif. FEBS Letters 1993; 330:323-328.

332. John DCA, Grant ME, Bulleid NJ. Cell-free synthesis and assembly of prolyl 4-hydroxylase: the role of the β-subunit (PDI) in preventing misfolding and segregation of the α-subunit. EMBO J 1993; 12:1587-1595.

333. Johnson DI, O'Brien JM, Jacobs CW. Isolation and sequence analysis of CDC43, a gene involved in the control of cell polarity in Saccharomyces cerevisiae. Gene 1991; 98:149-150.

334. Johnson DR, Cox AD, Solski PA et al. Functional analysis of protein N-myristoylation: metabolic labeling studies using three oxygen-substituted analogs of myristic acid and cultured mammalian cells provide evidence for protein-sequence-specific incorporation and analog-specific redistribution. Proc Natl Acad Sci 1990; 87:8511-8515.

335. Johnson LM, Bankaitis VA, Emr SD. Distinct sequence determinants direct intracellular sorting and modification of a yeast vacuolar protease. Cell 1987; 48:875-885.

336. Johnson KF, Chan W, Kornfeld S. Cation-dependent mannose 6-phosphate receptor contains two internalization signals in its cytoplasmic domain. Proc Natl Acad Sci 1990; 87:10010-10014.

337. Johnson KF, Kornfeld S. A His-Leu-Leu sequence near the carboxyl terminus of the cytoplasmic domain of the cation-dependent mannose 6-phosphate receptor is necessary for the lysosomal enzyme sorting function. J Biol Chem 1992; 267:17110-17115.

338. Johnson KF, Kornfeld S. The cytoplasmic tail of the mannose 6-phosphate/insulin-like growth factor-II receptor has two signals for lysosomal enzyme sorting in the Golgi. J Cell Biol 1992; 119:249-257.

339. Jones S, Litt RJ, Richardson CJ, Segev N. Requirement of nucleotide exchange factor for Ypt1 GTPase mediated protein transport. J Cell Biol 1995; 130:1051-1061.

340. Kahn RA, Clark J, Rulka C et al. Mutational analysis of sacchromyces cerevisiae ARF1. 1995; 270:143-150.

341. Kahn RA, Gilman AG. Purification of a protein cofactor involved required for ADP-ribosylation of the stimulatory regulatory component of adenylate cyclase by cholera toxin. J Biol Chem 1984; 259:6228-6234.

342. Kahn RA, Kern FG, Clark J et al. Human ADP-ribosylation factors. a functionally conserved family of GTP-binding proteins. J Biol Chem 1991; 266:2606-2614.

343. Kahn RA, Randazzo P, Serafini T et al. The amino terminus of ADP-ribosylation factor (ARF) is a critical determinant of ARF activities and is a potent and specific inhibitor of protein transport. J. Biol. Chem 1992; 267:13039-13046.

344. Kalies K-U, Gorlich D, Rapoport TA. Binding of ribosomes to the rough endoplasmic reticulum mediated by the Sec61p-complex. J Cell Biol 1994; 126:925-934.

345. Kao CY, Draper RK. Retention of secretory proteins in an intermediate compartment and disappearance of the Golgi complex in an END4 mutant of Chinese hamster ovary cells. J Cell Biol 1992; 117:701-715.

346. Kaplan JM, Varmus HE, Bishop JM. The src protein contains multiple domains for specific attachment to membranes. Mol Cell Biol 1990; 10:1000-1009.

347. Karaoglu D, Kelleher DJ, Gilmore R. Functional characterization of Ost3p. loss of the 34-kD subunit of the Saccharomyces cerevisiae oligosaccharyltransferase results in biased underglycosylation of acceptor substrates. J Cell Biol 1995; 130:567-577.

348. Katagiri H, Terasaki J, Murrata T, et al. A novel isoform of syntaxin-binding protein homologous to yeast Sec1 expressed ubiquitously in mammalian cells. J Biol Chem 1995; 270:4963-4966.

349. Deleted during revision.

350. Kee Y, Lin RC, Hsu SC, Scheller R. Distinct domains of syntaxin are required for synaptic vesicle fusion complex formation and dissociation. Cell 1995; 14:991-998.

351. Kehlenbach RH, Matthey J, Huttner WB. XLαs is a new type of G protein. Nature 1994; 372:804-809 (Correction in 1995; 375:253).

352. Kelleher DJ, Gilmore R. The Saccharomyces cerevisiae oligosaccharyltransferase is a protein complex composed of Wbp1p, Swp1p, and four additional

polypeptides. J Biol Chem 1994; 269: 12908-12917.

353. Kelleher DJ, Kreibich G, Gilmore R. Oligosaccharyltransferase activity is associated with a protein complex composed of ribophorins I and II and a 48 kD protein. Cell 1992; 69:55-65.

354. Kelly RB. Secretory granule and synaptic vesicle formation. Curr Opini Cell Biol 1991; 3:654-660.

355. Kelly RB. Storage and release of neurotransmitters. Cell 1993; 72:43-53.

356. Kim PS, Arvan P. Calnexin and BiP act as sequential molecular chaperones during thyroglobulin folding in the endoplasmic reticulum. J Cell Biol 1995; 128:29-38.

357. Kinsella BT, Maltese WA. Rab GTP-binding protiens implicated in vesicular transport are isoprenylated in vitro at cysteines within a novel carboxyl-terminal motif. J Biol Chem 1991; 266: 8540-8544.

358. Kirchhausen T. Identification of a putative yeast homolog of mammalian β chains of the clathrin-associated protein complexes. Mol Cell Biol 1990; 10: 6089-6090.

359. Kirchhausen T, Davis AC, Frucht S et al. AP17 and AP19, the mammalian small chains of the clathrin-associated protein complexes show homology to Yap17p, their putative homology in yeast. J Biol Chem 1991; 266: 11153-11157.

360. Kirchhausen T, Harrison SC, Chow EP et al. Clathrin heavy chain: molecular cloning and complete primary structure. Proc Natl Acad Sci 1987; 84:8805-8809.

361. Kirchhausen T, Nathanson KL, Matsui W et al. Structural and functional division into two domains of the large (100- to 115-kDa) chains of the clathrin-associated protein complex AP-2. Proc Natl Acad Sci 1989; 86:2612-1216.

362. Kishida S, Shirataki H, Sasaki T et al. Rab3A GTPase-activating protein-inhibiting activity of rabphilin-3A, a putative Rab3A target protein. J Biol Chem 1993; 268:22259-22261.

363. Klausner RD, Donaldson JG, Lippincott-Schwartz J. Brefeldin A: insights into the control of membrane traffic and organelle structure. J Cell Biol 1992; 116: 1071-1080.

364. Klionsky DJ, Banta LM, Emr SD. Intracellular sorting and processing of a yeast vacuolar hydrolase: proteinase A propeptide contains vacuolar targeting information. Mol Cell Biol 1988; 8: 2105-2116.

365. Klionsky DJ, Emr SD. Membrane protein sorting: Biogenesis, transport and processing of yeast vacuolar alkaline phosphatase. EMBO J 1989; 8: 2241-2250.

366. Klionsky DJ, Herman PK, Emr SD. The fungal vacuole: composition, function, and biogenesis. Microbiol Rev 1990; 54:266-292.

367. Knittler MR, Haas IG. Interaction of BiP with newly synthesized immunoglobulin light chain molecules: cycles of sequential binding and release. EMBO J 1992; 11:1573-1581.

368. Krijnse Locker J, Ericsson M, Rottier PJM, Griffiths G. Characterization of the budding compartment of mouse hepatitis virus: evidence that transport from the ER to the Golgi complex requires only one vesicular transport step. J Cell Biol 1994; 124:55-70.

369. Krijnse Locker J, Klumperman J, Oorschot V et al. The cytoplasmic tail of mouse hepatitis virus M protein is essential but not sufficient for its retention in the Golgi complex. J Biol Chem 1994; 269:28263-28269.

370. Krijnse Locker J, Opstelten D-J, Ericsson M et al. Oligomerization of a trans Golgi/trans Golgi network retained proteins occurs in the Golgi complex and may be part of its retention. J Biol Chem 1995; 270:8815-8821.

371. Kobayashi T, Storrie B, Simons K, Dotti CG. A functional barrier to movement of lipids in polarized neurons. Nature 1992; 359:647-650.

372. Koda T, Kakinuma M. Molecular cloning of a cDNA encoding a novel small

GTP-binding protein. FEBS Letters 1993; 328:21-24.

373. Kohl NE, Diehl RE, Schaber MD et al. Structural homology among mammalian and Saccharomyces cerevisiae isoprenyl-protein transferases. J Biol Chem 1991; 266:18884-18888.

374. Konrad RJ, Young RA, Record RD et al. The heterotrimeric G-protein Gi is localized to the insulin secretory granules of β-cells and is involved in insulin exocytosis. J Biol Chem 1995; 270:12869-12876.

375. Kornfeld R, Kornfeld S. Assembly of asparagine-linked oligosaccharides. Annu Rev Biochem. 1985; 54:631-664.

376. Kornfeld S. Mellman I. The biogenesis of lysosomes. Annu Rev Cell Biol 1989; 5:483-525.

377. Koster A, Saftig P, Matzner U et al. Targeted disruption of the Mr 46000 mannose 6-phosphate receptor gene in mice results in misrouting of lysosomal proteins. EMBO J 1993; 12:5219-5223.

378. Deleted during revision.

379. Kozutsumi Y, Segal M, Normington K et al. The presence of malfolded proteins in the endoplasmic reticulum signals the induction of glucose-regulated proteins. Nature 1988; 332:462-464.

380. Ktistakis NT, Brown HA, Sternweis PC, Roth MG. Phospholipase D is present on Golgi-enriched membranes and its activation by ADP-ribosylation factor is sensitive to brefeldin A. Proc Natl Acad Sci 1995; 92:4952-4956.

381. Ktistakis NT, Kao CY. Wang RH, Roth MG. A fluorescent lipid analogue can be used to monitor secretory activity and for isolation of mammalian secretion mutants. Mol Biol Cell 1995; 6:135-150.

382. Kuge O, Dascher C, Orci L et al. Sar1 promotes vesicle budding from the endoplasmic reticulum but not Golgi compartments. J Cell Biol 1994; 125:51-65.

383. Kuge O, Hara-Kuge S, Orci L et al. ζ-COP, a subunit of coatomer, is required for COP-coated vesicles assembly. J Cell Biol 1993; 123:1727-1734.

384. Kurzchalia TV, Dupree P, Parton RG, et al. VIP21, a 21-kD membrane protein is an integral component of trans Golgi-network-derived transport vesicles. J Cell Biol 1992; 118:1003-1014.

384a. Kutay U, Hartmann E, Rapoport TA. A class of membrane proteins with a C-terminal anchor. Trends Cell Biol, 1993; 3:72-75.

385. Kutay U, Ahnert-Hilger G, Hartmann E, et al. Transport route for synaptobrevin via a novel pathway of insertion into the endoplasmic reticulum membrane. EMBO J 1995; 14:217-223.

386. Kuznetsov G, Brostrom MA, Brostrom CO. Role of endoplasmic reticular calcium in oligosaccharide processing of α1-antitrypsin. J Biol Chem 1993; 268:2001-2008.

387. Lahtinen U, Dahllof B, Saraste J. Characterization of a 58 kDa cis-Golgi protein in pancreatic exocrine cells. J Cell Sci 1992; 103:321-333.

388. Lai MH, Bard M, Kirsch DR. Identification of a gene encoding a new Ypt/Rab-like monomeric G-protein in Saccharomyces cerevisia. Yeast 1994; 10:399-402.

389. Lambardi D, Soldati T, Riederer MA et al. Rab9 functions in transport from late endosomes to the trans Golgi network. EMBO J 1993; 12:677-682.

390. Lang J, Nishimoto I, Okamoto T et al. Direct control of exocytosis by receptor-mediated activation of the heterotrimeric GTPase G_i and G_0 or by the expression of their active Gα subunits. EMBO J 1995; 14:3635-3644.

391. Latterich M, Frohlich KU, Schekman R. Membrane fusion and the cell cycle: Cdc48p participates in the fusion of ER membranes. Cell 1995; 82:885-893.

392. Lauffer L, Garcia PD, Harkins RN, et al. Topology of signal recognition particle receptor in endoplasmic reticulum membrane. Nature 1985; 318:334-338.

393. Lazarovits J, Roth M. A single amino acid change in the cytoplasmic domain allows the influenza virus hemagglutinin

to be endocytosed through coated pits. Cell 1988; 53:743-752.

394. Le A, Steiner JL, Ferrell GA et al. Association between calnexin and a secretion-incompetent variant of human α1-antitrypsin. J Biol Chem 1994; 269: 7514-7519.

395. Le Biviv A, Quaroni A, Nichols B, Rodrigues-Boulan E. Biogenetic pathways of plasma membrane proteins in Caco-2, a human intestinal epithelial cell line. J Cell Biol 1990; 111:1351-1361.

396. Le Biviv A, Sambuy Y, Mostov K, Rodrigues-Boulan E. Vectorial targeting of an endogenous apical membrane sialoglycoprotein and uvomorulin in MDCK cells. J Cell Biol 1990; 110: 1533-1539.

397. Le Biviv A, Sambuy Y, Patzak A, et al. An internal deletion in the cytoplasmic tail reverses the apical localization of human NGF receptor in transfected MDCK cells. J Cell Biol 1991; 115:607-618.

398. Leberer E, Charuk JHM, Clarke DM et al. Molecular cloning and expression of cDNA encoding the 53,000-dalton glycoprotein of rabbit skeletal muscle sarcoplasmic reticulum. J Biol Chem 1989; 264:3484-3493.

399. Lee FJS, Stevens LA, Kao YL et al. Characterization of a glucose-repressible ADP-ribosylation factor 3 (ARF3) from Sacchromyces cerevisiae. J Biol Chem 1994; 269:20931-20937.

400. Lemmon SK, Pellicena-Palle A, Conley K, Freund CL. Sequence of the clathrin heavy chain from Saccharomyces cerevisiae and requirement of the COOH terminus for clathrin function. J Cell Biol 1991; 112:65-80.

401. Lenhard JM, Kahn RA, Stahl PD. Evidence for ADP-ribosylation factor (ARF) as a regulator of in vitro endosome-endosome fusion. J. Biol. Chem 1992; 267:13047-13052.

402. Lepage-Lezin A, Joseph-Bravo P, Devilliers G et al. Prosomatostatin is processed in the Golgi apparatus of rat neu-ronal cells. J Biol Chem 1991; 266: 1679-1688.

403. Letourneur F, Gaynor EC, Hennecke S, et al. Coatomer is essential for retrieval of dilysine-tagged proteins to the endoplasmic reticulum. Cell 1994; 79:1199-1207.

404. Letourneur F, Hennecke S, Demolliere C, Cosson P. Steric masking of a dilysine endoplasmic reticulum retention motif during assembly of the human high affinity receptor for immunoglobulin E. J Cell Biol 1995; 129:971-978.

405. Letourneur F, Klausner RD. A novel dileucine motif and a tyrosine-based motif independently mediate lysosomal targeting and endocytosis of CD3 chains. Cell 1992; 69:1143-1157.

406. Lewis MJ, Pelham HRB. A human homologue of the yeast HDEL receptor. Nature 1990; 348:162-163.

407. Lewis MJ, Pehham HRB. Ligand-induced redistribution of a human KDEL receptor from the Golgi complex to the endoplasmic reticulum. Cell 1992; 68: 353-364.

408. Lewis MJ, Sweet DJ, Pehham HRB. The ERD2 gene determine the specificity of the luminal ER protein retention system. Cell 1990; 61:1359-1363.

409. Li C, Takei K, Geppert M et al. Synaptic targeting of rabphilin-3A, a synaptic vesicle Ca²⁺/phospholipid-binding protein, depends on Rab3A/3C. Neuron 1994; 13:885-898.

410. Li C, Ullrich B, Zhang JZ et al. Ca²⁺-dependent and -independent activities of neural and non-neural synaptotagmins. Nature 1995; 375:594-599.

411. Li G, Barbieri MA, Colombo MI, Stahl PD. Structural features of the GTP-binding defective Rab5 mutants required for their inhibitory activity on endocytosis. J Biol Chem 1994; 269:14631-14636.

412. Li R, Havel C, Watson JA, Murray AW. The mitotic feedback control gene MAD2 encodes the α-subunit of a prenyltransferase. Nature 1993; 366:82-84.

413. Lian JP. Ferro-Novick S. Bos1p, an integral membrane protein of the endoplasmic reticulum to Golgi transport vesicles, is required for their fusion competence. Cell 1993; 73:735-745.

414. Lian JP, Stone S, Jiang Y et al. Ypt1p implicated in v-SNARE activation. Nature 1994; 372:698-701.

415. Lippincott-Schwartz J, Donaldson JG, Schweizer A et al. Microtubule-dependent retrograde transport of proteins into the ER in the presence of brefeldin A suggests an ER recycling pathway. Cell 1990; 60:821-836.

416. Lippincott-Schwartz J, Yuan LC, Bonifacino JS, Klausner RD. Rapid redistribution of Golgi proteins into the ER in cells treated with brefeldin A: evidence for membrane cycling from Golgi to ER. Cell 1989; 56:801-813.

417. Lisanti MP, Rodriguez-Boulan E. Glycophospholipid membrane anchoring provides clues to the mechanism of protein sorting in polarized epithelial cells. Trend Biochem Sci 1990; 15:113-118.

418. Lisanti MP, Caras IW, Davitz MA, Rodriguez-Boulan E. A glycophospholipid membrane anchor acts as an apical targeting signal in polarized epithelial cells. J Cell Biol 1989; 109:2145-2156.

419. Lisanti MP, Caras IW, Rodriguez-Boulan E. Fusion proteins containing a minimal GPI-attachment signal are apically expressed in transfected MDCK cells. J Cell Sci 1991; 99:637-640.

420. Lisanti MP, Le Bivic A, Sargiacomo M, Rodriguez-Boulan E. Steady-state distribution and biogenesis of endogenous Madin-Darby canine kidney glycoproteins: evidence for intracellular sorting and polarized cell surface delivery. J Cell Biol 1989; 109:2117-2127.

421. Lisanti MP, Sargiacomo M, Graeve L, Saltiel, AR, Rodriguez-Boulan E. Polarized apical distribution of glycosylphosphatidylinositol-anchored proteins in a renal epithelial cell line. Proc Natl Acad Sci 1988; 85:9557-9561.

422. Lisanti MP, Scherer PE, Tang Z, Sargiacomo M. Caveolae, caveolin and caveolin-rich membrane domains: a signalling hypothesis. Trends Cell Biol 1994; 4:231-235.

423. Littleton JT, Bellen HJ. Synaptotagmin controls and modulates synaptic-vesicle fusion in a Ca²⁺-dependent manner. Trends Neurosci 1995; 18:177-183.

424. Lledo P-M, Johnnne L, Vernier P et al. Rab3 proteins: Key players in the control of exocytosis. Trends Neurosci 1994; 17:426-432.

425. Lobel P, Dahms NM, Kornfeld S. Cloning and sequence analysis of the cation-independent mannose 6-phosphate receptor. J Biol Chem 1988; 263:2563-2570.

426. Lobel P, Fujimoto K, Ye RD et al. Mutations in the cytoplasmic domain of the 275 kd mannose 6-phosphate receptor differentially alter lysosomal enzyme sorting and endocytosis. Cell 1989; 57: 787-796.

426a. Lodish HF. Transport of secretory and membrane glycoproteins from the endoplasmic reticulum to the Golgi: a rate-limiting step in protein maturation and secretion. 1988; 263:2107-2110.

427. Lodish HF, Kong N. Cyclosporin A inhibits an initial step in folding of transferrin within the endoplasmic reticulum. J Biol Chem 1991; 266:14835-14838.

428. Lodish HF, Kong N. The secretory pathway is normal in dithiothreitol-treated cells, but disulfide-bonded proteins are reduced and reversibly retained in the endoplasmic reticulum. J Biol Chem 1993; 268:20598-20605.

429. Lodish HF, Kong N, Wikstrom L. Calcium is required for folding of newly made subunits of the asialoglycoprotein receptor within the endoplasmic reticulum. J Biol Chem 1992; 267: 12753-12760.

430. Losch A, Koch-Brandt C. Dithiothreitol treatment of Madin-Darby canine kidney cells reversibly blocks export from the endoplasmic reticulum but does not affect vectorial targeting of secretory proteins. J Biol Chem 1995; 270: 11543-11548.

431. Lotti LV, Torrisi M-R, Pascale MC, Bonatti S. Immunocytochemical analysis of the transfer of vesicular stomatitis virus G glycoprotein from the intermediate compartment to the Golgi complex. J Cell Biol 1992; 118:43-50.

432. Louvard D. Apical membrane aminopeptidase appears at site of cell-cell contact in cultured kidney epithelial cells. Proc Natl Acad Sci 1980; 77:4132-4136.

433. Low SH, Hong W. Protein trafficking in and out of the Golgi apparatus. Trends in GlysoSci and GlycoTechnol. 1994; 6:310-328.

434. Low SH, Tang BL, Wong SH, Hong W. Selective inhibition of protein targeting to the apical domain of MDCK cells by brefeldin A. J Cell Biol 1992; 118:51-62.

435. Low SH, Tang BL, Wong SH, Hong W. Golgi retardation in Madin-Darby canine kidney and Chinese hamster ovary cells of a transmembrane chimera of two surface proteins. J. Biol Chem 1994; 269:1985-1994.

436. Low SH, Tang BL, Wong SH, Hong W. Retardation of a surface protein chimera at the cis-Golgi. Biochemistry 1995; 34:5618-5626.

437. Low SH, Wong SH, Tang BL, Hong W. Involvement of both vectorial and transcytotic pathways in the preferential apical cell surface localization of rat dipeptidyl peptidase IV in transfected LLC-PK1 cells. J Biol Chem 1991; 266: 19710-19716.

438. Low SH, Wong SH, Tang BL, Subramaniam VN, Hong W. Apical cell surface expression of rat dipeptidyl peptidase IV in transfected Madin-Darby canine kidney cells. J Biol Chem 1991; 266:13391-13396.

439. Low SH, Wong SH, Tang BL, Tan P, Subramaniam VN, Hong W. Inhibition by brefeldin A of protein secretion from the apical cell surface of Madin-Darby canine kidney cells. J Biol Chem 1991; 266:17729-17732.

440. Low SH, Wong SH, Tang BL, Hong W. Effects of NH4Cl and nocodazole on polarized fibronectin secretion vary amongest different epithelial cell types. Mol Membr Biol 1994; 11:45-54.

441. Lowe SL, Wong SH, Hong W. The mammalian ARF-like protein 1 (Arl1) is associated with the Golgi complex. J Cell Sci 1996; 109:209-220.

442. Ludwig T, Ovitt CE, Bauer U et al. Targeted disruption of the mouse cation-dependent mannose 6-phosphate receptor results in partial missorting of multiple lysosomal enzymes. EMBO J 1993; 12:5225-5235.

443. Lutcke A, Jansson S, Parton RG et al. Rab17, a novel small GTPas, is specific for epithelial cells and is induced during cell polarization. J Cell Biol 1993; 121:553-564.

444. Lutcke A, Parton RG, Murphy C et al. Cloning and subcellular localization of novel rab proteins reveals polarized and cell type-specific expression. J Cell Sci 1994; 107:3437-3448.

445. Luzio JP, Banting G. Eukaryotic membrane traffic: retrieval and retention mechanisms to achieve organelle residence. Trends Biochem Sci 1993; 18: 395-398.

446. Luzio JP, Brake B, Banting G et al. Identification, sequencing and expression of an integral membrane protein of the trans Golgi-network (TGN38). Biochem J 1990; 270:97-102.

447. Machamer CE. Golgi retention ssignals: do membranes hold the key? Trends Cell Biol 1991; 1:141-144.

448. Machamer CE. Targeting and retention of Golgi membrane proteins. Cur Opi Cell Biol 1993; 5:606-612.

449. Machamer CE, Grim MG, Esquela A et al. Retention of a cis Golgi protein requires polar residues on one face of a predicted α-helix in the transmembrane domain. Mol Biol Cell 1993; 4:695-704.

450. Machamer CE, Mentone SA, Rose JK, Farquhar MG. The E1 glycoprotein of an avian coronavirus is targeted to the cis Golgi complex. Proc Natl Acad Sci 1990; 87:6944-6948.

451. Machamer CE, Rose JK. A specific transmembrane domain of a coronavirus E1 glycoprotein is required for its retention in the Golgi region. J Cell Biol 1987; 105:1205-1214.

452. Malhotra V, Orci L, Glick BS et al. Role of an N-ethylmaleimide-sensitive transport component in promoting fusion of transport vesicles with cisternae of the Golgi stack. Cell 1988; 54:221-227.

453. Malhotra V, Serafini T, Orci L et al. Purification of a novel class of coated vesicles mediating biosynthetic protein transport through the Golgi stack. Cell 1989; 58:329-336.

454. Mallabiabarrena A, Fresno M, Alarcon B. An endoplasmic reticulum retention signal in the CD3ε chain of the T-cell receptor. Nature 1992; 357:593-596.

455. Mallabiabarrena A, Jimenez MA, Rico M, Alarcon B. A tyrosine-containing motif mediates ER retention of CD3-ε and adopts a helix-turn structure. EMBO J 1995; 14:2257-2268.

456. Mann SA, Eichelsbacher U, Ungewickell E. Structural relationships between clathrin assembly proteins from the Golgi and the plasma membrane. EMBO J 1988; 7:919-929.

457. Marcusson EG, Horazdovsky BF, Cereghino JL et al. The sorting receptor for yeast vacuolar caboxylpeptidase Y is encoded by the VPS10 gene. Cell 1994; 77:579-586.

458. Marsh BJ, Alm RA, McIntosh SR, James DE. Molecular regulation of GLUT4 targeting in 3T3-L1 adipocytes. J Cell Biol 1995; 130:1081-1091.

459. Martinez O, Schmidt A, Salamero J et al. The small GTP-binding protein rab6 functions in intra-Golgi transport. J Cell Biol 1994; 127:1575-1588.

460. Martoglio B, Hofmann MW, Brunner J, Dobberstein B. The protein-conducting channel in the membrane of the endoplasmic reticulum is open laterally toward the lipid bilayer. Cell 1995; 81:207-214.

461. Masaki R, Yamamoto A, Tashiro Y. Microsomal aldhyde dehydrogenase is localized to the endoplasmic reticulum via its carboxyl-terminal 35 amino acids. J Cell Biol 1994; 126:1407-1420.

462. Masibay AS, Balaji PV, Boeggeman EE, Qasba PK. Mutational analysis of the Golgi retention signal of bovine β-1,4-galactosyltransferase. J Biol Chem 1993; 268:9908-9916.

463. Mathews PM, Martinie JB, Fambrough DM. The pathway and targeting signal for delivery of the integral membrane glycoprotein LEP100 to lysosomes. J Cell Biol 1992; 118:1027-1040.

464. Matlack KES, Walter P. The 70 carboxyl-terminal amino acids of nascent secretory proteins are protected from proteolysis by the ribosome and the protein translocation apparatus of the endoplasmic reticulum membrane. J Biol Chem 1995; 270:6170-6180.

465. Matsui Y, Kikuchi A, Araki S et al. Molecular cloning and characterization of a novel type of regulatory protein (GDI) for smg p25A, a ras p21-like GTP-binding protein. Mol Cell Biol 1990; 10: 4116-4122.

466. Matter K, Brauchbar M, Bucher K, Hauri H-P. Sorting of endogenous plasma membrane proteins occurs from two sites in cultured human intestinal epithelial cells (Caco-2). Cell 1990; 60:429-437.

467. Matteoli M, Takai K, Cameron R et al. Association of Rab3A with synaptic vesicles at late stage of the secretory pathway. J Cell Biol 1991; 115:625-633.

468. Matter K, Yamamoto EM, Mellman I. Structural requirements and sequence motifs for polarized sorting and endocytosis of LDL and Fc receptors in MDCK cells. J Cell Biol 1994; 126:991-1004.

469. Matter K, Whitney JA, Yamamoto EM, Mellman I. Common signals control low density lipoprotein receptor sorting in endosomes and the Golgi complex of MDCK cells. Cell 1993; 74:1053-1064.

470. Matter K, Hunziker W, Mellman I. Basolateral sorting of LDL receptor in MDCK cells: the cytoplasmic domain

contains two tyrosine-dependent targeting determinants. Cell 1992; 71:741-753.

471. McGee TP, Skinner HB, Whitters EA, et al. A phosphatidylinositol transfer protein controls the phosphatidylcholine content of yeast Golgi membranes. J Cell Biol 1994; 124: 273-287.

472. McIntyre GF, Erickson AH. The lysosomal proenzyme receptor that binds procathepsin L to microsomal membrane at pH5 is a 43-kDa integral membrane protein. Proc Natl Acad Sci 1993; 90: 10588-10592.

473. McMahon HT, Sudhof TC. Synaptic core complex of synaptobrevin, syntaxin, and SNAP25 forms high affinity α-SNAP binding sites. J Biol Chem 1995; 270: 2213-2217.

474. McMahon HT, Ushkaryov YA, Edelmann L et al. Cellubrevin is a ubiquitous tetanus toxin substrate homologous to a putative synaptic vesicle fusion protein. Nature 1993; 364:346-349.

475. Melancon P, Glick BS, Malhotra V et al. Involvement of GTP-binding "G" proteins in transport through the Golgi stack. Cell 1987:51:1053-1062.

476. Mellman I. Enigma variations: protein mediators of membrane fusion. Cell

477. Mellman I, Simon K. The Golgi complex: in vitro veritas? Cell 1992; 68: 829-840.

478. Melnick J, Dul JL, Argon Y. Sequential interaction of the chaperones BiP and GRP94 with immunoglobulin chains in the endoplasmic reticulum. Nature 1994; 370:373-375.

479. Meyer DI. Protein translocation into the endoplasmic reticulum: a light at the end of the channel. Trend Cell Biol 1991; 1:154-159.

480. Miesenbock G, Rothman JE. The capacity to retrieve escaped ER proteins extends to the trans-most cisternae of the Golgi stack. J Cell Biol 1995; 129: 309-319.

481. Migliaccio G, Nicchitta CV, Blobel G. The signal sequence receptor, unlike the signal recognition particle receptor, is not essential for protein translocation. J Cell Biol 1992; 117:15-25.

482. Miller JD, Bernstein HD, Walter P. Interaction of E coli Ffh/4.5S ribonucleoprotein and FtsY mimics that of mammalian signal recognition particle and its receptor. Nature 1994; 367:557-659.

483. Miller JD, Wilhelm H, Gierasch L, et al.. GTP binding and hydrolysis by the signal recognition particle during initiation of protein translocation. Nature 1993; 366:351-354.

484. Miller JD, Tajima S, Lauffer L, Walter P. The β subunit of the signal recognition particle receptor is a transmembrane GTPase that anchors the a subunit, a peripheral membrane GTPase, to the endoplasmic reticulum membrane. J Cell Biol 1995; 128:273-282.

485. Miller SG, Moore HPH. Reconstitution of constitutive secretion using semi-intact cells: regulation by GTP but not calcium. J Cell Biol 1991; 112:39-54.

486. Miyake S, Yamamoto M. Identification of ras-related, YPT family genes in Schizosaccharomyces pombe. EMBO J 1990; 9:1417-1422.

486a. Mizuno M, Singer SJ. A soluble secretory protein is first concentrated in the endoplasmic reticulum before transfer to the Golgi apparatus. Proc Natl Acad Sci 1993; 90:5732-5736.

487. Molenaar CMT, Prange R, Gallwitz D. A carboxyl-terminal cysteine residue is required for palmitic acid binding and biological activity of the ras-related yeast YPT1 protein. EMBO J 1988; 7: 971-976.

488. Molloy SS, Thomas L, Vanslyke JK et al. Intracellular trafficking and activation of the furin proprotein convertase: localization to the TGN and recycling from the cell surface. EMBO J 1994; 13: 18-33.

489. Moore HP, Kelly RB. Re-routing of a secretory protein by fusion with human growth hormone sequences. Nature 1986; 321:443-446.

490. Moores SL, Schaber MD, Mosser SD et al. Sequence dependence of protein isoprenylation. J Biol Chem 1991; 266: 14603-14610.

491. Morgan A, Burgoyne RD. A role for soluble NSF attachment proteins (SNAPs) in regulating exocytosis in adrenal chromaffin cells. EMBO J 1995; 14:232-239.

492. Morgan A, Dimaline R, Burgoyne RD. The ATPase activity of N-ethylmaleimide-sensitive fusion protein (NSF) is regulated by soluble NSF attachment proteins. J Biol Chem 1994; 269: 29347-29350.

493. Morgan A, Burgoyne RD. A role for soluble NSF attachment proteins (SNAPs) in regulated exocytosis in adrenal chromaffin cells. EMBO J 1995; 14:232-239.

494. Deleted during revision.

495. Mori K, Ma W, Gething M-J, Sambrook J. A transmembrane protein with a cdc2+/CDC28-related kinase activity is required for signalling from the ER to the nucleus. Cell 1993; 74;743-756.

496. Moss J, Vaughan M. Structure and function of ARF proteins: activators of cholera toxin and critical components of intracellular vesicular transport process. J Biol Chem 1995; 270:12327-12330.

497. Mostov K, Apodaca G, Aroeti B, OKamoto,C. Plasma membrane protein sorting in polarized epithelial cell. J. Cell Biol. 1992 116:577-583.

498. Mostov K, Deitcher DL. Polymeric immunoglobulin receptor expressed in MDCK cells transcytoses IgA. Cell 1986; 46:613-621.

499. Mostov K, de Bruyn Kops A, Deitcher DL. Deletion of the cytoplasmic domain of the polymeric immunoglobulin receptor prevents basolateral localization and endocytosis. Cell 1986; 47:359-364.

500. Moya M, Roberts D, Novick P. DSS4-1 is a dominant suppressor of sec4-8 that encodes a nucleotide exchange protein that aids Sec4p function. Nature 1993; 361:460-463.

501. Mumby SM, Heukeroth RO, Gordon JI, Gilman AG. G-protein α-subunit expression, myristoylation, and membrane association in COS cells. Proc Natl Acad Sci 1990; 87:728-732.

502. Munro S. Sequences within and adjacent to the transmembrane segment of α-2,6-sialyltransferase specify Golgi retention. EMBO J 1991; 10:3577-3588.

503. Munro S, Pelham HRB. A C-terminal signal prevents secretion of luminal ER proteins. Cell 1987; 48:899-907.

504. Musch A, Wiedmann M, Rapoport TA. Yeast Sec proteins interact with polypeptides traversing the endoplasmic reticulum membrane. Cell 1992; 69:343-352.

505. Nabi IR, Le Bivic A, Fambrough D, Rodriguez-Boulan E. An endogenous MDCK lysosomal membrane glycoprotein is targeted basolaterally before delivery to lysosomes. J Cell Biol 1991; 115:1573-1584.

506. Nair J, Muller H, Peterson M, Novick P. Sec2 protein contains a coiled-coil domain essential for vesicular transport and a dispensable carboxy terminal domain. J Cell Biol 1990; 110:1897-1909.

507. Nakai M, Takada T, Endo T. Cloning of the YAP19 gene encoding a putative yeast homolog of AP19, the mammalian small chain of the clathrin-assembly proteins. Biochem Biophys Acta 1993; 1174:282-284.

508. Nakajima H, Hirata A, Ogawa Y et al. A cytoskeleton-related gene, USO1, is required for intracellular protein transport in Saccharomyces cerevisiae. J Cell Biol 1991; 113:245-260.

509. Nakano A, Brada D, Schekman R. A membrane glycoprotein, Sec12p, required for protein transport from the endoplasmic reticulum to the Golgi apparatus in yeast. J Cell Biol 1988; 107:851-863.

510. Nakano A, Muramatsu M. A novel GTP-binding protein, Sar1p, is involved in transport from the endoplasmic reticulum to the Golgi apparatus. J Cell Biol 1989; 109:2677-2691.

511. Nakayama Y, Gobel M, O'Brine Greco B et al. The medium chains of the mammalian clathrin-associated proteins have a homolog in yeast. Eur J Biochem 1991; 202:569-574.

512. Narula N, McMorrow I, Plopper G et al. Identification of a 200-kD, brefeldin-sensitive protein on Golgi membranes. J Cell Biol 1992; 117:27-38.

513. Narula N, Stow JL. Distinct coated vesicles labeled for p200 bud from trans Golgi network membranes. Proc Natl Acad Sci 1995; 92:2874-2878.

514. Neefjes J, Schumacher TNM, Ploegh HL. Assembly and intracellular transport of major histocompatibility complex molecules. Cur Opi Cell Biol 1991; 3:601-609.

515. Neer EJ, Schmedt CJ, Nambudripad R, Smith TF. The ancient regulatory-protein family of WD-repeat proteins. Nature 1994; 371:297-300.

516. Neher E, Penner R. Mice sans synaptotagmin. Nature 1994; 372:316-317.

517. Nelson WJ. Regulation of cell surface polarity from bacteria to mammals. Science 1992; 258:948-955.

518. Neufeld EF. Lysosomal storage diseases. Annu Rev Biochem 1991; 60:257-280.

519. Newman AP, Ferro-Novick S. Characterization of new mutants in the early part of the yeast secretory pathway isolated by a [³H]mannose suicide selection. J Cell Biol 1987; 105:1587-1594.

520. Newman AP, Graf J, Rossi G, et al. SEC22 and SLY2 are identical. Mol Cell Biol 1992; 12: 3663-3664.

521. Newman AP, Groesch ME, Ferro-Novick S. Bos1p, a membrane protein required for ER to Golgi transport in yeast, co-purifies with carrier vesicles and with Bet1p and the ER membrane. EMBO J 1992; 11:3609-3617.

522. Newman AP, Shim J, Ferro-Novick S. BET1, BOS1, and SEC22 are members of a group of interacting yeast genes required for transport from the endoplasmic reticulum to the Golgi complex. Mol Cell Biol 1990; 10:3405-3414.

523. Newman LS, McKeever MO, Okano HJ, Darnell RB. β-NAP, a cerebellar degeneration antigen, is a neuron-specific vesicle coat protein. Cell 1995; 82: 773-783.

524. Ngsee JK, Fleming AM, Scheller RH. A rab protein regulates the localization of secretory granules in AtT-20 cells. Mol Biol Cell 1993; 4:747-756.

525. Nicchitta CV, Migliaccio G, Blobel G. Biochemical fractionation and assembly of the membrane components that mediate nascent chain targeting and translocation. Cell 1991; 65:587-598.

526. Nickel W, Huber LA, Kahn RA et al. ADP-ribosylation factor and a 14-kD polypeptide are associated with haparin sulfate-carrying post-trans Golgi network secretory vesicle in rat hepatocytes. J Cell Biol 1994; 125:721-732.

527. Niemann H, Blasi J, Jahn R. Clostridial neurotoxins: new tools for dissecting exocytosis. Trends Cell Biol 1994; 4: 179-185.

528. Nilsson I, von Heijne G. Determination of the distance between the oligosaccharyltransferase active site and the endoplasmic reticulum membrane. J Biol Chem 1993; 268:5798-5801.

529. Nilsson T, Hoe MH, Slusarewicz P et al. Kin recognition between medial Golgi enzymes in Hela cells. EMBO J 1994; 13:562-574.

530. Nilsson T, Jackson M, Peterson PA. Short cytoplasmic sequences serve as retention signals for transmembrane proteins in the endoplasmic reticulum. Cell 1987; 58:707-718.

531. Nilsson T, Lucocq JM, Mackay D, Warren G. The membrane spanning domain of β-1,4-galactosyltransferase specifies trans Golgi localization. EMBO J 1991; 10:3567-3575.

532. Nilsson T, Slusarewicz P, Hoe MH, Warren G. Kin recognition: a model for the retention of Golgi enzymes. FEBS Letters 1993; 330:1-4.

533. Nishimura N, Nakamura H, Takai Y, Sano K. Molecular cloning and charac-

terization of two rab GDI species from rat brain: brain-specific and ubiquitous types. J Biol Chem 1994; 269: 14191-14198.

534. Nothwehr SF, Stevens TH. Sorting of membrane proteins in the yeast secretory pathway. J Biol Chem 1994; 269: 10185-10188.

535. Novick P, Brennwald P. Friends and family: the role of the Rab GTPases in vesicular traffic. Cell 1993; 75:597-601.

536. Novick P, Ferro S, Schekman R. Orders of events in the yeast secretory pathway. Cell 1981; 25:461-469.

537. Novick P, Field C, Schekman R. Identification of 23 complementation groups required for post-translational events in the yeast secretory pathway. Cell 1980; 21:205-215.

538. Novick P, Schekman R. Secretion and cell-surface growth are blocked in a temperature-sensitive mutant of Saccharomyces cerevisiae. Proc Natl Acad Sci 1979; 76:1858-1862.

539. Nuoffer C, Balch WE. GTPases: Multifunctional molecular switches regulating vesicular traffic. Annu Rev Biochem 1994; 63:949-990.

540. Nuoffer C, Davidson HW, Matteson J et al. A GTP-bound form of Rab1 inhibits protein export from the endoplasmic reticulum and transport between Golgi compartments. J Cell Biol 1994: 125:225-237.

541. Oberhauser AF, Monck JR, Balch WE, Fernandez JM. Exocytotic fusion is activated by Rab3a peptides. Nature 1992; 360:270-273.

542. Ogata S, Fukuda M. Lysosomal targeting of Limp II membrane glycoprotein requires a novel Leu-Ile motif at a particular position in its cytoplasmic tail. J Biol Chem 1994; 269:5210-5217.

543. Ohno H, Stewart J, Fournier MC et al. Interaction of tyrosine-based sorting signals with clathrin-associated proteins. Science 1995; 269:1872-1873.

544. Oka T, Nakano A. Inhibition of GTP hydrolysis by Sar1p causes accumulation of vesicles that are a functional intermediate of the ER-to-Golgi transport in yeast. J Cell Biol 1994; 124:425-434.

545. Oka T, Nishikawa S-i, Nakano A. Reconstitution of GTP-binding Sar1 protein function in ER to Golgi transport. J Cell Biol 1991; 114:671-679.

546. Okamoto CT, Shia SP, Bird C, et al.. The cytoplasmic domain of the polymeric immunoglobulin receptor contains two internalization signals that are distinct from its basolateral sorting signal. J Biol Chem 1992; 267: 9925-9932.

547. Olkkonen VM, Dupree P, Killisch I et al. Molecular cloning and subcellular localization of three GTP-binding proteins of the rab subfamily. J Cell Sci 1994; 106:1249-1261.

548. Olkkonen VM, Peterson JR, Dupree P et al. Isolation of a mouse cDNA encoding Rab23, a small novel GTPase expressed predominantly in the brain. Gene 1994; 138:207-211.

549. Oprins A, Duden R, Kreis TE et al. β-COP localizes mainly to the cis-Golgi side in exocrine pancreas. J Cell Biol 1993; 121:49-59.

550. Orci L, Glick BS, Rothman JE. A new type of coated vesicular carrier that appears not to contain clathrin: its possible role in protein transport within the Golgi stack. Cell 1986:46:171-184.

551. Orci L, Malhotra V, Amherdt M et al. Dissection of a single round of vesicular transport: sequential intermediates for intercisternal movement in the Golgi stack. Cell 1989; 56:357-368.

552. Orci L, Palmer DJ, Amherdt M, Rothman JE. Coated vesicle assembly in the Golgi requires only coatomer and ARF proteins from the cytosol. Nature 1993; 364:732-734.

553. Orci L, Palmer DJ, Ravazzola M et al. Budding from Golgi membranes requires the coatomer complex of non-clathrin coat protein. Nature 1993; 362:648-652.

554. Orci L, Perrelet A, Ravazzola M et al. Coatomer-rich endoplasmic reticulum Proc Natl Acad Sci 1994; 91:11924-11928.

555. Orci L, Ravazzola M, Amherdt M et al. The trans-most cisternae of the Golgi complex: a compartment for sorting of secretory and plasma membrane proteins. Cell 1987; 51:1039-1051.

556. Orci L, Ravazzola M, Meda P et al. Mammalian Sec23p homologue is restricted to the endoplasmic reticulum transitional cytoplasm. Proc Natl Acad Sci 1991; 88:8611-8615.

557. Oshima A, Nolan CM, Kyle JW et al. The human cation-independent mannose 6-phosphate receptor: cloning and sequence of the full-length cDNA and expression of functional receptor in Cos cells. J Biol Chem 1988; 263:2553-2562.

558. Ossig R, Dascher C, Trepte H-H et al. The yeast SLY genes products, suppressors of defects in the essential GTP-binding Ypt1 protein, may act in endoplasmic reticulum-to-Golgi transport. Mol Cell Biol 1991; 11:2980-2993.

559. Ostermann J, Orci L, Tani K et al. Stepwise assembly of functional active transport vesicles. Cell 1993; 75: 1015-1025.

560. Ou W-J, Cameron PH, Thomas DY, Bergeron JJM. Association of folding intermediates of glycoproteins with calnexin during protein maturation. Nature 1993; 364:771-776.

561. Oyler GA, Higgins GA, Hart RA et al. The identification of a novel synaptosomal-associated protein, SNAP-25, differentially expressed by neuronal subpopulations. J Cell Biol 1989; 109: 3039-3052.

562. Padfield PJ, Balch WE, Jamieson JD. A synthetic peptide of the rab3a effector domain stimulates amylase release from permeabilized pancreatic acini. Proc Natl Acad Sci 1992; 89:1656-1660.

562a. Palade, GE. Intracellular aspects of the processing of protein synthesis. Science 1975; 189:347-354.

563. Pallanck, L, Ordway RW, Ramaswami M et al. Distinct roles for N-ethylmaleimide-sensitive fusion protein (NSF) suggested by the identification of a second Drosophila NSF homolog. J Biol Chem 1995; 270:18742-18744.

564. Palmer DJ, Helms JB, Beckers CJM et al. Binding of coatamer to Golgi membranes requires ADP-ribosylation factor. J Biol. Chem 1993; 268:12083-12089.

565. Panzner S, Dreier L, Hartmenn E et al. Posttranslational protein transport in yeast reconstituted with a purified complex of Sec proteins and Kar2p. Cell 1995; 81:561-570.

566. Parlati F, Dignard D, Bergeron JJM, Thomas DY. The calnexin homologue cnx1⁺ in Schizosaccharomyces pombe, is an essential gene which can be complemented by its soluble ER domain. EMBO J 1995; 14:3064-3072.

567. Parlati F, Dominguez M, Bergeron JJM, Thomas DY. Saccharomyces cerevisiae CEN1 encodes a endoplasmic reticulum (ER) membrane protein with sequence similarity to calnexin and calreticulum and functions as a constituent of the ER quality control apparatus. J Biol Chem 1995; 270:244-253.

568. Payne GS, Schekman R. Clathrin: a role in the intracellular retention of a Golgi membrane protein. Science 1989; 245: 1358-1365.

569. Pearse BMF. Receptors compete for adaptors found in plasma membrane coated pits. EMBO J 1988; 7: 3331-3336.

570. Pearse BMF, Robinson MS. Clathrin, adaptors and sorting. Annu Rev Cell Biol 1990; 151-171.

571. Pelham HRB. Evidence that luminal proteins are sorted from secreted proteins in a post-ER compartment. EMBO J 1988; 7:913-918.

572. Pelham HRB. The retention signal for soluble proteins of the endoplasmic reticulum. Trends Biochem Sci 1990; 15:483-486.

573. Pelham HRB. Is epimorphin involved in vesicular transport. Cell 1993; 73: 425-428.

574. Pelham HRB. About turn for the COPs? Cell 1994; 79:1125-1127.

575. Pelham HRB, Hardwick KG, Lewis MJ. Sorting of soluble ER proteins in yeast. EMBO J 1989; 7:1757-1762.

576. Pepperkok R, Scheel J, Horstmann H et al. β-COP is essential for biosynthetic membrane transport from the endoplasmic reticulum to the Golgi complex in vivo. Cell 1993; 74:71-82.

577. Perin MS, Brose N, Jahn R, Sudhof TC. Domain structure of synaptotagmin (p65). J Biol Chem 1991; 266:623-629.

578. Perin MS, Fried VA, Mignery GA et al. Phospholipid binding by a synaptic vesicle protein homologous to the regulatory region of protein kinase C. Nature 1990; 345:260-263.

579. Perin MS, Johnston PA, Ozcelik T et al. Structural and functional conservations of synaptotagmin (p65) in Drosophila and human. J Biol Chem 1991; 266:615-622.

580. Peter F, Nuoffer C, Pind SN, Balch WE. Guanine nucleotide dissociation inhibitor is essential for Rab1 function in budding from the endoplasmic reticulum and transport through the Golgi stack. J Cell Biol 1994; 126:1393-1406.

581. Peter F, Plutner H, Zhu H et al. β-COP is essential for transport of protein from the endoplasmic reticulum to the Golgi in vitro. J Cell Biol 1993; 122: 1155-1167.

582. Peters C, Braun M, Weber B et al. Targeting of a lysosomal membrane protein: a tyrosine-containing endocytosis signal in the cytoplasmic tail of lysosomal acid phosphatase is necessary and sufficient for targeting to lysosomes. EMBO J 1990; 9:3497-3506.

583. Peters JM, Walsh MJ, Franke WW. An abundant and ubiquitous homo-oligomeric ring-shaped ATPase particle related to the putative vesicle fusion protein Sec18p and NSF. EMBO J 1990; 9: 1757-1767.

584. Peters PJ, Hsu VW, Ooi CE et al. Overexpression of wild-type and mutant ARF1 and ARF6: distinct perturbations of nonoverlapping membrane compartments. J Cell Biol 1995; 128:1003-1017.

585. Petrenko AG, Perin MS, Davletov BA et al. Binding of synaptotagmin to the α-latrotoxin receptor implicates both in synaptic vesicle exocytosis. Nature 1991; 353:65-68.

586. Pevsner J, Hsu SC, Braun JEA, et al. Specificity and regulation of a synaptic vesicle docking complex. Neuron 1994; 13:353-361.

587. Pevsner J, Hsu SC, Scheller RH. n-Sec1: a neural-specific syntaxin-binding protein. Proc Natl Acad Sci 1994; 91:1445-1449.

588. Pevsner J, Volknandt W, Wong BR, Scheller RH. Two rat homologs of clathrin-associated adaptor proteins. Gene 1994; 146:279-283.

589. Pfanner N, Glick BS, Arden SR, Rothman JE. Fatty acylation promotes fusion of transport vesicles with Golgi cisternae. J Cell Biol 1990; 111:955-961.

590. Pfeffer SR, Dirac-Svejstrup AB, Soldati T. Rab GDP dissociation inhibitors: putting Rab GTPases in the right place. J Biol Chem 1995; 270:17057-17059.

590a. Pfeffer SR, Rothman JE. Biosynthetic protein transport and sorting by the endoplasmic reticulum and Golgi. Annu Rev Biochem 1987; 56:829-852.

591. Phan, HL, Finlay J, Chu DS et al. The saccharomyces cerevisiae APS1 gene encodes a homolog of the small subunit of the mammalian clathrin AP-1 complex: evidence for functional interaction with clathrin at the Golgi complex. EMBO J 1994; 13:1706-1717.

592. Pimplikar SW, Ikonen E, Simons K. Basolateral protein transport in streptolysin O-permeabilized MDCK cells. J Cell Biol 1994; 125:1025-1035

593. Pimplikar SW, Simons K. Regulation of apical transport in epithelial cells by a Gs class of heterotrimeric G protein. Nature 1993; 362:456-458.

594. Pimplikar SW, Simons K. Activators of protein kinase A stimulates apical but not basolateral transport in epithelial Madin-Darby canine kidney cells. J Biol Chem 1994; 269:19054-19059.

595. Pind SN, Nuoffer C, McCaffery JM et al. Rab1 and Ca²⁺ are required for the fusion of carrier vesicles mediating endoplasmic reticulum to Golgi transport. J Cell Biol 1994; 125:239-252.

596. Piper RC, Tai, C, Kulesza P. GLU-4 NH2 terminus contains a phenylalanine-based targeting motif that regulates intracellular sequestration. J Cell Biol 1993; 121:1221-1232.

596a. Piper RC, Whitters EA, Stevens TH. Yeast Vps45p is a Sec1p-like protein required for the consumption of vacuole-targeted, post-Golgi vesicles. Eur J Cell Biol 1994; 65:305-318.

597. Plutner H, Cox AD, Pind S et al. Rab1b regulates vesicular transport between the endoplasmic reticulum and successive Golgi compartments. J. Cell Biol 1991; 115:31-43.

598. Plutner H, Davidson HW, Saraste J, Balch WE. Morphological analysis of protein transport from the ER to the Golgi membrane in digitonin-permeabilized cells: role of the p58 containing compartment. J Cell Biol 1992; 119: 1097-1116.

599. Podos SD, Reddy P, Ashkenas J, Krieger M. LDLC encodes a brefeldin A-sensitive, peripheral Golgi protein required for normal Golgi function. J Cell Biol 1994; 127:679-691.

600. Ponnambalam S, Rabouille C, Luzio P et al. The TGN38 glycoprotein contains two non-overlapping signals that mediate localization to the trans Golgi network. J Cell Biol 1994; 125:253-268.

601. Ponnambalam S, Robinson MS, Jackson AP et al. Conservation and diversity in families of coated vesicle adaptins. J Biol Chem 1990; 265:4814-4820.

602. Pousi B, Hautala T, Heikkinen J et al. Alu-Alu recombination results in a duplication of seven exons in the lysyl hydroxylase gene in patient with the type VI variant of Ehlers-Danlos syndrome. Am J Hum Genet 1994; 55:899-906.

603. Potenza M, Bowser R, Muller H, Novick P. SEC6 encodes an 85 kDa soluble protein required for exocytosis in yeast. Yeast 1992; 8:549-558.

603a. Powers T, Walter P. Reciprocal stimulation of GTP hydrolysis by two directly interacting GTPases. Science 1995; 269:1422-1424.

604. Price SR, Nightingale M, Tsai SC et al. Guanine nucleotide-binding proteins that enhance choleragen ADP-ribosyltransferase activity: nucleotide and deduced amino acid sequence of an ADP-ribosylation factor cDNA. Proc Natl Acad Sci 1988; 85:5488-5491.

605. Prill V, Lehmann L, von Figura K, Peters C. The cytoplasmic tail of lysosomal acid phosphatase contains overlapping but distinct signals for basolateral sorting and rapid internalization in polarized MDCK cells. EMBO J 1993; 12:2181-2193.

606. Protopopov V, Govindan B, Novick P, Gerst JE. Homologs of the synaptobrevin/VAMP family of synaptic vesicle proteins function on the late secretory pathway in *S. cerevisiae*. Cell 1993; 74:855-861.

607. Prydz K, Brandli AW, Bomsel M, Simons K. Surface distribution of the mannose 6-phosphate receptors in epithelial Madin-Darby canine kidney cells. J Biol Chem 1990; 265:12629-12635.

608. Pryer NK, Salama NR, Schekman R, Kaiser CA. Cytosolic Sec13p complex is required for vesicle formation from the endoplasmic reticulum in vitro. J Cell Biol 1993; 120:865-875.

609. Pryer NK, Wuestehube LJ, Schekman R. Vesicle-mediated protein sorting. Annu Rev Biochem 1992; 61:471-516.

610. Quemeneur E, Guthapfel R, Gueguen P. A major phosphoprotein of the endoplasmic reticulum is protein disulfide isomerase. J Biol Chem 1994; 269:5485-5488.

611. Rabouille C, Levine TP, Peters JM, Warren G. An NSF-like ATPase, p97, and NSF mediate cisternal regrowth from mitotic Golgi fragments. Cell 1995; 82:905-914.

612. Rad MR, Phan HL, Kirchrath L et al. Saccharomyces cerevisiae Apl2p, a homologue of the mammalian clathrin AP β subunit, plays a role in clathrin-dependent Golgi functions. J Cell Sci 1995; 108:1605-1615.

613. Rajasekaran AK, Humphrey JS, Wagner M et al. TGN38 recycles basolaterally in polarized Madin-Darby canine kidney cells. Mol Cell Biol 1994; 5:1093-1103.

614. Randazzo PA, Yang YC, Rulka C, Kahn RA. Activation of ADP-ribosylation factor by Golgi membranes. evidence for a brefeldin A- and protease-sensitive activating factor on Golgi membrane. J Biol Chem 1993; 268:955-9563.

615. Rapiejko PJ, Gilmore R. Protein translocation across the ER requires a functional GTP binding site in the alpha subunit of the signal recognition particle receptor. J Cell Biol 1992; 117:493-503.

616. Rapiejko PJ, Gilmore R. Signal sequence recognition and targeting of ribosomes to the endoplasmic reticulum by the signal recognition particle do not require GTP. Mol Biol Cell 1994; 5:887-897.

617. Rapoport TA. Protein transport across the endoplasmic reticulum membrane: facts, models, mysteries. FASEB J 1991; 5:2792-2798.

618. Rapoport TA. Transport of proteins across the endoplasmic reticulum membrane. Science 1992; 258:931-936.

619. Raymond CK, Howald-Stevenson I, Vater CA, Stevens TH. Morphological classification of the yeast vacuolar protein sorting mutants: evidence for a prevacuolar compartment in class E vps mutants. Mol Biol Cell 1992; 3: 1389-1402.

620. Raymond CK, Roberts CJ, Moore KE, et al. Biogenesis of the vacuole in Saccharomyces cerevisia. Internatl Rev Cytol 1992; 139:59-120.

621. Reaves B, Horn M, Banting G. TGN38/ 41 recycles between the cell surface and the TGN: brefeldin A affects its rate of return to the TGN. Mol Cell Biol 1993; 4:93-105.

622. Reaves B, Horn M, Banting G. Identification, molecular characterization and immunolocalization of an isoform of the trans Golgi-network (TGN)-specific integral membrane protein TGN38. Biochem J 1992; 283:313-316.

623. Regazzi R, Wollheim CB, Lang J et al. VAMP-2 and cellubrevin are expressed in pancreatic β-cells and are essential for Ca^{2+} but not for GTPγS-induced insulin secretion. EMBO J 1995; 14:2723-2730.

624. Reiss Y, Stradley SJ, Gierasch LM et al. Sequence requirement for peptide recognition by rat brain p21^{ras} protein farnesyltransferase. Proc Natl Acad Sci 1991; 88:732-736.

625. Resh MD, Ling HP. Identification of a 32K plasma membrane protein that binds to the myristylated amino-terminal sequence of p60^{v-src}. Nature 1990; 346: 84-86.

626. Rexach MF, Latterich M, Schekman RW. Characterization of endoplasmic reticulum-derived transport vesicles. J Cell Biol 1994; 1133-1148.

627. Rexach MF, Schekman RW. Distinct biochemical requirements for budding, targeting, and fusion of ER-derived transport vesicles. J Cell Biol 1991; 114: 219-229.

628. Riederer MA, Soldati T, Shapiro AD et al. Lysosome biogenesis requires Rab9 function and receptor recycling form endosomes to the trans Golgi network. J Cell Biol 1994; 125:573-582.

629. Rindler MJ, Traber MG. A specific sorting signal is not required for the polarized secretion of newly-synthesized proteins from cultured intestinal epithelial cells. J Cell Biol 1988; 107:471-479.

630. Roa M, Cornet V, Yang C, Goud B. The small GTP-binding protein rab6 is redistributed in the cytosol by brefeldin A. J Cell Sci 1993; 106:789-802.

631. Robinson JS, Klionsky DJ, Banta LM, Emr SD. Protein sorting in Saccharomyces cerevisiae: isolation of mutants defective in delivery and processing of multiple vacuolar hydrolases. Mol Cell Biol 1988; 8:4936-4948.

632. Robinson MS. Cloning of cDNAs encoding two related 100-kD coated vesicle proteins (α-adaptins). J Cell Biol 1989; 108:833-842.

633. Robinson MS. Cloning and expression of gamma-adaptin, a component of clathrin-coated vesicles associated with the Golgi apparatus. J Cell Biol 1990; 111: 2319-2326.

634. Robinson MS. Adaptins. Trends Cell Biol 1992; 2:293-297.

635. Robinson MS, Kreis TE. Recrutiment of coat proteins onto Golgi membranes in intact and permeabilized cells: effects of brefeldin A and G protein activators. Cell 1992; 69:129-138.

636. Robinson PJ, Liu J-P, Powell KA et al. Phosphorylation of dynamin I and synaptic-vesicle recycling. Trends Neurosci 1994; 17:348-353.

637. Rocque WJ, Mcwherter CA, Wood DC, Gordon JI. A comparative analysis of the kinetic mechanism and peptide substrate specificity of human and Saccharomyces cerevisiae myristoyl-CoA:protein N-myristoyltransferase. J Biol Chem 1993; 268:9964-9971.

638. Rodgers W, Crise B, Rose JK. Signals determining protein tyrosine kinase and glycosyl-phosphatidylinositol-anchored protein targeting to a glycolipid-enriched membrane fractions. Mol Cell Biol 1994:14:5384-5391.

639. Rodriguez-Boulan E, Nelson WJ. Morphogenesis of the polarized epithelial cell phenotype. Science 1989; 245:718-725.

640. Rohrer J, Schweizer A, Johnson KF, Kornfeld S. A determinant in the cytoplasmic tail of the cation-dependent mannose-6-phospate receptor prevents trafficking to lysosomes. J Cell Biol 1995; 130:1297-1306.

641. RomischK, Webb J, Lingelbach K, et al. The 54-kD protein of signal recognition particle contains a methionine-rich RNA binding domain. J Cell Biol 1990; 111: 1793-1802.

642. Rosa P, Weiss U, Pepperkok R et al. An antibody against secretogranin I (chromogranin A) is packaged into secretory granules. J Cell Biol 1989; 109:17-34.

643. Rosahl TW, Spillane D, Missler M et al. Essential functions of synapsins I and II in synaptic vesicle regulation. Nature 1995; 375:488-493.

644. Rose JK, Doms RW. Regulation of protein export from the endoplasmic reticulum. Ann Rev Cell Biol 1988; 4: 257-288.

645. Rossi G, Jing Y, Newman AP, Ferro-Novick S. Dependence of Ypt1p and Sec4 membrane attachment on Bet2. Nature 1991; 351:158-161.

646. Rothberg KG, Heuser JE, Donzell WC, et al. Caveolin, a protein component of caveolae membrane coats. Cell 1992; 68: 673-682.

647. Rothman JE, Orci L. Molecular dissection of the secretory pathway. Nature 1992; 355: 409-415.

648. Rothman JE. Transport of the vesicular stomatitis glycoprotein to trans Golgi membrane in a cell-free system. J Biol Chem 1987; 262:12502-12510.

649. Rothman JE. Mechanism of intracellular protein transport. Nature 1994; 372: 55-63.

650. Rothman JE, Warren G. Implications of the SNARE hypothesis for intracellular membrane topology and dynamics. Curr Biol 1994; 4:220-233.

651. Rothman JH, Raymond CK, Gilbert T, et al. A putative GTP binding protein homologous to interferon-inducible Mx protein performs an essential function in yeast protein sorting. Cell 1990; 61:1063-1074.

652. Ruohola H, Kabcenell AK, Ferro-Novick S. Reconstitution of protein transport from the endoplasmic reticulum to the Golgi complex in yeast: the acceptor Golgi compartment is defective in the sec23 mutant. J Cell Biol 1988; 107: 1465-1476.

653. Russo RN, Shaper NL, Taatjes DJ, Shaper JH. β1,4-galactosyltransferase: a short NH2-terminal fragment that include the cytoplasmic and transmembrane domain is sufficient for Golgi retention. J Biol Chem 1992; 267: 9241-9247.

654. Sadoul K, Lang J, Montecucco C et al. SNAP-25 is expressed in islets of Langerhans and is involved in insulin release. J Cell Biol 1995;128:1019-1028.

655. Salama NR, Yeung T, Schekman RW. The Sec13p complex and reconstitution of vesicle budding from the ER with purified cytosolic proteins. EMBO J 1993; 12:4073-4082.

656. Salminen A, Novick PJ. A ras-like protein is required for a post-Golgi event in yeast secretion. Cell 1987; 49:527-538.

657. Salminen A, Novick PJ. The Sec15 protein responds to the function of the GTP-binding protein, Sec4, to control vesicular traffic in yeast. J Cell Biol 1989; 109:1023-1036.

658. Sandoval IV, Arredondo JJ, Alcalde J et al. The residues Leu(Ile)[475]-Ile(Leu, Val, Ala)[476], contained in the extended carboxyl cytoplasmic tail, are critical for targeting of resident lysosomal membrane protein LIMP II to lysosomes. J Biol Chem 1994; 269:6622-6631.

659. Sandoval IV, Bakke O. Targeting of membrane proteins to endosomes and lysosomes. Trends Cell Biol 1994; 4: 292-297.

660. Sapperstein SK, Walter, DM, Grosvenor AR, et al. p115 is a general vesicular transport factor related to the yeast endoplasmic reticulum to Golgi transport factor Uso1p. Proc Natl Acad Sci 1995; 92:522-526.

661. Saraste J, Svensson K. Distribution of the intermediate elements operating in ER to Golgi transport. J Cell Sci 1991; 100: 415-430.

662. Sargiacomo M, Lisanti M, Graeve L, et al. Integral and peripheral protein composition of the apical and basolateral membrane domains in MDCK cells. J Membr Biol 1989; 107:277-286.

663. Sargiacomo M, Sudol M, Tang ZL, Lisanti MP. Signal transducing molecules and glycosyl-phosphatidylinositol-linked proteins from a caveolin-rich insoluble complex in MDCK cells. J Cell Biol 1993; 122:789-807.

664. Schafer W, Stroh A, Berghofer S et al. Two independent targeting signals in the cytoplasmic domain determine trans Golgi network localization and endosomal trafficking of the proprotein convertase furin. EMBO J 1995; 14: 2424-2435.

665. Schell MJ, Maurice M, Stieger B, Hubbard AL. 5' nucleotide is sorted to the apical domain of hepatocytes via an indirect route. J Cell Biol 1992; 119: 1173-1182.

666. Scheller RH. Membrane trafficking in the presynaptic nerve terminal. Neuron 1995; 14:893-897.

667. Schiavo G, Rossetto, O, Montecucco C. Clostridial neurotoxins as tools to investigate the molecular events of neurotransmitter release. Seminars Cell Biol 1994; 5:221-229.

668. Schiavo G, Shone CC, Bennett MK et al. Botulinum neurotoxin type C cleaves a single Lys-Ala bond within the carboxyl-terminal region of syntaxins. J Biol Chem 1995; 270:10566-10570.

669. Schimmoller F, Riezman H. Involvement of Ypt7p, a small GTPase, in traffic from late endosome to the vacuole in yeast. J Cell Sci 1993; 106:823-830.

670. Schimmoller F, Singer-Kruger B, Schroder S et al. The absence of Emp-24p, a component of ER-derived COPII-coated vesicles, causes a defect in transport of selected proteins to the Golgi. EMBO J 1995; 14:1329-1339.

671. Schindler R, Itin C, Zerial M et al. ERGIC-53, a membrane protein of the ER-Golgi intermediate compartment, carries an ER retention motif. Eur J Cell Biol 1993; 61:1-9.

672. Schmitt HD, Pfaff WE, Gallwitz D. The ras-related YPT1 gene product in yeast: a GTP-binding protein that might be involved in microtubule organization. Cell 1986; 47:401-412.

673. Schmitt HD, Puzicha M, Gallwitz D. Study of a temperature-sensitive mutant of the ras-related YPT1 gene product in yeast suggests a role in the regulation of intracellular calcium. Cell 1988, 53: 635-647.

674. Schu PV, Takegawa K, Fry MJ, et al. Phosphatidylinositol 3-kinase encoded by yeast VPS34 gene essential for protein sorting. Science 1993; 260:88-91.

675. Schurmann A, Breiner M, Becker W et al. Cloning of two novel ADP-ribosylation factor-like proteins and characterization of their differential expression in 3T3-L1 cells. J Biol Chem 1994; 269:15683-15688.

676. Schutze M-P, Peterson PA, Jackson MR. An N-terminal double-arginine motif maintains type II membane proteins in the endoplasmic reticulum. EMBO J 1994; 13:1696-1705.

677. Schwaninger R, Beckers CJM, Balch WE. Sequential transport of protein between the endoplasmic reticulum and succesive Golgi compartments in semi-intact cells. J Biol Chem 1991; 266:13055-13063.

678. Schwaninger R, Plutner H, Bokoch GM, Balch WE. Multiple GTP-binding proteins regulate vesicular transport from the ER to Golgi membranes. J Cell Biol 1992; 119:1077-1096.

679. Schweizer A, Fransen JAM, Bachi T et al. Identification, by a monoclonal antibody, of a 53-kD protein associated with a tubule-vesicular compartment at the cis-side of the Golgi apparatus. J Cell Biol 1988; 107:1643-1653.

680. Schweizer A, Fransen JAM, Matter K et al. Identification of an intermediate compartment involved in protein transport from endoplasmic reticulum to Golgi apparatus. Eur J Cell Biol 1990; 53:185-196.

681. Schweizer A, Matter K, Ketcham CM, Hauri H-P. The isolated ER-Golgi intermediate compartment exhibits properties that are different from ER and cis-Golgi. J Cell Biol 1991; 113:45-54.

682. Schweizer FE, Betz H, Augustine GJ. From vesicle docking to endocytosis: intermediate reactions of exocytosis. Neuron 1995; 14:689-696.

683. Seabra MC, Goldstein JL, Sudhof TC, Brown MS. Rab geranylgeranyltransferase: a multisubunit enzyme that prenylates GTP-binding proteins terminating in Cys-X-Cys or Cys-Cys. J Biol Chem 1992; 267:14497-14503.

684. Seabra MC, Reiss Y, Casey PJ et al. Protein farnesyltransferase and geranylgeranyltransferease share a common α subunit. Cell 1991; 65:429-434.

685. Seeger M, Payne GS. Selective and immediate effects of clathrin heavy chain mutations on Golgi membrane protein retention in Saccharomyces cerevisiae. J Cell Biol 1992; 118:531-540.

686. Seeger M, Payne GS. A role for clathrin in the sorting of vacuolar proteins in the Golgi complex of yeast. EMBO J 1992; 11:2811-2818.

687. Segev N. Mediation of the attachment or fusion step in vesicular transport by the GTP-binding Ypt1 protein. Science 1991; 252:1553-1556.

688. Segev N, Mulholland J, Botstein D. The yeast GTP-binding YPT1 protein and a mammalian counterpart are associated with the secretion machinery. Cell 1988; 52:915-924.

689. Semenza JC, Hardwick KG, Dean N, Pelham HRB. ERD2, a yeast gene required for the receptor-mediated retrieval of luminal ER proteins from the secretory pathway. Cell 1990; 61:1349-1357.

690. Serafini T, Orci L, Amherdt M et al. ADP-ribosylation factor is a subunit of the coat of Golgi-derived COP-coated vesicles: A novel role for a GTP-binding protein. Cell 1991; 67:239-253.

691. Serafini T, Stenbeck G, Brecht A et al. A coat subunit of Golgi-derived non-clathrin-coated vesicles with homology to clathrin-coated vesicle coat protein β-adaptin. Nature 1991; 349:215-220.

692. Sevarino KA, Stork P, Ventimiglia R et al. Amino-terminal sequences of prosomatostatin direct intracellular targeting but not processing specificity. Cell 1989; 57:11-19.

693. Sewell JL, Kahn RA. Sequences of the bovine and yeast ADP-ribosylation factor and comparison to other GTP-binding proteins. Proc. Natl. Acad. Sci 1988; 85:4620-4624.

694. Shamu CE, Cox JS, Walter P. The unfolded-protein-response pathway in yeast. Trends Cell Biol 1994; 4:56-60.

695. Shapiro AD, Riederer MA, Pfeffer SR. Biochemical analysis of rab9, a ras-like GTPase involved in protein transport from late endosomes to the trans Golgi network. J Biol Chem 1993; 268: 6925-6931.

696. Shaywitz DA, Orci L, Ravazzola M, et al. Human SEC13Rp functions in yeast and is localized on transport vesicles budding from the endoplasmic reticulum. J Cell Biol 1995; 128:769-777.

697. Shelness GS, Blobel G. Two subunits of the canine signal peptidase complex are homologous to yeast SEC11 protein. J Biol Chem 1990; 265:9512-9519.

698. Shelness GS, Lin L, Nicchitta CV. Membrane topology and biogenesis of enkaryotic signal peptidase. J Biol Chem 1993; 268:5201-5208.

699. Shen KA, Hammond CM, Moore H-PH. Molecular analysis of SAR1-related cDNAs from a mouse pituitary cell line. FEBS Letters 1993; 335:380-385.

700. Shim J, Newman AP, Ferro-Novick S. The BOS1 gene encodes an essential 27-kD putative membrane protein that is required for vesicular transport from the ER to the Golgi complex in yeast. J Cell Biol 1991; 113:55-64.

701. Shin J, Dunbrack Jr RL, Lee S, Strominger JL. Signals for retention of transmembrane proteins in the endoplasmic reticulum studied with CD4 truncation mutants. Proc Natl Acad Sci 1991; 88:1918-1922.

702. Shirataki H, Kaibuchi K, Sakoda T et al. Rabphilin-3A, a putative target protein for smg p25A/Rab3A p25 small GTP-binding protein related to synaptotagmin. Mol Cell Biol 1993; 13: 2061-2068.

703. Shirataki H, Kaibuchi K, Yamaguchi T et al. A possible target protein for smg-25A/rab3A small GTP-binding protein. J Biol Chem 1992; 267:10946-10949.

704. Shirataki H, Yamamato T, Hagi S et al. Rabphilin-3A is asssociated with synaptic vesicles through a vesicle protein in a manner independent of Rab3A. J Biol Chem 1994; 269:32717-32720.

705. Shisheva A, Buxton J, Czech MP. Differential intracellular localizations of GDP dissociation inhibitor isoforms. J Biol Chem 1994; 269:13865-23868.

706. Shisheva A, Sudhof TC, Czech MP. Cloning, characterization, and expression of a novel GDP dissociation inhibitor isoform from skeletal muscle. Mol Cell Biol 1994; 14:3459-3468.

707. Silberstein S, Collins PG, Kelleher DJ et al. The α subunit of the saccharomyces cerevisiae oligosaccharyltransferase complex is essential for vegetative growth of yeast and is homologous to mammalian ribophorin I. J Cell Biol 1995; 128: 525-536.

707a. Silberstein S, Collins PG, Kelleher DJ, Gilmore R. The essential OST2 gene encodes the 16-kD subunit of the yeast oligosaccharyltransferase, a highly conserved protein expressed in diverse eukaryotic organisms. J Cell Biol 1995; 371-383.

708. Silberstein S, Kelleher DJ, Gilmore R. The 48-kDa subunit of the mammalian oligosaccharyltransferase complex is homologous to the essential yeast protein Wbp1. J Biol Chem. 1992; 267: 23658-23663.

709. Silveira LA, Wong DH, Masiarz FR, Schekman R. Yeast clathrin has a distinctive light chain that is important for cell growth. J Cell Biol 1990; 111: 1439-1449.

710. Deleted during revision.

711. Simons K, Fuller SD. Cell surface polarity in epithelia. Annu Rev Cell Biol 1985; 1:243-288.

712. Simons K, Virta H. Perforated MDCK cells support intracellular transport. EMBO J 1987:8:2241-2247.

713. Simons K, Wandinger-Ness A. Polarized sorting in epithelia. Cell 1990; 62:207-210.

714. Simons, K, Zerial, M. Rab proteins and the road maps for intracellular transport. Neuron 1993; 11:789-799.

715. Singer, SJ. The structure and insertion of integral proteins in membranes. Annu Rev Cell Biol 1990; 6:247-296.

716. Singer-Kruger B, Stenmark H, Dusterhoft A et al. Role of three Rab5-like GTPases, Ypt51p, Ypt52p and Ypt53p, in the endocytotic and vacuolar sorting pathways of yeast. J Cell Biol 1994; 125:283-298.

717. Singh P, Tang BL, Wong SH, Hong W. Transmembrane topology of the mammalian KDEL receptor. Mol Cell Biol 1993; 13:6435-6441.

718. Skehel, PA, Martin KC, Kandel ER, Bartsch D. A VAMP-binding protein from Aplysia required for neurotransmitter release. Science 1995; 269:1580-1583.

719. Slusarewicz P, Nilsson T, Hui N et al. Isolation of a matrix that binds medial Golgi enzymes. J Cell Biol 1994; 124: 405-413.

720. Sogaard M, Tani K, Ye RB, et al. A rab protein is required for the assembly of SNARE complexes in the docking of transport vesicles. Cell 1994; 78:937-948.

721. Soldatt T, Riederer MA, Pfeffer SR. Rab GDI: a solubilizing and recycling factor for rab9 protein. Mol Biol Cell 1993; 4:425-434.

722. Soller T, Whiteheart SW, Brunner M et al. SNAP receptors implicated in vesicle targeting and fusion. Nature 1993; 362:318-324.

723. Soller T, Bennett MK, Whiteheart SW et al. A protein assembly-disassembly pathway in vitro that may correspond to sequential steps of synaptic vesicle docking, activation, and fusion. Cell 1993; 75:409-418.

724. Somons JF, Ferro-Novick S, Rose MD, Helenius A. BiP/Kar2p serves as a molecular chaperone during carboxylpeptidase Y folding in yeast. J Cell Biol 1995; 130:41-49.

725. Sonnichsen B, Fullekrug J Van PN et al. Retention and retrieval: both mechanisms cooperates to maintain calreticulin in the endoplasmic reticulum. J Cell Sci 1994; 107:2705-2717.

726. Deleted during revision.

727. Soldati T, Shapiro AD, Dirac Svejstrup AB, Pfeffer SR. Membrane targeting of small GTPase Rab9 is accompanied by nucleotide exchange. Nature 1994; 369: 76-78.

728. Sosa MA, Schmidt B, von Figura K, Hille-Rehfeld A. In vitro binding of plasma membrane-coated vesicle adaptors to the cytoplasmic domain of lysosomal acid phosphatase. J Biol Chem 1993; 268:12537-12543.

729. Sossin WS, Fisher JM, Scheller RH. Cellular and molecular biology of neuropeptide processing and packaging. Neuron 1989; 2:1407-1417.

730. Sossin WS, Fisher JM, Scheller RH. Sorting within the regulated secretory pathway occurs in the trans Golgi network. J Cell Biol 1990; 110:1-12.

731. Sousa M, Parodi AJ. The molecular basis for the recognition of misfolded glycoprotein by the UDP-Glc:glycoprotein glucosyltransferase. EMBO J 1995; 14:4196-4203.

732. Stack JH, DeWald DB, Takegawa K, Emr SD. Vesicle-mediated protein sorting: regulatory interactions between the Vps15 protein kinase and the Vps34 PtdIns 3-kinase essential for protein sorting to the vacuole in yeast. J Cell Biol 1995; 129:321-334.

733. Stack JH, Emr SD. Genetic and biochemical studies of protein sorting to the yeast vacuole. Cur Opi Cell Biol 1993; 5:641-646.

734. Stack JH, Emr SD. Vps34p required for yeast vacuolar protein sorting is a multiple specificity kinase that exhibits both protein kinase and phosphatidylinositol-specific PI 3-kinase activities. J Biol Chem 1994; 269:31552-31562.

735. Stack JH, Herman PK, Schu PV, Emr SD. A membrane-associated complex containing the Vps15 protein kinase and the Vps34 PI 3-kinase is essential for protein sorting to the yeast lysosome-like vacuole. EMBO J 1993; 12:2195-2204.

736. Stahl B, Fischer von Mollard G, Walch-Solimena C, Jahn R. GTP cleavage by the small GTP-binding protein rab3A is associated with exocytosis of synaptic vesicles induced by α-Latrotoxin. J Biol Chem 1994; 269:24770-24776.

737. Stamnes MA, Craighead MW, Hoe MH et al. An integral membrane component of coatomer-coated transport vesicles defines a new family of proteins in budding. Proc Natl Acad Sci 1995; 92: 8011-8015.

738. Stamnes MA, Rothman JE. The binding of AP-1 clathrin adaptor particles to Golgi membranes requires ADP-ribosylation factor, a small GTP-binding protein. Cell 1993; 73;999-1005.

739. Stanley KK, Howell KE. TGN38/41: a molecule on the move. Trends Cell Biol 1993; 3:252-255.

740. Steele-Mortimer O, Clague MJ, Huber LA et al. The N-terminal domain of a rab protein is involved in membrane-membrane recognition and/or fusion. EMBO J 1994; 13:34-41.

741. Steele-Mortimer O, Gruenberg J, Clague MJ. Phosphorylation of GDI and membrane cycling of rab proteins. FEBS Letters 1993; 329:313-318.

742. Stenbeck G, Harter C, Brecht A et al. β'-COP, a novel subunit of coatomer. EMBO J 1993; 12:2841-2845.

743. Stenbeck G, Schreiner R, Hermann D et al. γ-COP, a coat subunit of non-clathrin-coated vesicle with homology to Sec21p. FEBS Letters 1992; 314: 195-198.

744. Stenmark H, Valencia A, Martinez O et al. Distinct structural elements of Rab5 define its functional specificity. EMBO J 1994; 13:575-583.

745. Stenmark H, Parton RG, Steele-Mortimer O et al. Inhibition of rab5 GTPase activity stimulates membrane fusion in endocytosis. EMBO J 1994; 13:1287-1296.

746. Sterns T, Kahn RA, Bostein D, Hoyt MA. ADP ribosylation factor is an essential protein in Sacchromyces cerevisiae and is encoded by by two genes. Mol Cell Biol 1990; 10:6690-6699.

747. Sterns T, Willingham MC, Bostein D, Kahn RA. ADP-ribosylation factor is functionally and physically associated with the Golgi complex. Proc Natl Acad Sci 1990; 87:1238-1242.

748. Stoller TJ, Shields D. The propeptide of preprosomatostatin mediates intracellular transport and secretion of α-globin from mammalian cells. J Cell Biol 1989; 108:1647-1655.

749. Stow JL, de Almeida JB, Narula N et al. A heterotrimeric G protein, Ga_{i-3}, on Golgi membranes regulates the secretion of a heparan sulfate proteoglycan in LLC-PK_1 epithelial cells. J Cell Biol 1991; 114:1113-1124.

750. Strom M, Vollmer P, Tan TJ, Gallwitz D. A yeast GTPase-activating protein that interacts specifically with a member of the Ypt-Rab family. Nature 1993; 361:736-738.

751. Subramaniam VN, Krijnse-Locker J, Tang BL et al. Monoclonal antibody HFD9 identifies a novel 28 kDa integral membrane protein on the cis-Golgi. J Cell Sci 1995; 108:2405-2414.

752. Subramaniam VN, Peter F, Phil R, Hong W. A 28 kDa cis-Golgi integral protein involved in ER-Golgi transport. Science, submitted.

753. Sudhof TC. The synaptic vesicle cycle: a cacades of protein-protein interactions. Nature 1995; 375:645-653.

754. Sudhof TC, Baumert M, Perin MS, Jahn R. A synaptic vesicle membrane protein is conserved from mammals to Drosophila. Neuron 1989; 2:1475-1481.

755. Sudhof TC, Jahn R. Proteins of synaptic vesicles involved in exocytosis and membrane recycling. Neuron 1991; 6:665-677.

756. Sumida M, Hong RM, Tagaya M. Role of two nucleotide-binding regions in an N-ethylmaleimide-sensitive factor involved in vesicle-mediated protein transport. J Biol Chem 1994; 269: 20636-20641.

757. Swaroop A, Yang-Feng TL, Liu W et al. Molecular characterization of a novel human gene, Sec13R, related to the yeast secretory pathway gene SEC13, and mapped to a conserved linkage group on human chromasome 3p24-p25 and mouse chromosome 6. Human Mol Gen 1994; 3:1281-1286.

758. Sweet DJ, Pelham HRB. The Saccharomyces cerevisiae SEC20 gene encodes a membrane glycoprotein which is sorted by the HDEL retrieval system. EMBO J 1992; 11:423-432.

759. Sweet DJ, Pelham HRB. The TIP1 gene of Saccharomyces cerevisiae encodes an 80 kDa cytoplasmic protein that interacts with the cytoplasmic domain of Sec20p. EMBO J 1993; 12:2831-2840.

760. Swift AM, Machamer CE. A Golgi retention signal in a membrane-spanning domain of coronavirus E1 protein. J Cell Biol 1991; 115:19-30.

761. Sykes K, Gething M-J, Sambrook J. Proline isomerases function during heat shock. Proc Natl Acad Sci 1993; 90: 5853-5857.

762. Tagaya M, Wilson DW, Brunner M et al. Domain structure of an N-ethylmaleimide-sensitive fusion protein involved in vesicular transport. J Biol Chem 1993; 268:2662-2666.

763. Tajima S, Lauffer L, Rath VL, Walter P. The signal recognition particle receptor is a complex that contains two polypeptide chains. J Cell Biol 1986; 103:1167-1178.

764. Takai Y, Kaibuchi K, Kikuchi A, Kawata M. Small GTP-binding proteins. Internatl Rev Cytol 1992; 133:187-230.

765. Takei K, McPherson PS, Schmid SL, De Camilli P. Tubular membrane invaginations coated by dynamin rings are induced by GTP-γS in nerve terminals. Nature 1995; 374:186-190.

766. Takizawa PA, Malhotra V. Coatomers and SNAREs in promoting membrane traffic. Cell 1993; 75:593-596.

767. Tam WWH, Andreasson KI, Loh YP. The amino-terminal sequence of pro-opiomelanocortin directs intracellular targeting to the regulated secretory pathway. Eur J Cell Biol 1993; 62:294-306.

768. Tamkun JW, Kahn RA, Kissinger M et al. The arflike gene encodes an essential GTP-binding protein in Drosophila. Proc Natl Acad Sci 1991; 88:3120-3124.

769. Tang BL, Low SH, Wong SH, Hong W. Cell type differences in Golgi retention signals for transmembrane proteins. Eur J Cell Biol 1995; 66:365-374.

770. Tang BL, Low SH, Hauri HP, Hong W. Segregation of ERGIC53 and the mammalian KDEL receptor upon exit from the 15°C compartment. Eur J Cell Biol, in press

771. Tang BL, Wong SH, Low SH, Hong W. The transmembrane domain of N-glucosamyltransferase I contains a Golgi retention signal. J Biol Chem 1992; 267: 10122-10126.

772. Tang BL, Wong SH, Qi XL, et al. Molecular cloning, characterization, subcellular localization and dynamics of p23, the mammalian KDEL receptor. J Cell Biol 1993; 120:325-338.

773. Tang BL, Wong SH, Low SH et al. Cytosolic factors block antibody binding to the C-terminal cytoplasmic tail of the KDEL receptor. Eur J Cell Biol 1994; 65:298-304.

774. Tanigawa C, Orci L, Amherdt M et al. Hydrolysis of bound GTP by ARF protein triggers uncoating of Golgi-derived COP-coated vesicles. J. Cell Biol 1993; 123:1365-1371.

775. Taylor TC, Kanstein M, Weidman P, Melancon P. Cytosolic ARFs are required for vesicle formation but not for cell-free intra-Golgi transport: evidence for coated vesicle-independent transport. Mol Biol Cell 1994; 5:237-252.

776. Tatu U, Braakman, I, Helenius A. Membrane glycoprotein folding, oligomerization and intracellular transport: effects of dithiothreitol in living cells. EMBO J 1993; 12:2151-2157.

777. Teal SB, Hsu VW, Peters PJ et al. An activating mutation in ARF1 stabilizes

coatomer binding to Golgi membrane. J Biol Chem 1994; 269:3135-3138.

778. Teasdale R, D'Agostaro G, Gleeson PA. The signal for Golgi retention of β1,4-galactosyltransferase is in the transmembrane domain. J Biol Chem 1992; 267: 4084-4096.

779. Tellam JT, McIntosh S, James DE. Molecular identification of two novel Munc-18 isoforms expressed in non-neuronal tissues. J Biol Chem 1995; 270: 5357-5863.

780. TerBush DR, Novick P. Sec6, Sec8, Sec15 are components of a multisubunit complex which localizes to small bud tips in Saccharomyces cerevisiae. J Cell Biol 1995; 130:299-312.

781. Teyton L, Peterson PA. Invariant chain-a regulator of antigen presentation. Trends Cell Biol 1992; 2:52-56.

782. Thastrup O, Cullen PJ, Drobak BK et al. Thapsigargin, a tumor promoter, discharges intracellular Ca^{2+} stores by specific inhibition of the endoplasmic reticulum Ca^{2+}-ATPase. Proc Natl Acad Sci 1990; 87:2466-2470.

783. Thomas DC, Roth MG. The basolateral targeting signal in the cytoplasmic domain of glycoprotein G from vesicular stomatitis virus resembles a variety of intracellular targeting motifs related by primary sequence but having diverse targeting activities. J Biol Chem 1994; 269:15732-15739.

784. Thomas DC, Brewer CB, Roth MG. Vesicular stomatitis virus glycoprotein contains a dominant cytoplasmic basolateral sorting signal critically dependent upon a tyrosine. J Biol Chem 1993; 268:3313-3320.

785. Thurieau C, Brosius J, Burne P et al. Molecular cloning of complete amino acid sequence of AP50, an assembly protein associated with clathrin-coated vesicles. DNA 1988; 7:663-669.

786. Tisdale EJ, Bourne JR, Khosravi-Far R et al. GTP-binding mutants of Rab1 and Rab2 are potent inhibitors of vesicular transport from the Endoplasmic Reticu-lum to the Golgi Complex. J. Cell Biol 1992; 119:749-761.

787. Tjoelker LW, Seyfried CE, Eddy Jr RL et al. Human, mouse, and rat calnexin cDNA cloning: identification of potential calcium binding motifs and gene localization to human chromosome 5. Biochemistry 1994; 33:3229-3236.

788. Tooze SA, Huttner WB. Cell-free protein sorting to the regulated and constitutive secretory pathway. Cell 1990; 60:837-847.

789. Tooze SA, Weiss U, Huttner WB. Requirement for GTP hydrolysis in the formation of secretory vesicles. Nature 1990; 347:207-208.

790. Townsley FM, Frigerio G, Pelham HRB. Retrieval of HDEL proteins is required for growth of yeast cells. J Cell Biol 1994; 127:21-28.

791. Townsley FM, Pelham HRB. The KKXX signal mediates retrieval of membrane proteins from the Golgi to the ER in yeast. Eur J Cell Biol 1994; 64:211-216.

792. Townsley FM, Wilson DW, Pelham HRB. Mutational analysis of the human KDEL receptor: distinct structural requirements for Golgi retention, ligand binding and retrograde transport. EMBO J 1993; 12:2821-2829.

793. Traber MG, Kayden HJ, Rindler MJ. Polarized secretion of newly-synthesized lipoproteins by the Caco-2 human intestinal cell line. J Lipid Res 1987; 28: 1350-1363.

794. Trimble WS, Cowan DM, Scheller RH. VAMP-1: A synaptic vesicle associated integral membrane protein. Proc Natl Acad Sci 1988; 85:4538-4542.

795. Tsuchiya M, Price SR, Tsai SC et al. Molecular identification of ADP-ribosylation factor mRNAs and their expression in mammalian cells. J Biol Chem 1991; 266:2772-2777.

796. Ui M, Okada T, Hazeki K, Hazeki O. Wortmannin as a unique probe for an intracellular signalling protein, phosphoinositide 3-kinase. Trends Biochem Sci. 1995; 20:303-307.

797. Ullrich O, Horiuchi H, Bucci C, Zerial M. Membrane association of Rab5 mediated by GDP-dissociation inhibitor and accompanied by GDP/GTP exchange. Nature 1994; 368:157-160.

798. Ullrich O, Stenmark H, Alexandrov K et al. Rab GDI dissociation inhibitor as a general regulator for the membrane association of Rab proteins. J Biol Chem 1993; 268:18143-18150.

799. Urban J., Parczyk K, Leutz A, Kayne M, Kondor-koch C. Constitutive apical secretion of an 80-kD sulfated glycoprotein complex in polarized epithelial Madin-Darby canine kidney cell line. J Cell Biol 1987; 105:2735-2743.

800. Urbe S, Huber LA, Zerial M et al. Rab11, a small GTPase associated with both constitutive and regulated secretory pathways in PC12 cells. FEBS Letters 1993; 334:175-182.

801. Ushkaryov YA, Petrenko AG, Geppert M, Sudhof TC. Neurexins: synaptic cell surface proteins related to the α-latrotoxin receptor and laminin. Science 1992; 257:50-56.

802. Vallee RB, Okamoto PM. The regulation of endocytosis: identifying dynamin's binding partners. Trends Cell Biol 1995; 5:43-47.

803. Valls LA, Hunter CP, Rothman JH, Stevens TH. Protein sorting in yeast: the localization determinant of yeast vacuolar carboxylpeptidase Y resides in the propeptide. Cell 1987; 48:887-897.

804. Valls LA, Winther JR, Stevens TH. Yeast carboxylpeptidase Y vacuolar targeting signal is defined by four propeptide amino acids. J Cell Biol 1990; 111: 361-368.

805. van der Sluijs P, Hull M, Huber LA et al. Reversible phosphorylation-dephosphorylation determines the localization of rab4 during cell cycle. EMBO J 1992; 11:4379-4389.

806. van der Sluijs P, Hull M, Webster P et al. The small GTP-binding protein rab4 controls an early sorting event on the endocytotic pathway. Cell 1992; 70: 729-740.

807. van der Sluijs P, Hull M, Zahraoui A et al. The small GTP-binding protein rab4 is associated with early endosomes. Proc Natl Acad Sci 1991; 88:6313-6317.

808. Vater CA, Raymond CK, Ekena K et al. The VPS1 protein, a homolog of dynamin required for vacuolar protein sorting in Saccharomyces cerevisiae, is a GTPase with two functionally separable domains. J Cell Biol 1992; 119:773-786.

809. Vega MA, Rodriguez F, Segui B et al. Targeting of lysosomal integral membrane protein LIMP II: the tyrosine-lacking carboxyl cytoplasmic tail of LIMP II is sufficient for direct targeting to lysosomes. J Biol Chem 1991; 266: 16269-16272.

810. Verhey KJ, Birnbaum MJ. A leu-leu sequence is essential for COOH-terminal targeting signal of GLU4 glucose transporter in fibroblasts. J Biol Chem 1994; 269:2353-2356.

811. Verhey KJ, Yeh JI, Birnbaum MJ. Distinct signals in the GLUT4 glucose transporter for internalization and for targeting to an insulin-responsible compartment. J Cell Biol 1995; 130:1071-1079.

812. Vey M, Schafer W, Berghofer S et al. Maturation of the trans Golgi network protease furin: compartmentalization of propeptide removal, substrate cleavage, and COOH-trminal truncation. J Cell Biol 1994; 127:1829-1842.

813. Vida TA, Graham TR, Emr SD. In vitro reconstitution of intercompartmental protein transport to the yeast vacuole. J Cell Biol 1990; 111:2871-2884.

814. Vida TA, Huyer G, Emr SD. Yeast vacuolar proenzymes are sorted in the late Golgi complex and transported to the vacuole via a prevacuolar endosome-like compartment. J Cell Biol 1993; 121: 1245-1256.

815. Volinia S, Dhand R, Vanhaesebroeck B, et al. A human phosphatidylinositol 3-kinase complex related to the yeast Vps34p-Vps15p protein sorting system. EMBO J 1995; 14:3339-3348.

816. von Mollard GF, Stahl B, Walch-Solimena C et al. Localization of Rab5 to synaptic vesicles identifies endosomal intermediate in synaptic vesicle recycling pathway. Eur J Cell Biol 1994; 65: 319-326.

817. Wada I, Ou W-J, Liu M-C, Scheele G. Chaperone function of calnexin for the folding intermediates of gp80, the major secretory protein in MDCK cells. J Biol Chem 1994; 269:7464-7472.

818. Wada I, Rindress D, Cameron PH et al. SSRa and associated calnexin are major calcium binding proteins of the endoplasmic reticulum membrane. J Biol Chem 1991; 266:19599-19610.

818a. Wada Y, Kitamoto K, Kanbe T et al. The SLP1 gene of Saccharomyces cerevisiae is essential for vacuolar morphogenesis and function. Mol Cell Biol 1990; 10:2214-2223.

819. Wadhwa R, Kaul SC, Komatsu Y, et al. Identification and differential expression of yeast SEC23-related gene (Msec23) in mouse tissues. FEBS Letters 1993; 315:193-196.

820. Wagner ACC, Strowski MZ, Goke B, Williams JA. Molecular cloning of a new member of the Rab protein family, Rab26, from rat pancreas. Biochem Biophys Res Commun 1995; 207: 950-956.

821. Wahlberg JM, Geffen I, Reymond F et al. Trans Golgi retention of a plasma membrane protein: mutations in the cytoplasmic domain of the asialoglycoprotein receptor subunit 1H result in trans Golgi retetnion. J Cell Biol 1995; 130:285-297.

822. Walch-Solimena C, Blasi J, Edelmann L et al. The t-SNAREs syntaxin 1 and SNAP-25 are present in organelles that participate in synaptic vesicle recycling. J Cell Biol 1995; 128:637-645.

823. Walter P, Blobel G. Purification of a membrane-associated protein complex required for protein translocation across the endoplasmic reticulum. Proc Natl Acad Sci 1980; 77:7112-7116.

824. Walter P, Blobel G. Translocation of protein across the endoplasmic reticulum III. signal recognition protein (SRP) causes signal sequence-dependent and site-specific arrest of chain elongation that is release by microsomal membrane. J Cell Biol 1981; 91:557-561.

825. Walter P, Blobel G. Signal recognition particle contains a 7S RNA essential for protein translocation across the endoplasmic reticulum. Nature 1982; 299: 691-698.

826. Walter P, Johnson AE. Signal sequence recognition and protein targeting to the endoplasmic reticulum. Annu Rev Cell Biol 1994; 10:87-119.

827. Walter P, Lingappa VR. Mechanism of protein translocation across the endoplasmic reticulum membrane. Annu Rev Cell Biol 1986; 2:499-516.

828. Walworth NC, Brennwald P, Kabcehell AK et al. Hydrolysis of GTP by Sec4 protein plays an important role in vesicular transport and is stimulated by a GTPase-activating protein in Saccharomyces cerevisiae. Mol Cell Biol 1992; 12:2017-2028.

829. Waters MG, Clary DO, Rothman JE. A novel 115-kD peripheral membrane protein is required for intercisternal transport in the Golgi stack. J Cell Biol 1992; 118:1015-1026.

830. Waters MG, Serafini T, Rothman JE. Coatomer: a cytosolic protein complex containing subunits of non-clathrin-coated Golgi transport vesicles. Nature 1991; 349:248-251.

831. Weidman PJ, Melancon P, Block MR, Rothman JE. Binding of N-ethylmaleimide-sensitive fusion protein to Golgi membrane requires both a soluble protein(s) and an integral membrane receptor. J Cell Biol 1989; 108:1589-1596.

832. Weisz OA, Swift AM, Machamer CE. Oligomerization of a memrbane protein correlates with its retention in the Golgi complex. J Cell Biol 1993; 122: 1185-1196.

833. Wendinger-Ness A, Bennett MK, Antony C, Simons K. Distinct transport vesicles mediates the delivery of plasma membrane proteins to the apical and basolateral domains of MDCK cells. J Cell Biol 1990; 111:987-1000.

834. Wessels HP, Geffen, I, Spiess M. A hepatocyte-specific basolateral membrane protein is targeted to the same domain when expressed in Madin-Darby canine kidney cells. J Biol Chem 1989; 264:17-20.

835. Westlund B, Dahms NM, Kornfeld S. The bovine mannose 6-phosphate/insulin-like growth factor II receptor. localization of mannose 6-phosphate binding sites to domains 1-3 and 7-11 of the extracellular region. J Biol Chem 191; 266:23233-23239.

836. Whiteheart SW, Brunner M, Wilson DW et al. Soluble N-thylmaleimide-sensitive fusion attachment proteins (SNAPs) bind to a multi-SNAP receptor complex in Golgi membrane. J Biol Chem 1992; 267:12239-12243.

837. Whiteheart SW, Griff IC, Brunner M et al. SNAP family of NSF attachment proteins includes a brain-specific isoform. Nature 1993; 362:353-355.

838. Whiteheart SW, Kubalek EW. SNAPs and NSF: general members of the fusion apparatus. Trends Cell Biol 1995; 5:6 4-69.

839. Whiteheart SW, Rossnagel K, Buhrow SA et al. N-ethylmaleimide-sensitive fusion protein: A trimeric ATPase whose hydrolysis of ATP is required for membrane fusion. J Cell Biol 1994; 126: 945-954.

840. Wichmann H, Hengst L, Gallwitz D. Endocytosis in yeast: evidence for the involvement of a small GTP-binding protein (Ypt7p). Cell 1992; 71:1131-1142.

841. Wide A, Dempsey C, Banting G. The tyrosine-containing internalization motif in the cytoplasmic domain of TGN38/41 lies within a nascent helix. J Biol Chem 1994; 269:7131-7136.

842. Wiedmann B, Sakai H, Davis TA, Wiedmann M. A protein complex required for signal-sequence-specific sorting and translocation. Nature 1994; 370: 434-440.

842a. Wieland FT, Gleason ML, Serafini TA, Rothman JE. The rate of bulk flow from the endoplasmic reticulum to the cell surface. Cell 1987; 50:289-300.

843. Williams MA, Fukuda M. Accumulation of membrane glycoproteins in lysosomes requires a tyrosine residue at a particular postion in the cytoplamic tail. J Cell Biol 1990; 111:955-966.

844. Wilsbach K, Payne GS. Dynamic retention of TGN membrane proteins in Saccharomyces cerevisiae. Trends Cell Biol 1993; 3:426-431.

845. Wilsbach K, Payne GS. Vps1p, a member of dynamin GTPase family, is necessary for Golgi membrane protein retention in Saccharomyces cerevisiae. EMBO J 1993; 12:3049-3059.

846. Wilson AL, Maltese WA. Isoprenylation of Rab1b is impaired by mutations in its effector domain. J Biol Chem 1993; 268:14561-14564.

847. Wilson BS, Nuoffer C, Meinkoth JL et al. A Rab1 mutant affecting guanine nucleotide exchange promotes disassembly of the Golgi apparatus. J Cell Biol 1994; 125:557-571.

848. Wilson BS, Palade GE, Farquhar, MG. Endoplasmic reticulum-through-Golgi transport assay based on O-glycosylation of native glycophorin in permeabilized erythroleukemia cells: role of G_{i3}. Proc Natl Acad Sci 1993; 90:1681-1685.

849. Wilson DW, Lewis MJ, Pelham HRB. pH-dependent binding of KDEL to its receptor in vitro. J Biol Chem 1993; 268:7465-7468.

850. Wilson DW, Whiteheart SW, Wiedmann M et al. A multisubunit particle implicated in membrane fusion. J Cell Biol 1992; 117:531-538.

851. Deleted during revision.

852. Wilson DW, Wilcox CA, Flynn GC et al. A fusion protein required for vesicle-mediated transport in both mammalian cells and yeast. Nature 1989; 339: 355-359.

853. Wilson JM, Fasel N, Kraehenbuhl J-P. Polarity of endogenous and exogenous glycosyl-phosphatidylinositol-anchored membrane proteins in Madin-Darby canine kidney cells. J Cell Sci 1990; 96:143-149.

854. Wittinghofer A, Pai E. The structure of ras protein: a model for a universal molecular switch. Trends Biochem Sci 1991; 16:382-387.

855. Wolin SL. From the elephant to E. coli: SRP-dependent protein targeting. Cell 1994; 77:787-790.

856. Wong SH, Low SH, Hong W. The 17-residue transmembrane domain of β-galactoside α2,6-sialyltransferase is sufficient for Golgi retention. J Cell Biol 1992; 117:245-258.

857. Wong SH, Hong W. The SXYQRL sequence in the cytoplasmic domain of TGN38 plays a major role in trans Golgi network localization. J Biol Chem 1993; 268:22853-22862.

858. Xu H, Shields D. Prohormone processing in the trans Golgi network: endoproteolytic cleavage of prosomatostatin and formation of nascent secretory vesicles in permeabilized cells. J Cell Biol 1993; 122:1169-1184.

859. YaDeau JT, Klein C, Blobel G. Yeast signal peptidase contains a glycoprotein and the Sec11 gene product. Proc Natl Acad Sci 1991; 88:517-521.

860. Yamaguchi N, Fukuda, MN. Golgi retention mechanism of β1,4-galactosyltransferase: membrane-spanning domain-dependent homodimerization and association with α- and β-tubulins. J Biol Chem 1995; 270:12170-12176.

861. Yamasaki S, Hu Y, Binz T et al. Synaptobrevin/vesicle-associated membrane protein (VAMP) of Aplysia californica: structure and proteolysis by tetanus toxin and butalinal neurotoxins type D and F. Proc Natl Acd Sci 1994; 91:4688-4692.

862. Yang C, Slepnev VI, Goud B. Rab proteins form in vivo complexes with two isoforms of the GDP-dissociation inhibitor protein (GDI). J Biol Chem 1994; 269:31891-31899.

863. Yeowell HN, Ha V, Clark WL et al. Sequence analysis of a cDNA for lysyl hydroxylase isolated from human skin fibroblasts from a normal donor: differences from human lysyl hydroxylase cDNA. J Invest Dermatol 1994; 102:382-384.

864. Yoshihisa T, Barlowe C, Schekman R. Requirement for a GTPase-activating protein in vesicle budding from the endoplasmic reticulum. Science 1993; 259:1466-1468.

865. Yu H, Leaf DS, Moore HPH. Gene cloning and characterization of a GTP-binding RAB protein from mouse pituitary AtT-20 cells. Gene 1993; 132:273-278.

866. Zahraoui A, Joberty G, Arpin M et al. A small rab GTPase is distributed in cytoplasmic vesicles in non polarized cells but colocalizes with the tight junction marker ZO-1 in polarized epithelial cells. J Cell Biol 1994; 124:101-115.

867. Zerial M, Stenmark H. Rab GTPases in vesicular transport. Curr Opi Cell Biol 1993; 5:613-620.

868. Zhang CJ, Rosenwald AG, Willingham MC et al. Expression of a dominant mutant allele of human ARF1 inhibits membrane traffic in vivo. J Cell Biol 1994; 124:289-300.

869. Zhang FL, Diehl RE, Kohl NE et al. cDNA cloning and expression of rat and human protein geranylgeranyltransferase type-I. J Biol Chem 1994; 269:3175-3180.

870. Zhang JZ, Davletov BA, Sudhof TC, Anderson RGW. Synaptotagmin I is a high affinity receptor for clathrin AP-2: implications for membrane recycling. Cell 1994; 78:751-760.

871. Zopf D, Bernstein HD, Johnson AE, Walter P. The methionine-rich domain of the 54 kd protein subunit of the signal recognition particle contains an RNA binding site and can be crosslinked to a signal sequence. EMBO J 1990; 9: 4511-4517.

872. Zurzolo C, Polistina C, Saini M, et al.. Opposite polarity of virus budding and of viral envelope glycoprotein distribution in epithelial cells derived from different tissues. J Cell Biol 1992; 117:551-564.

873. Zurzolo C, van't Hof W, van Meer G, Rodriguez-Boulan E. VIP21/caveolin, glycosphingolipid clusters and the sorting of glycosylphosphatidylinositol-anchored proteins in epithelial cells. EMBO J 1994; 13:42-53.

INDEX

Page numbers in italics denote figures (f) or tables (t).